Management and Environmental Factor Contributions to Maize Yield

Management and Environmental Factor Contributions to Maize Yield

Special Issue Editors

Frederick E. Below
Juliann R. Seebauer

MDPI • Basel • Beijing • Wuhan • Barcelona • Belgrade

MDPI

Special Issue Editors
Frederick E. Below
University of Illinois at Urbana-Champaign
USA

Juliann R. Seebauer
University of Illinois at Urbana-Champaign
USA

Editorial Office
MDPI
St. Alban-Anlage 66
4052 Basel, Switzerland

This is a reprint of articles from the Special Issue published online in the open access journal *Agronomy* (ISSN 2073-4395) from 2018 to 2019 (available at: https://www.mdpi.com/journal/agronomy/special_issues/management_environmental_maize_yield)

For citation purposes, cite each article independently as indicated on the article page online and as indicated below:

LastName, A.A.; LastName, B.B.; LastName, C.C. Article Title. *Journal Name* **Year**, *Article Number*, Page Range.

ISBN 978-3-03897-612-7 (Pbk)
ISBN 978-3-03897-613-4 (PDF)

Cover image courtesy of Juliann R. Seebauer

Contents

About the Special Issue Editors

Frederick E. Below, Dr., is a Professor of Crop Physiology in the Department of Crop Sciences at the University of Illinois. He received his B.S, M.S, and Ph.D. degrees in Agronomy all from the University of Illinois, and he has been in his current position since 1985. His research is focused on understanding factors limiting crop productivity, particularly corn and soybean. He has been an advisor to dozens of graduate students, teaches advanced courses, and has given numerous presentations at national conferences and local meetings. In 2016, he was elected Fellow of the American Society of Agronomy. He developed the 'Seven Wonders of The Corn Yield World' and the 'Six Secrets of Soybean Success' as tools to teach farmers and agricultural professionals the value of their individual crop management decisions, and he is actively using these concepts to develop systems capable of sustainably producing high yield corn and soybean.

Juliann R. Seebauer is a Principal Research Specialist in Agriculture in the Department of Crop Sciences at the University of Illinois. She received her B.S and M.S. degrees in Agricultural Sciences and Agronomy, respectively, from the University of Illinois, and has been in her current position as Laboratory Manager since 1990. She has received research awards at the college and departmental levels. Her research has centered around whole plant growth and development in regards to nutrient use efficiency. Grain composition has been a particular interest, especially using in vitro kernel culture to alter metabolism and growth. Most recently, she has focused on how the weather influences final grain quality.

Preface to "Management and Environmental Factor Contributions to Maize Yield"

Agricultural production must increase substantially to meet the increasing per capita demand for food, feed, fuel, and fiber of a rising human census. The amount of arable land is limited due to soil type, weather, and ecosystem considerations; therefore it is necessary to increase yields on current fields. To obtain the greatest maize (*Zea mays* L.) yield, a farmer needs to nurture the crop as much as possible. Weather and nitrogen availability are well-known as two factors that normally have the greatest influence on maize yields and grain quality. Some management factors a producer may need to consider while growing a maize crop are mineral fertilization, genotype, plant population, and protection from insects and diseases. Additionally, there are numerous biological and chemical compounds that can stimulate plant growth, such as in-furrow mixes and foliar fungicides. Field management also plays a role in final grain yield, including crop rotation, tillage, soil pH and nutrient levels, weed control, and drainage.

This special issue highlights research that focuses on weather and other crop agronomic management factors and their relative independent and/or interactive influence on maize growth and yield. As maize is grown world-wide, this issue has contributions from Africa, Asia, and Europe, as well as North and South America. Global collaboration is essential for the most efficient progress towards the goal of sustainable increased agricultural production.

The editors wish to thank the contributors, reviewers, and the support of the *Agronomy* editorial staff, whose professionalism and dedication have made this issue possible.

Frederick E. Below, Juliann R. Seebauer
Special Issue Editors

agronomy

MDPI

Article

Weather During Key Growth Stages Explains Grain Quality and Yield of Maize

Carrie J. Butts-Wilmsmeyer [1], Juliann R. Seebauer [1], Lee Singleton [2] and Frederick E. Below [1,*]

[1] Department of Crop Sciences, University of Illinois at Urbana-Champaign, 1102 S. Goodwin Ave., Urbana, IL 61801, USA; cjbutts2@illinois.edu (C.J.B.-W.); jzzz@illinois.edu (J.R.S.)
[2] Centrec Consulting Group, 3 College Park Ct., Savoy, IL 61874, USA; lsingleton@centrec.com
* Correspondence: fbelow@illinois.edu; Tel.: +1-217-333-9745

Received: 14 November 2018; Accepted: 25 December 2018; Published: 2 January 2019

Abstract: Maize (*Zea mays* L.) grain yield and compositional quality are interrelated and are highly influenced by environmental factors such as temperature, total precipitation, and soil water storage. Our aim was to develop a regression model to account for this relationship among grain yield and compositional quality traits across a large geographical region. Three key growth periods were used to develop algorithms based on the week of emergence, the week of 50% silking, and the week of maturity that enabled collection and modeling of the effect of weather and climatic variables across the major maize growing region of the United States. Principal component analysis (PCA), stepwise linear regression models, and hierarchical clustering analyses were used to evaluate the multivariate relationship between weather, grain quality, and yield. Two PCAs were found that could identify superior grain compositional quality as a result of ideal environmental factors as opposed to low-yielding conditions. Above-average grain protein and oil levels were favored by less nitrogen leaching during early vegetative growth and higher temperatures at flowering, while greater oil than protein concentrations resulted from lower temperatures during flowering and grain fill. Water availability during flowering and grain fill was highly explanatory of grain yield and compositional quality.

Keywords: maize; grain quality; yield; climate; temperature; precipitation; data mining; principal component analysis; crop models; corn

1. Introduction

Maize (*Zea mays* L.) is among the United States' most valuable economic exports. In 2017, the United States exported over $10.1 billion in maize alone [1]. Of all grains produced in the U.S., corn is the major feed grain and constitutes greater than 95% of feed grain production and use [2]. Given its frequent use as animal feed, exported maize grain quality is of utmost importance to international buyers. Ideally, feed grain should have a relatively high protein concentration, should be relatively free of broken kernels and foreign matter, and should have minimal levels of mycotoxins. In response to the desires of their international stakeholders, the U.S. Grains Council has published a short annual data summary report since 2011 [3]. However, these reports are typically not available until a few months after the majority of the U.S. maize crop has been harvested. The ability to predict maize grain quality prior to harvest would be of benefit to both international buyers and domestic exporters. Furthermore, grain composition traits are known to be strongly intercorrelated and responsive to weather conditions [4–6], but those relationships have not been explored on a multi-state production basis. Rather, many models are state-specific [5,7]. This unique, comprehensive dataset, when used in conjunction with weather, climatic, and yield databases, provides an opportunity to build multivariate, multi-state predictive models which consider not just grain yield, but also grain quality.

Concurrent yield and compositional grain quality improvement has proven difficult in the past. For the purposes of this study, we define compositional grain quality, or chemical compositional quality, as maize grain having a superior concentration of protein and/or oil relative to other currently, commercially produced, grain. Of the three main chemical components of maize grain, namely starch, protein, and oil, starch is the most prevalent, with a typical range in values between 700 and 750 g/kg in the U.S. corn belt [3]. Maize yield, the trait which has traditionally been given the utmost priority in U.S. corn production, is most closely related to the starch concentration [8]. Oil is a valuable nutrient because of its relatively high energy density. Although previous studies, including the Illinois Long Term Selection Experiments, have examined the potential for creating maize cultivars with high concentrations of protein and oil, these studies have also shown that maize grain yield decreases if the protein concentration is increased beyond approximately 110–120 g/kg, with efforts to increase the grain oil concentration exhibiting a similar limitation [8–10]. Since only 14% of U.S. maize is exported [3,11], grain composition traits often have been neglected in maize improvement research [8,12]. The intercorrelated relationship among yield and these grain quality variables suggests that any predictive models should use multivariate approaches to account for this relationship.

Final yield and grain quality in maize are a result of the interaction of genetic, environmental, and agronomic management factors. Although the genotype has a large influence on final grain composition [13], the temperature and available moisture throughout development, but especially during key physiological growing periods, also plays a role. Specifically, this study focused on the following three key periods: the three weeks following emergence (early growth), the week before to two weeks after silking (flowering), and from five weeks after silking until physiological grain maturity (grain fill). Early plant growth encompasses the time when the photosynthetic potential initiates and the earshoot (panicle) is forming the ovules of the potential future grain [14]. A second critical growth stage borders pollination, when temperature or water availability have a great influence on final numbers of kernels per ear [15–17]. The third critical stage is when the grain is accumulating storage materials, primarily starch and protein, which are sensitive to weather factors affecting photosynthesis, including temperature and soil moisture [18–20].

Throughout the growing season, nitrogen (N) is necessary for optimum maize growth, photosynthesis, grain formation and protein accumulation [21–25], and since maize plants require more N than soils typically supply, it is common practice to apply nitrogen fertilizer. Nitrogen availability in the soil, however, is dynamic, and varies due to many factors, including temperature and water status [26,27].

Additionally, due to different growing season lengths, planting dates, harvest dates, etc., these key growth periods vary in time across the major corn-growing states. In the past, separate predictive models have been built for each state, irrespective of the fact that seed companies market the same hybrids in large, multi-state regions [28]. As such, knowledge of these critical growth periods, as a function of emergence, silking, and maturity, could enable the construction of multi-state predictive models for grain quality and yield.

The premise of this study is that while grain quality is of utmost importance to international buyers, quality traits such as protein and oil composition are frequently secondary considerations in domestic U.S. maize production. Subsequently, more research efforts have been devoted to the development of predictive models for maize yield than for grain quality, and efforts to predict grain quality and yield simultaneously are even more rare. However, the increasing willingness of international buyers to pay a premium for improved grain quality, particularly higher protein quality, suggests that grain composition should be a greater consideration in U.S. maize production. The overall goal of this research was to identify general weather conditions during key points in the maize growing season which influence grain quality. To accomplish this on a multi-state basis, a new standardization technique was developed to quantitatively define weather conditions during the early vegetative, flowering, and grain fill stages. Principal component analysis (PCA) was conducted to understand the multivariate relationship in grain composition variables. These standardized weather variables and PCAs, along with yield, were then employed in predictive models to delineate the most important weather factors that influence grain quality and yield simultaneously.

2. Materials and Methods

2.1. Data Compilation

From 2011 to 2017, a random set of grain elevators which were geographically representative of the maize grain exported from the United States each year were selected for grain composition analyses. Each elevator randomly sampled incoming truckloads, noted the location of origin, and sent 1100 g samples to the Illinois Crop Improvement Association's Identity Preserved Grain Laboratory (IPG Lab) in Champaign, Illinois, U.S.A. After arriving, samples were dried to a suitable moisture content, if needed, and analyzed for grain compositional characteristics by near infrared transmission (NIT) (Infratrec 1241; Foss, Hillerod, Denmark) and reported on a dry basis [29]. Test weight was also determined as a measure of grain weight per standardized volume [29]. There were 2654 samples total, comprised of 360, 160, 132, 629, 527, 624, and 222 samples from 2011, 2012, 2013, 2014, 2015, 2016, and 2017, respectively (Figure S1). Yield data (15.5% moisture) by county of origin was collected from USDA NASS [30].

Latitude and longitude coordinates for the centroid of each county of sample origin were calculated using the coordinate data provided in the R maps package [31]. The code for this data collection step can be found in the Supplementary Materials. The county centroid latitude and longitude coordinates were submitted to the Nutrient Star TED Framework Tool [32]. This returned information regarding the soil water storage (SWS), the aridity index (AI), and the typical number of growing degree days (GDD) accumulated in a region. Briefly, each of these climatic variables are ordinal in nature. The SWS has 7 classes ranging from 0–50 mm to greater than 300 mm by increasing intervals of 50 mm. The greater the value for SWS, the greater the soil water storage capacity of the soil. The AI is a unitless index calculated as a function of the ratio of typical annual precipitation to evapotranspiration. It has 10 classes ranging from 0–2695 to greater than 12,877, with a smaller value indicating a more arid environment. Lastly, GDD is the typical annual growing degree days (sum of daily mean temperature above 0 °C) recorded for a region. It has 10 classes ranging from 0–2670 units to greater than 9851 units. The SWS, AI, and GDD were recorded for the county centroid GPS coordinate. Typically, the TED framework tool returned three sets of SWS, AI, and GDD values per county centroid. Since each ordinal class consists of a range of values, the median of these values was recorded for each of these variables (e.g., the first class for SWS, that being 0–50 mm, was given a median value of 25 mm). The modes of the SWS, AI, and GDD median values were recorded by county, and these are the values that were used in the linear regression models.

The week of maize emergence, week of silking, and week of maturity for each growing season were obtained from USDA NASS [30]. These weeks were defined as follows. The week of emergence was recorded as the week of the year at which a given geographical location [state or Agricultural Statistics District (ASD) [33] first exceeded 50% corn emergence. Likewise, the week of silking and week of maturity were recorded as the week of the year in which 50% of the fields sampled in a geographic area first exceeded 50% silking or full maturity, respectively. For the states of Iowa, Kansas, Kentucky, Minnesota, Missouri, Nebraska, North Dakota, South Dakota, and Wisconsin, these dates were recorded by state. The great difference in climatic conditions in the northern versus southern counties of Illinois and Indiana dictated that these dates be recorded for individual ASDs. These data were available by ASD for Illinois, but it was necessary to interpolate the emergence, silking, and maturity dates for Indiana using the following algorithm. Sections of Indiana were broken into three latitudes: northern (ASDs 10, 20, and 30), central (ASDs 40, 50, and 60), and southern (ASDs 70, 80, and 90) [34]. If ASDs occurred in the same latitude group, they were given the same emergence, silking, and maturity dates. The northern region of Indiana was assigned values based on the average of Illinois ASDs 20 and 50. The central region of Indiana was assigned values based on the average of Illinois ASDs 50 and 70. Lastly, the southern region of Indiana was assigned values based on the average of Illinois ASDs 70 and 90. In the case of a non-integer mean for the week of emergence, silking, or maturity, the mean was truncated (e.g., a mean of 39.5 would be truncated to 39 weeks). This process was repeated annually for years 2011 through 2017.

Three critical growth intervals were established: early growth, flowering, and grain fill:

$$EG \in [E, E + 3 \text{ weeks}]$$

$$F \in [S - 1 \text{ week}, S + 2 \text{ weeks}]$$

$$GF \in [S + 5 \text{ weeks}, M]$$

where *EG* is the set of dates contained within the early growth stage, *F* is the set of dates contained within the flowering growth stage, and *GF* is the set of dates contained within the grain fill growth stage. In the sets above, *E* is the week of emergence, *S* is the week of 50% silking, and *M* is the week of maturity, as defined previously. By specifying the three critical growth periods this way, no dates overlapped between the critical growth periods (e.g., if the week of 50% silking was recorded as the 30th week of the year, *F* would contain the weather information between the 29th and the 32nd weeks, and *GF* would contain the weather information between the 35th week and the week of maturity.

For each county sampled, the total precipitation and the average mean temperature of each of the three growth intervals as well as the average minimum temperature during grain fill were obtained from the National Weather Service in Lincoln, IL (NWSLI) through the Midwestern Regional Climate Center (MRCC) Application Tools Environment (cli-MATE) [35]. In the instance that data were not available for a particular county, data from a neighboring county, preferably to the east or west and no closer to a large body of water than the county of question, were used. In the case that data were recorded for multiple locations within the same county, the median of the locations was used. In the instance that the county information for a sample was unknown, the median of all the counties in the same ASD was used to impute the weather data.

2.2. Correlation and Principal Component Analyses

Once the database was assembled, Pearson correlation coefficients between all response and between all putative explanatory variables were calculated using PROC CORR of SAS (version 9.4; SAS Institute, Inc. Cary, NC, USA). Since the correlation coefficients among the explanatory variables were very weak in most cases, stepwise regression analyses were conducted as described below to account for the rare correlation among explanatory variables. Given the large number of samples, the *p*-value associated with the correlation coefficient is nearly meaningless (i.e., the power to detect even slight differences from $r = 0$ is extraordinary). Thus, the following thresholds for the absolute value of the correlation coefficient were used to describe the strength of the relationship between variables:

$$0.0 < |r| \le 0.3 \text{ indicated a weak relationship}$$

$$0.3 < |r| \le 0.7 \text{ indicated a moderate relationship}$$

$$0.7 < |r| \le 1.0 \text{ indicated a strong relationship}$$

Values of $|r| \ge 0.5$ indicated a potential multicollinearity issue may arise between two predictor variables. This was also used as the threshold for inclusion in the PCA of the response variables.

The PCA of the response variables exhibiting $|r| \ge 0.5$ was conducted using PROC PRINCOMP of SAS (version 9.4). The PCAs were calculated based on the correlation matrix. Only PCAs with eigenvalues greater than 1 were maintained [36]. The PCA scores were output using the Output Delivery System (ODS) in SAS. The PCAs were interpreted based upon their vector loadings.

2.3. Stepwise Linear Regression and Remedial Measures

Separate models were built for PCA_1, PCA_2, and yield. Each model was an additive multiple regression model such that:

$$Y_i = \beta_0 + \sum_{k=1}^{p-1} \beta_k X_{ik}$$

where X_{ik} is the *k*th weather or climatic predictor variable.

A total of $p - 1 = 11$ possible weather and climatic predictor variables potentially could have been entered into the model, although one of these predictor variables is a covariate that was identified through the PCA. This covariate is described in more detail in the results and discussion section. A general description of all predictor variables is provided in Table 1.

Table 1. Description of weather and climatic predictor variables and their utilization in models.

X_{ik}	Acronym	General Description	Models Where Included		
			PCA_1	PCA_2	Yield
X_{i1}	EGP	The total precipitation during the early vegetative growth stage in inches	Y	N	Y
X_{i2}	EGT	The average daily temperature during the early vegetative growth stage in °F	N	N	N
X_{i3}	FP	The total precipitation during the flowering growth stage in inches	N	Y	Y
X_{i4}	FT	The average daily temperature during flowering in °F	Y	Y	Y
X_{i5}	GFP	The total precipitation during grain fill in inches	Y	N	Y
X_{i6}	GFT	The average daily temperature during grain fill in °F	Y	Y	Y
X_{i7}	GFMT	The average minimum daily temperature during grain fill in °F	N	Y	Y
X_{i8}	SWS	Soil water storage, more positive values indicate a greater soil water storage capacity	N	Y	Y
X_{i9}	AI	The aridity index, smaller values indicate a more arid environment as a function of average annual precipitation and rate of evapotranspiration	Y	Y	Y
X_{i10}	GDD	The average growing degree days for an area	N	N	N
X_{i11}	D	A qualitative covariate accounting for the greater protein content typical of hybrids grown in the Dakotas. This variable was assigned a value of 0 if the sample in question came from either ND or SD and a value of 1 otherwise.	Y	Y	Y

Stepwise selection methods were used to build all three models in PROC REG. An entry rate of 0.10 and a retention rate of 0.15 were used. Added variable plots were used in remedial measure analysis to ensure the addition of interaction terms was not warranted. Assumptions of normality were validated using QQ-plots produced in the diagnostics output of PROC REG. Assumptions of homogeneity of variance were validated by examining plots of the semi-studentized residuals versus the predicted values and versus the individual regressors. In the case that an issue with homogeneity of variance presented itself, iterative weighted least squares (WLS) regression was used in order to estimate the regression parameter values. Iterative WLS was continued until additional iterations converged to the same parameter estimates within 5% for each of the previous parameter estimates. Extreme outliers were removed based on semi-studentized residual values and leverage values and thresholds calculated in PROC REG. Extremely influential points, as measured by Cook's D, were removed.

2.4. Cluster Analyses and Imputation Methods

Due to sampling limitations, particularly in the earlier years of the study, not all ASDs were represented each year. However, as a result of the extremely different growing conditions encountered from 2011 to 2017, not including all years was initially found to penalize some ASDs more than others. In particular, ASDs that were able to produce enough grain to be sampled in adverse years, such as the drought conditions encountered in 2012, were more heavily penalized than ASDs that were not able to provide samples during such conditions. Thus, it was necessary to impute certain Year-by-ASD combinations before clustering analyses could be conducted.

Imputation was completed as follows. Each Year-by-ASD combination that was not measured by the U.S. Grains Council was recorded. A typical county from that ASD that had been sampled in multiple other years in the U.S. Grains Council dataset was identified. Yield data (wet basis) from these counties were recorded from USDA NASS [30]. The SWS, AI, and GDD values had already been recorded for those counties in a different year, and these values were reused for the imputation dataset. Emergence, silking, and maturity dates were available for all states and ASDs, as previously described, and these dates were matched to the counties in the imputation set. The precipitation and temperature data were recorded for these counties as previously described. Then, PCA_1 and PCA_2 scores were calculated for each Year-by-ASD combination in the imputation set using the regression parameters estimated from the stepwise multiple linear regression models.

The observed values from the U.S. Grains Council database and the imputed dataset were combined. The LSMEAN PCA_1, PCA_2, and yield values were calculated by first using PROC MEANS of SAS 9.4 to take the mean values of each of these response values for each Year-by-ASD combination and then again using PROC MEANS to take the mean of the resulting values by ASD. As an example,

$$LSMEAN_{PCA1,\ ASD_j} = \frac{\sum_{i=1}^{7} \overline{Y}_{Year_i,ASD_j}}{7}$$

where $\overline{Y}_{Year_i,ASD_j}$ corresponds to the mean PCA_1 value in the *i*th year and the *j*th ASD.

The LSMEANs were then standardized using PROC STDIZE of SAS (version 9.4). The standardized values were used to conduct a hierarchical clustering analysis, this being a form of machine learning which identifies groups based on their level of dissimilarity. The approach used is a slight modification of the approach presented in Butts-Wilmsmeyer et al. [37]. Briefly, the cluster analysis was conducted in PROC CLUSTER of SAS using Ward's Minimum Variance Approach. When Ward's method is employed, the number of clusters selected is left to the discretion of the scientist. The following two guidelines were used. First, the number of clusters selected corresponded with an R^2 value greater than 80%. Second, if a large increase in the between cluster sums of squares occurred when two clusters were joined, then clustering ceased and the number of clusters used prior to the large increase in the between cluster sums of squares was selected.

3. Results and Discussion

3.1. Correlation and Principal Component Analysis

A moderate correlation existed between the average flowering temperature and both of the grain fill temperature variables, the minimum and average temperature during the grain fill period being strongly correlated (Table 2). The GDD were correlated with the average temperature during early vegetative growth ($r = 0.52$), the average temperature during grain fill ($r = 0.70$), and the minimum temperature during grain fill ($r = 0.69$). The presence of correlated predictor variables, while somewhat infrequent, suggested that multicollinearity issues may arise. The use of PCAs as predictor variables was considered, but only four of the possible $\binom{10}{2} = 45$ correlations exhibited values above the threshold established as an indicator of multicollinearity. As such, it is not surprising that an exploratory PCA (results not shown) was only capable of reducing the number of explanatory variables from ten to seven. Therefore, stepwise linear regression models were used to account for the occasional intercorrelation between predictor variables, as previously described in the materials and methods.

The starch concentration was negatively correlated with both the protein and oil concentrations, with Pearson correlation coefficients of −0.54 and −0.60, respectively. Yield was not correlated with any of the chemical composition traits above the established threshold, although it was moderately negatively correlated with the protein concentration ($r = -0.43$; Table 3).

Correlations between test weight and the chemical composition variables, as well as between test weight and yield, changed considerably depending on the year (Table S1). Given that these correlations between test weight and the other response variables were not stable and that only 5.7% of the samples had test weight values less than 69.9 kg/hL (56 lb/bushel), test weight was not included in the subsequent analyses.

Table 2. Pearson correlation coefficients between weather and climate predictor variables. Correlations which surpassed the threshold for multicollinearity concerns ($|r| \geq 0.50$) are highlighted in orange. Other correlations of moderate strength are shown in blue.

EGP [†]	EGT	FP	FT	GFP	GFT	GFMT	SWS	GDD	AI	
	0.132	−0.175	−0.110	−0.019	−0.243	−0.193	0.006	−0.025	−0.004	EGP
		−0.260	0.169	−0.058	0.221	0.213	−0.086	0.520	0.442	EGT
			0.019	0.153	0.150	0.237	−0.060	−0.003	0.099	FP
				−0.175	0.496	0.420	0.098	0.373	0.004	FT
					0.084	0.203	0.024	0.049	0.207	GFP
						0.953	0.029	0.701	0.261	GFT
							0.001	0.693	0.346	GFMT
								0.011	−0.178	SWS
									0.479	GDD

[†] EGP, early growth precipitation; EGT, early growth temperature; FP, flowering period precipitation; FT, flowering period daily average temperature; GFP, precipitation during grain fill; GFT, average temperature during grain fill; GFMT, Average minimum temperature during grain fill; SWS, soil water storage capacity; GDD, average growing degree days for an area; AI, aridity index.

Table 3. Pearson correlation coefficients between response variables. Correlations that surpassed the threshold for inclusion in PCA ($|r| \geq 0.50$) are highlighted in orange. Other correlations of moderate strength are shown in blue.

	Grain Concentration			Test			
	Protein	Starch	Oil	Weight	PCA$_1$	PCA$_2$
Yield	−0.431	0.063	0.248	0.176	−0.087	−0.488
Protein		−0.544	−0.001	−0.018	NA [†]	NA
Starch			−0.599	0.176	NA	NA
Oil				−0.070	NA	NA
Test Weight					−0.126	0.034
PCA$_1$						0.000

[†] NA, Not applicable.

The PCA indicated that greater than 93.6% of the variability in the chemical composition measures could be explained using two PCAs, both of which had eigenvalues greater than 1. The vector loadings for these PCAs can be found in Table S2. Generally, PCA$_1$ can be described as a contrast between the amount of protein and oil in a maize kernel in comparison to the starch concentration. Furthermore, PCA$_2$ can be described as a contrast between protein and oil concentration. Yield was not correlated with PCA$_1$, but the correlation between yield and PCA$_2$ was moderate at $r = -0.49$ (Table 3). These results suggested that these two PCAs might be capable of distinguishing the difference between higher protein concentration as a result better compositional quality versus as a result of reduced starch deposition and lower yields.

When the starch-to-protein ratio was plotted by state (Figure S2), it was noted that both North Dakota (ND) and South Dakota (SD) consistently had higher than average protein concentrations than the other ten states included in the study. To adapt to an inherently shorter growing season than the majority of the U.S. Corn Belt, the hybrids grown in the Dakotas purportedly were derived with a greater proportion of flint germplasm [38]. Flint germplasm is characterized by early maturing hybrids which are more resistant to the molds and adversely cooler temperatures encountered in the northern United States, and it is also noted for its harder kernels in comparison to dent varieties [39]. Flint germplasm has a higher ratio of horny to floury endosperm, and a higher protein concentration but less yield than other germplasm sources, even under identical weather conditions [40–42]. To account for this genetic difference between hybrids, a covariate was included in all stepwise models such that:

$$X_{i11} = \begin{cases} 0 & \text{if state} \in D \\ 1 & \text{otherwise} \end{cases}$$

where $D = \{\text{ND, SD}\}$

Average yields, based on collected county information, were calculated for each of the seven years included in the study (Table 4). Generally, 2014–2017 were high yielding years, with average yields greater than 10.67 metric tons/hectare (170 bushels/acre) each. The year 2013 can be characterized as moderate to moderately-high yielding, with an average yield of 10 metric tons/hectare (159 bushels/acre). The year 2011 was less ideal, with severe flooding across much of the U.S. Corn Belt during the early growing season and drought during flowering, but yields were still acceptable at an average of 8.93 metric tons/hectare (142 bushels/acre). The year 2012, which was characterized by prolonged drought and exceptionally high temperatures during much of the growing season, was the worst yielding year among the seven years included in the study. The average yield in 2012 was 7.19 metric tons / hectare (114.5 bushels/acre), with some counties recording an average of 0 metric tons / hectare yield to the USDA [30].

Table 4. Average maize grain chemical composition, PCA, and yield values between 2011 to 2017 for U.S. Corn Belt samples with and without the Dakota states.

Year	Grain Concentration			PCA$_1$	PCA$_2$	Yield
	Protein	Starch	Oil			
		—g/kg—				T/ha
All States Included						
2011	87.2	734.7	36.7	−0.40	0.49	8.93
2012	94.4	731.6	37.5	0.42	1.03	7.19
2013	85.8	734.1	38.5	−0.17	0.03	10.00
2014	84.6	735.0	37.6	−0.43	0.07	10.86
2015	81.9	736.9	37.7	−0.75	−0.19	10.86
2016	85.7	724.7	40.4	0.84	−0.32	10.97
2017	86.2	723.2	41.2	1.09	−0.42	10.75
Excluding Dakotas						
2011	86.8	734.8	36.8	−0.41	0.43	9.41
2012	94.3	731.6	37.5	0.41	1.01	7.46
2013	85.8	734.1	38.5	−0.17	0.03	10.00
2014	83.8	735.6	37.9	−0.50	−0.05	11.51
2015	81.0	737.4	37.8	−0.82	−0.29	11.00
2016	84.9	725.6	40.2	0.70	−0.36	11.22
2017	85.9	723.6	41.1	1.04	−0.43	11.45

The years 2011 and 2012 were the two years studied which had the highest grain protein concentration but reduced yields and grain starch deposition as a result of extremely adverse weather conditions, especially during flowering [18,20]. The negative relationship between yield and protein concentration in the adverse weather conditions did not extend to years characterized by moderate or optimal weather conditions (2013–2017). Quite to the contrary, 2016 and 2017, both high-yielding years, were also characterized by protein concentrations that were comparable to 2013, a moderate year in terms of weather and, consequently, yield (Table 4). This observation remained true even after accounting for the greater number of samples from the Dakotas in 2015–2017 as opposed to 2013 (data not shown). Furthermore, the grain oil concentration was also relatively high in 2016 and 2017, but it was at relatively similar levels in 2012, 2014, and 2015.

Collectively, these observations suggest that these two PCAs can be used as indices to distinguish apparent improved chemical composition quality as a result of reduced yield and lower starch deposition (unfavorable) from actual improved chemical composition in conjunction with higher yields (favorable). Arithmetic means of the PCAs showed that positive mean values for PCA_1 occurred in 2012, 2016, and 2017, whereas positive mean values for PCA_2 occurred in 2011–2013 (Table 4; Figure 1). Four outliers from 2011 with extreme PCA_2 values were discovered in the scatterplot and were removed prior to stepwise regression analyses (Figure 1).

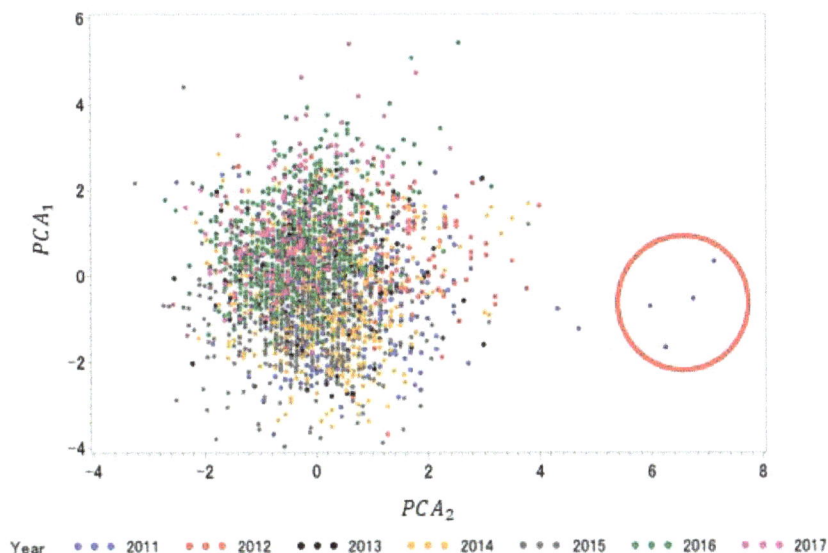

Figure 1. Scatterplot of PCAs by year. Different years are represented by different colors. Four outliers were identified for removal based on the PCA, these being circled in red in the figure above. Years 2016 and 2017, represented by green and magenta points, respectively, were both high yielding years characterized by higher protein and oil concentrations. These two years separate from the other years in the scatterplot, suggesting that these two PCAs could be used to characterize improved compositional grain quality and yield simultaneously.

3.2. Stepwise Regression with Weather and Climatic Variables

Linear regression models were fit for PCA_1, PCA_2, and yield. A summary of all variables included in each of these three models can be found in Table 1. All three models included the covariate that accounted for the protein-rich germplasm grown in the Dakotas, the AI, the average temperature at flowering, and the average temperature during grain fill. None of the models included GDD or the average temperature during early vegetative growth. It stands to reason that GDD would not likely be included in the regression models due to its collinearity with the average and minimum temperature during grain fill, these latter variables often being included in the regression model. Given that the covariate accounting for the Dakotas is somewhat correlated with the other weather and climatic predictors, it was not interpreted in analyses below [43]. Rather, it is included in the model only to improve the model's predictive ability.

3.2.1. PCA$_1$—High Grain Protein and Oil

More positive values of PCA$_1$ were the result of higher protein and oil concentrations as opposed to starch concentration, irrespective of whether that increase was due to actual grain quality improvement or a reduced starch concentration and lower yields. More positive values of PCA$_1$ are ideal if attempting to determine which weather conditions lead to more favorable concentrations of protein and oil. The most important predictor in explaining PCA$_1$ was the total precipitation during early vegetative growth, with a partial R^2 of 5.1%. The addition of five other predictor variables, namely the average temperature during flowering, the AI, the total precipitation during grain fill, the covariate accounting for the Dakotas, and the average temperature during grain fill (in order of addition to the model using stepwise selection), led to a final model R^2 of 12.7%. Given that nothing is known of the specific production management strategies employed or the specific hybrids used, this is a reasonably accurate model. Wet conditions during early growth resulted in reduced PCA$_1$ values, most likely due to nitrogen fertilizer leaching or denitrification and a reduced grain protein concentration [44]. Hot mean temperatures during flowering and grain fill as well as more arid climates resulted in more positive PCA$_1$ values, likely due to drought and heat stress reducing photosynthesis, resulting in reduced starch deposition [45–47]. However, PCA$_1$ is a function of both protein and oil, and both of these constituents were found to be at higher concentrations in the grain during favorable-yielding years. More positive values of PCA$_1$ were also observed when sufficient water was available during grain fill. Having an optimal balance of N availability and photoassimilates in a non-water-limiting environment can lead to larger maize kernels with a concurrent higher level of protein [48].

3.2.2. PCA$_2$—High Grain Protein Over Oil

More positive values of PCA$_2$ are the result of higher grain protein as opposed to oil concentration, having already accounted for the chemical composition differences captured by PCA$_1$. Thus, PCA$_2$ is instrumental in describing stressful conditions which influence compositional grain quality. More positive values of PCA$_2$ are indicative of a higher protein concentration as a result of stressful conditions, either drought or heat stress, that decrease starch and oil deposition in the grain. Heat stress during grain fill has been found to decrease kernel oil concentration in semi-dent hybrids [49]. As such, more negative values of PCA$_2$ are ideal for greater yield and oil, but this measure alone will not capture favorable protein concentrations without also examining PCA$_1$. The final regression model for PCA$_2$ had a model R^2 of 18.9%, which is moderate (multiple correlation coefficient = 0.453). More negative values of PCA$_2$ were the result of less arid environments where the SWS was also greater. More negative values of PCA$_2$ were also observed in environments with lower temperatures during flowering and grain fill and with greater precipitation during flowering. The average minimum temperature during grain fill was also included in the regression model to improve the predictive ability of the model, but was unnecessary due to the high degree of multicollinearity between the two temperature variables during grain fill.

3.2.3. Yield

Interestingly, even though nothing is known of the specific production management strategies used during the growing season of these samples, the regression of yield against the climatic and weather predictor variables explained 47.7% of the total variability in yield, which is fairly high (multiple correlation coefficient = 0.69). Nine of the eleven possible predictor variables were included in the model, the two that were not included being the average temperature during early vegetative growth and the GDD. In general, yield was higher under growing conditions where ample moisture was available during flowering and grain fill, and where drought was less likely to be a limiting factor due to SWS, AI, or hot temperatures during flowering and grain fill. Too much precipitation early in the growing season was found to decrease yield, likely due to the loss of nitrogen fertilizer from the soil environment. The final model was capable of predicting the average county yields to within 0.89 metric tons/hectare (14 bushels/acre), as a median (Table S3). An alternative measure of model accuracy, the root mean square error (RMSE), was found to be 1.44 metric tons/hectare (23 bushels/acre) in this study. By comparison, the USDA WASDE model, a computationally intensive model that makes use of weather data and satellite imagery to compute multivariate non-linear predictive models for grain yield, was recently shown to have an RMSE of 1.11 metric tons per hectare (18 bushels/acre) early in the growing season [5].

Thus, even though the model we show here is computationally simple, it is similar in accuracy to much more complex models such as the USDA WASDE. Furthermore, it highlights the importance of minimizing drought stress at flowering and grain fill. Otherwise, both yield and grain quality will suffer. Our linear models serve as a foundation for more complex models in the future by indicating (i) maize yield and maize quality are dependent on a shared set of conditions during critical growth periods, and (ii) these critical growth periods should be given greater weight in complex predictive models for the multivariate prediction of yield and compositional quality. As a second consideration, the more complex nonlinear models that are characterized by a higher predictive ability are also characterized by predictor variables that are all highly intercorrelated, meaning that their parameter estimates should not be interpreted [43]. Given that one of our goals was to identify which of the putative critical growth stages are important influencers of grain yield and chemical composition, it was imperative to build models that were characterized by both low multicollinearity and adequate predictive ability. The models presented here, particularly those for PCA$_2$ and yield, accomplish that goal.

3.3. Multivariate Clustering Analysis by ASD

Cluster analysis using Ward's Minimum Variance Approach indicated that the 76 ASDs used in this study could be subdivided into 10 clusters based on their standardized average PCA$_1$, PCA$_2$, and yield values. These clusters are indicated by different colors in Figure 2.

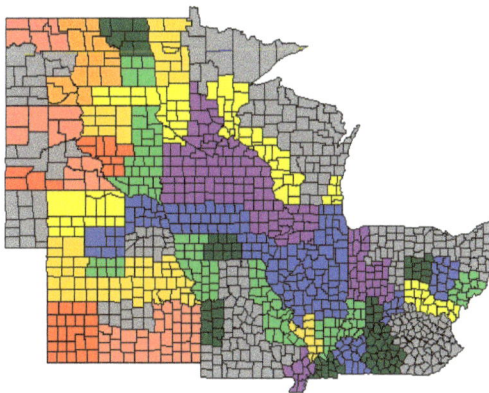

Figure 2. Multivariate clustering analysis by ASD. The ASDs with the same color fall in the same cluster and have similar maize yield and compositional quality. A color-spectrum approach was used to represent the clusters, with purple being high yielding ASDs with lower protein content. Blue is high yielding with decent compositional quality. The greens and yellows are used to describe ASDs with moderate yield and compositional quality values. Lastly, the orange and progressively red ASDs represent areas where protein concentration is higher, but at the expense of yield.

In Figure 2, a color-spectrum approach was used to visualize the multivariate presentation of yield and the PCAs, as is described in more detail in Table 5. No one cluster of ASDs was ideal; all had their advantages and disadvantages (Table 5). While Cluster 1 (purple) and Cluster 2 (blue) undeniably had the highest yielding averages, the samples from Cluster 2 had somewhat better chemical composition quality overall but slightly less yield.

Table 5. Means of response variables and number of ASDs included for each cluster of maize grain quality and quantity relationship to weather from 2011–2017. Blue cells represent more desirable means and orange cells represent less desirable means for PCA_1 (relatively high protein and oil concentrations), PCA_2 (more protein than oil), and yield. Yield is presented in metric tons per hectare with bushels per acre in parentheses.

Cluster	Color	ASD [†] Count	PCA_1	PCA_2	Yield
1	Purple	13	−0.44397	0.09525	11.214 (178.644)
2	Blue	14	0.1441	−0.17026	10.840 (172.674)
3	Green	12	0.19138	0.33464	9.186 (146.339)
4	Dark Green	7	−0.07588	−0.10005	9.063 (144.381)
5	Yellow	9	−0.73882	0.43401	10.184 (162.237)
6	Orange	8	0.72065	0.59974	8.638 (137.604)
7	Gold	3	1.52549	0.54987	6.980 (111.195)
8	Salmon	6	0.4815	0.82565	6.564 (104.568)
9	Brick Red	3	1.30486	1.15527	9.190 (146.394)
10	Red	1	−1.78707	0.90478	6.362 (101.340)

† ASD—Agricultural Statistics District.

Overall, ASDs clustered together as one might expect based on similar weather and climatic conditions. The historically high-yielding regions of Iowa, Illinois, and Southern Minnesota fell into Clusters 1 and 2, and ASDs from clusters described by moderate values of all three response variables (green and yellow clusters) falling adjacent to these regions. The ASDs in the Plains States in the west typically fell into more protein rich clusters, but at the expense of reduced yield and oil concentration. Given the aridity of these regions and the frequency of drought conditions, this is to be expected. However, there were three ASDs (NE 30, NE 50, and MO 90) that fell into clusters that were somewhat different than might be expected given their geographical location and the cluster assignments of the neighboring ASDs. Upon further examination, it was noted that these three ASDs all lie in regions where cropland is heavily irrigated (Figure S3). Therefore, it is probable that the improved yields and chemical composition of the grain sampled from these ASDs is due in at least part to the presence of irrigation [50]. Other ASDs of interest are KY 20, IN 90, and OH 50. These ASDs fell into Cluster 2, this cluster typically being reserved for ASDs in the major maize growing regions of Iowa and Illinois. All three of these ASDs are located in areas with a greater presence of rivers than is typical of most of the ASDs included in this study [51]. Thus, these observations lead us to conclude that grain yield and grain chemical composition can be modeled and improved simultaneously, and the key factor involved is non-limited water conditions during flowering and grain fill.

Based on these results, it is apparent that water availability as a function of total rainfall, temperature, AI, and SWS is a major predictor of grain compositional quality and yield. Too much rainfall during early vegetative growth leads to reduced protein concentration and yield, most likely as a result of nitrogen leaching or denitrification. On the other hand, water availability during the two critical growth stages of flowering and grain fill is largely responsible for both grain yield and compositional quality, as indicated by both the multiple regression models and clustering analyses used in this study. Previous studies have also found that irrigation has a greater impact on maize yield than temperature over the season [50]. These findings may be used to predict when weather conditions may hinder yield and/or compositional quality of the grain and could also be used to build more sophisticated models (e.g., nonlinear multivariate models, spatial error models, etc.) that have a stronger weight on the weather conditions at the identified critical growth stages. Ultimately, these findings indicate that both yield and grain compositional quality can be monitored and improved simultaneously, that improved maize grain chemical composition as a result of favorable environmental conditions can be distinguished from superficial, apparent improvement as the result of low-yielding environmental conditions, and that the key limiter to improving grain yield and compositional quality is access to water.

Supplementary Materials: The following are available online at http://www.mdpi.com/2073-4395/9/1/16/s1, Figure S1: Sampling information by ASD, Figure S2: Plot of starch-to-protein ratio by state, Figure S3: Map of irrigation prevalence in the United States, Table S1: Pearson correlation coefficients between test weight and other response variables across seven years of study, Table S2: Vector loadings of PCAs, Table S3: Absolute differences in observed versus predicted, Supplemental Code: R code for latitude and longitude coordinates.

Author Contributions: Conceptualization, J.R.S. and F.E.B.; methodology, L.S.; statistical methodology and formal analysis, C.J.B.-W.; investigation, J.R.S.; resources, F.E.B.; data curation, L.S.; cross-database dataset compilation, C.J.B.-W.; writing—original draft preparation, C.J.B.-W. and J.R.S.; writing—review and editing, J.R.S.; visualization, C.J.B.-W.; supervision, F.E.B.; project administration, J.R.S.; funding acquisition, F.E.B.

Funding: This research was made possible with partial funding from the United States Department of Agriculture National Institute of Food and Agriculture project NC-1200 "Regulation of Photosynthetic Processes" and the Illinois Agriculture Experiment Station project 802-906. This project also would not have been possible without the data collection and analysis support of the U.S. Grains Council (http://grains.org).

Acknowledgments: The authors with to thank Centrec Consulting Group, LLC. for coordinating sample acquisition, analysis, and data compilation.

Conflicts of Interest: The authors declare no conflict of interest.

References

1. U.S. Census Bureau. U.S. Exports to World Total by 5-Digit End-Use Code. Available online: https://www.census.gov/foreign-trade/statistics/product/enduse/exports/c0000.html (accessed on 20 August 2018).
2. USDA ERS. Corn and Other Feedgrains: Background. Available online: https://www.ers.usda.gov/topics/crops/corn-and-other-feedgrains/background/ (accessed on 20 August 2018).
3. U.S. Grains Council. Corn Reports. Available online: https://grains.org/corn_report/ (accessed on 31 July 2018).
4. Wilhelm, E.P.; Mullen, R.E.; Keeling, P.L.; Singletary, G.W. Heat stress during grain filling in maize: Effects on kernel growth and metabolism. *Crop Sci.* **1999**, *39*, 1733–1741. [CrossRef]
5. Peng, B.; Guan, K.; Pan, M.; Li, Y. Benefits of seasonal climate prediction and satellite data for forecasting US maize yield. *Geophys. Res. Lett.* **2018**, *45*, 9662–9671. [CrossRef]
6. Lecerf, R.; Ceglar, A.; López-Lozano, R.; Van Der Velde, M.; Bauruth, B. Assessing the information in crop model and meteorological indicators to forecast crop yield over Europe. *Agric. Syst.* **2019**, *168*, 191–202. [CrossRef]
7. Warren, F.B. Forecasting corn ear weights from daily weather data. In Proceedings of the Conference Applied Statistics in Agriculture, Manhattan, KS, USA, 30 April–2 May 1989. [CrossRef]
8. Alexander, D.E. Breeding special nutritional and industrial types. In *Corn and Corn Improvement*, 3rd ed.; Sprague, G., Dudley, J., Eds.; ASA-CSSA-SSSA: Madison, WI, USA, 1988.
9. Thomison, P.R.; Geyer, A.B.; Lotz, L.D.; Siegrist, H.J.; Dobbels, T.L. Topcross high oil corn production: Select grain quality attributes. *Agron. J.* **2003**, *95*, 147–154. [CrossRef]
10. Uribelarrea, M.; Below, F.E.; Moose, S.P. Grain composition and productivity of maize hybrids derived from the Illinois Protein Strains in response to variable nitrogen supply. *Crop Sci.* **2004**, *44*, 1593–1600. [CrossRef]

11. USDA ERS. Corn and Other Feedgrains: Trade. Available online: https://www.ers.usda.gov/topics/crops/corn-and-other-feedgrains/trade/ (accessed on 20 August 2018).

12. Butts-Wilmsmeyer, C.J.; Mumm, R.H.; Bohn, M.O. Concentration of beneficial phytochemicals in harvested grain of US yellow dent maize (*Zea mays* L.) germplasm. *J. Agric. Food Chem.* **2017**, *65*, 8311–8318. [CrossRef] [PubMed]

13. Cook, J.P.; McMullen, M.D.; Holland, J.B.; Tian, F.; Bradbury, P.; Ross-Ibarra, J.; Buckler, E.S.; Flint-Garcia, S.A. Genetic architecture of maize kernel composition in the nested association mapping and inbred association panels. *Plant Physiol.* **2012**, *158*, 824–834. [CrossRef] [PubMed]

14. Lejeune, P.; Bernier, G. Effect of environment on the early steps of ear initiation in maize (*Zea mays* L.). *Plant Cell Environ.* **1996**, *19*, 217–224. [CrossRef]

15. Cantarero, M.G.; Cirilo, A.G.; Andrade, F.H. Night temperature at silking affects kernel set in maize. *Crop Sci.* **1999**, *39*, 703–710. [CrossRef]

16. Edreira, J.I.R.; Carpici, E.B.; Sammarro, D.; Otegui, M.E. Heat stress effects around flowering on kernel set of temperate and tropical maize hybrids. *Field Crops Res.* **2011**, *123*, 62–73. [CrossRef]

17. Prasad, P.V.V.; Bheemanahalli, R.; Jagadish, S.V.K. Field crops and the fear of heat stress-opportunities, challenges and future directions. *Field Crops Res.* **2017**, *200*, 114–121. [CrossRef]

18. Jones, R.J.; Gengenbach, B.G.; Cardwell, V.B. Temperature effects on in vitro kernel development of maize. *Crop Sci.* **1981**, *21*, 761–766. [CrossRef]

19. Seebauer, J.R.; Singletary, G.W.; Krumpelman, P.M.; Ruffo, M.L.; Below, F.E. Relationship of source and sink in determining kernel composition of maize. *J. Exp. Bot.* **2010**, *61*, 511–519. [CrossRef] [PubMed]

20. Singletary, G.W.; Banisadr, R.; Keeling, P.L. Heat-stress during grain filling in maize—effects on carbohydrate storage and metabolism. *Aust. J. Plant Physiol.* **1994**, *21*, 829–841. [CrossRef]

21. Abe, A.; Menkir, A.; Moose, S.P.; Adetimirin, V.O.; Olaniyan, A.B. Genetic variation for nitrogen-use efficiency among selected tropical maize hybrids differing in grain yield potential. *J. Crop Improv.* **2013**, *27*, 31–52. [CrossRef]

22. Blackmer, A.M.; Pottker, D.; Cerrato, M.E.; Webb, J. Correlations between soil nitrate concentrations in late spring and corn yields in Iowa. *J. Prod. Agric.* **1989**, *2*, 103–109. [CrossRef]

23. Jeong, H.; Bhattarai, R. Exploring the effects of nitrogen fertilization management alternatives on nitrate loss and crop yields in tile-drained fields in Illinois. *J. Environ. Manag.* **2018**, *213*, 341–352. [CrossRef]

24. Mastrodomenico, A.T.; Hendrix, C.C.; Below, F.E. Nitrogen use efficiency and the genetic variation of maize expired plant variety protection germplasm. *Agriculture* **2018**, *8*, 3. [CrossRef]

25. Seebauer, J.R.; Moose, S.P.; Fabbri, B.J.; Crossland, L.D.; Below, F.E. Amino acid metabolism in maize earshoots. Implications for assimilate preconditioning and nitrogen signaling. *Plant Physiol.* **2004**, *136*, 4326–4334. [CrossRef]

26. Archontoulis, S.V.; Miguez, F.E.; Moore, K.J. Evaluating APSIM maize, soil water, soil nitrogen, manure, and soil temperature modules in the midwestern United States. *Agron. J.* **2014**, *106*, 1025–1040. [CrossRef]

27. Yang, X.L.; Lu, Y.L.; Tong, Y.A.; Yin, X.F. A 5-year lysimeter monitoring of nitrate leaching from wheat-maize rotation system: Comparison between optimum N fertilization and conventional farmer N fertilization. *Agric. Ecosyst. Environ.* **2015**, *199*, 34–42. [CrossRef]

28. Troyer, A.F. Adaptedness and heterosis in corn and mule hybrids. *Crop Sci.* **2006**, *46*, 528–543. [CrossRef]

29. Illinois Crop Improvement Association. Grain Laboratory Services. Available online: https://www.ilcrop.com/labservices/grainservices (accessed on 20 January 2018).

30. USDA NASS. Quick Stats. Available online: https://quickstats.nass.usda.gov/ (accessed on 1 August 2018).

31. Becker, R.A.; Wilks, A.R.; Brownrigg, R.; Minka, T.P.; Deckmyn, A. Maps: Draw Geographical Maps. R Package Version 3.3.0. Available online: https://CRAN.R-project.org/package=maps (accessed on 15 August 2018).

32. Chapman, K.; McGuire, J. NutrientStar TED Framework Tool. Available online: http://nutrientstar.org/ted-framework/ (accessed on 31 July 2018).

33. USDA NASS. County Data Frequently Asked Questions. Available online: https://www.nass.usda.gov/Data_and_Statistics/County_Data_Files/Frequently_Asked_Questions/index.php# (accessed on 20 July 2018).

34. USDA NASS. Charts and Maps. Available online: https://www.nass.usda.gov/Charts_and_Maps/Crops_County/boundary_maps/indexgif.php (accessed on 20 July 2018).

35. Midwestern Regional Climate Center. Cli-Mate: Daily County Data between Two Dates. Available online: https://mrcc.illinois.edu/CLIMATE/ (accessed on 20 July 2018).
36. Johnson, D.E. Principal component analysis. In *Applied Multivariate Methods for Data Analysts*; Brooks/Cole Publishing Company: Pacific Grove, CA, USA, 1998; pp. 110–111.
37. Butts-Wilmsmeyer, C.J.; Mumm, R.H.; Rausch, K.; Kandhola, G.; Yana, N.A.; Happ, M.M.; Ostezan, A.; Wasmund, M.; Bohn, M.O. Changes in phenolic acid content in maize during food product processing. *J. Agric. Food Chem.* **2018**, *66*, 3378–3385. [CrossRef] [PubMed]
38. Labate, J.A.; Lamkey, K.R.; Mitchell, S.E.; Kresovich, S.; Sullivan, H.; Smith, J.S.C. Molecular and historical aspects of corn belt dent diversity. *Crop Sci.* **2003**, *43*, 80–91. [CrossRef]
39. Li, P.X.P.; Hardacre, A.K.; Campanella, O.H.; Kirkpatrick, K.J. Determination of endosperm characteristics of 38 corn hybrids using the Stenvert Hardness test. *Cereal Chem.* **1996**, *73*, 466–471.
40. Brown, W.L.; Zuber, M.S.; Darrah, L.L.; Glover, D.V. Origin, adaptation, and types of corn. In *National Corn Handbook*; Cooperative Extension Service, Iowa State Univ.: Ames, IA, USA, 1985; pp. 1–6.
41. Gayral, M.; Gaillard, C.; Bakan, B.; Dalgalarrondo, M.; Elmorjani, K.; Delluc, C.; Brunet, S.; Linossier, L.; Morel, M.H.; Marion, D. Transition from vitreous to floury endosperm in maize (*Zea mays* L.) kernels is related to protein and starch gradients. *J. Cereal Sci.* **2016**, *68*, 148–154. [CrossRef]
42. Kereliuk, G.R.; Sosulski, F.W. Properties of corn samples varying in percentage of dent and flint kernels. *Food Sci. Technol. Leb.-Wiss. Technol.* **1995**, *28*, 589–597. [CrossRef]
43. Kutner, M.H.; Nachtsheim, C.J.; Neter, J.; Li, W. Applied linear statistical models. In *Applied Linear Statistical Models*, 5th ed.; Gordon, B., Hercher, R.T., Eds.; McGraw-Hill: New York, NY, USA, 2005.
44. Dinnes, D.L.; Karlen, D.L.; Jaynes, D.B.; Kaspar, T.C.; Hatfield, J.L.; Colvin, T.S.; Cambardella, C.A. Nitrogen management strategies to reduce nitrate leaching in tile-drained midwestern soils. *Agron. J.* **2002**, *94*, 153–171. [CrossRef]
45. Dwyer, L.M.; Stewart, D.W.; Tollenaar, M. Analysis of maize leaf photosynthesis under drought stress. *Can. J. Plant Sci.* **1992**, *72*, 477–481. [CrossRef]
46. Lobell, D.B.; Field, C.B. Global scale climate—Crop yield relationships and the impacts of recent warming. *Environ. Res. Lett.* **2007**, *2*, 014002. [CrossRef]
47. Nielson, B.L. Warm Nights & High Yields of Corn: Oil & Water? Available online: https://www.agry.purdue.edu/ext/corn/news/timeless/WarmNights.html (accessed on 22 August 2018).
48. Below, F.E.; Cazetta, J.O.; Seebauer, J.R. Carbon/nitrogen interactions during ear and kernel development of maize. In *Physiology and Modeling Kernel Set in Maize*; Westgate, M., Boote, K., Eds.; Crop Science Society of America and American Society of Agronomy: Madison, WI, USA, 2000; pp. 15–24. [CrossRef]
49. Mayer, L.I.; Edreira, J.I.R.; Maddonni, G.A. Oil yield components of maize crops exposed to heat stress during early and late grain-filling stages. *Crop Sci.* **2014**, *54*, 2236–2250. [CrossRef]
50. Carter, E.K.; Melkonian, J.; Riha, S.J.; Shaw, S.B. Separating heat stress from moisture stress: Analyzing yield response to high temperature in irrigated maize. *Environ. Res. Lett.* **2016**, *11*, 094012. [CrossRef]
51. Glelck, P.; Heberger, M. American Rivers: A Graphic. Available online: http://pacinst.org/american-rivers-a-graphic/ (accessed on 24 October 2018).

agronomy

MDPI

Article

Carbohydrate Dynamics in Maize Leaves and Developing Ears in Response to Nitrogen Application

Peng Ning [1,2] , Yunfeng Peng [2,3] and Felix B. Fritschi [2,*]

[1] College of Natural Resources and Environment, Northwest A&F University, Yangling 712100, Shaanxi, China; ningp@nwafu.edu.cn

[2] Division of Plant Sciences 1-31 Agriculture Building, University of Missouri, Columbia, MO 65211, USA; pengyf@ibcas.ac.cn

[3] Key Laboratory of Vegetation and Environmental Change, Institute of Botany, The Chinese Academy of Sciences, Beijing 100093, China

* Correspondence: fritschif@missouri.edu; Tel.: +1-573-882-3023

Received: 4 November 2018; Accepted: 12 December 2018; Published: 15 December 2018

Abstract: Maize grain yield is considered to be highly associated with ear and leaf carbohydrate dynamics during the critical period bracketing silking and during the fast grain filling phase. However, a full understanding of how differences in N availability/plant N status influence carbohydrate dynamics and processes underlying yield formation remains elusive. Two field experiments were conducted to examine maize ear development, grain yield and the dynamics of carbohydrates in maize ear leaves and developing ears in response to differences in N availability. Increasing N availability stimulated ear growth during the critical two weeks bracketing silking and during the fast grain-filling phase, consequently resulting in greater maize grain yield. In ear leaves, sucrose and starch concentrations exhibited an obvious diurnal pattern at both silking and 20 days after silking, and N fertilization led to more carbon flux to sucrose biosynthesis than to starch accumulation. The elevated transcript abundance of key genes involved in starch biosynthesis and maltose export, as well as the sugar transporters (SWEETs) important for phloem loading, indicated greater starch turnover and sucrose export from leaves under N-fertilized conditions. In developing ears, N fertilization likely enhanced the cleavage of sucrose to glucose and fructose in the cob prior to and at silking and the synthesis from glucose and fructose to sucrose in the kernels after silking, and thus increasing kernel setting and filling. At the end, we propose a source-sink carbon partitioning framework to illustrates how N application influences carbon assimilation in leaves, transport, and conversions in developing reproductive tissues, ultimately leading to greater yield.

Keywords: grain yield; nitrogen; nonstructural soluble carbohydrate (NSC); starch; sucrose; *Zea mays* L.

---

1. Introduction

Maize (*Zea mays* L.) is one of the world major crops and an important food source for the growing population [1]. Generally, maize yields are strongly correlated with amounts of nitrogen (N) fertilizer applied to the crop. The grain yield of maize is considered to be greatly dependent on ear growth during the critical period bracketing silking [2,3], that is, around 1 week before to 2 weeks after silking [4–6]. At this stage, carbon assimilation in leaves, photoassimilate (mainly sucrose) transport in the phloem, and carbohydrate interconversion and transport between the maternal tissues and developing kernels are important steps which determine grain filling and yield formation [7,8]. Therefore, a better understanding of N effects on carbohydrate dynamics in leaves and the developing ear is pivotal to improve maize N use efficiency as well as yields under N-deficient conditions.

Over the past decades, numerous studies have been conducted to examine the effects of N supply on leaf growth and photosynthetic capacity [9–12], leaf carbohydrate concentrations [13–15], sucrose loading and transport in the phloem [8], and ear growth and carbohydrate dynamics [13–16], and how these factors influence yield and yield components. For instance, increased grain yield in response to N application was associated with greater leaf area [9–11], ear dry weight (DW) [13–15], and developing kernel number per ear [17]. Enhanced maize yield with N fertilization has been positively associated with increased leaf total and soluble protein concentrations [9,14,18,19], as well as kernel soluble protein [16,20] and starch concentrations [16,21], but yield improvement was negatively related to increasing soluble carbohydrate concentrations in the developing kernels [16,21]. Although these studies have advanced our knowledge of the mechanism underlying yield increases in maize in response to N applications, questions remain about how N status influences the carbohydrate dynamics in leaves, sucrose transport from source to developing sink and carbohydrate conversion in the maternal tissues and growing kernels, and their correlations with yield production.

To address some of these questions, we conducted two field experiments to examine carbohydrate dynamics at early stages of ear development and the fast grain filling phase in response to three N application levels (zero-N, medium and high N levels). In Experiment I, we measured ear growth and nonstructural soluble carbohydrate (NSC) dynamics in developing ears from 10 days before silking (DBS) to physiological maturity. In Experiment II, we examined the diurnal dynamics of the NSC in maize ear-leaves at silking and 20 days after silking (DAS), and the associated expression of genes involved in sucrose export and starch turnover. Based on these results, we present a framework of how N application affects leaf carbohydrate assimilation, sugar export from leaves, and carbohydrate conversion and transport in maternal tissues and developing kernels, ultimately leading to enhanced grain yield.

2. Materials and Methods

2.1. Site Description

Field experiments were performed at the University of Missouri Bradford Research Center, Columbia, MO, USA (38°53′ N; 92°12′ W) in 2010–2011 and 2015 on a Mexico silt loam (Fine, smectitic, mesic Aeric Vertic Epiaqualf) soil. Experiments in the three years were performed in different field plots. The soil chemical properties in the three years were as follows: pH 6.7, 6.6 and 6.5, organic matter 21, 28 and 31 g kg^{-1}, total-N 1.3 g kg^{-1} (2010) and 1.6 g kg^{-1} (2011), mineral N 12 mg kg^{-1} (2015), Bray-I P 11.5, 14.5 and 6.4 mg kg^{-1}, NH_4OAc extractable K 124, 93 and 78 mg kg^{-1}, respectively. Precipitation during the maize growing season in May, June, July, August and September was 108, 84, 204, 105 and 176 mm in 2010, 136, 83, 59, 61 and 46 mm in 2011, and 140, 129, 204, 106 and 21 mm in 2015, respectively.

Prior to sowing, fields were disked to approximately 0.15 m depth, followed by a single pass with a harrow. Maize hybrid 'Pioneer 32D79' was selected for the study and was sown in rows 0.76 m apart on 25 May 2010, 9 May 2011, and 8 May 2015 to achieve a stand density of 78,000, 72,000, and 78,000 plants ha^{-1}, respectively. A randomized complete block design with four replications was used each year. Weeds were controlled by pre-emergence herbicide application (Atrazine plus S-metolachlor) followed by manually hoeing as needed. Other aspects specific to each of the two experiments are described below.

2.2. Experiment I

Phosphorus and K fertilizer was applied before sowing in 2010 (105 kg ha^{-1} P_2O_5 and 105 kg ha^{-1} K_2O) and in 2011 (85 kg ha^{-1} P_2O_5 and 110 kg ha^{-1} K_2O). Three N treatments were imposed in 2010 and 2011, i.e., (i) N0, 0 kg N ha^{-1} as control; (ii) N150, 150 kg N ha^{-1} applied at planting (N150); and (iii) N300, 150 kg N ha^{-1} applied at planting, 75 kg N ha^{-1} applied at V6 (6th leaf fully expanded), and 75 kg N ha^{-1} applied at V10 (10th leaf fully expanded). Nitrogen was broadcast applied as urea with an over-the-shoulder broadcast spreader. Individual plots were 9.14 m long and 6.1 m wide in both years.

Developing ears were sampled at 10 and 6 days before silking (DBS), silking, 8 days after silking (DAS) and 63 DAS (R6) in 2010, and 6 DBS, silking, 8, 18, and 57 DAS (R6) in 2011. All plant samplings were performed between 11:00 AM and 1:00 PM. At each sampling, whole ears and ear leaves of eight consecutive plants in a row were removed from each plot and four of them were immediately flash-frozen in liquid N and stored at −80 °C. Ear DW was determined based on the remaining four ears. Ears were separated into cobs and kernels (after silking). Prior to carbohydrate analyses, frozen samples were lyophilized and ground with a coffee grinder and mortar and pestle to obtain a fine powder for carbohydrate assay.

The NSC (glucose, fructose, sucrose) and starch concentrations were assayed as previously described [14,15,22]. Briefly, approximately 30 mg freeze-dried and ground tissue was extracted three times with 1 ml 80% EtOH at 80 °C in a water bath for 15 min. The three supernatants were combined in a test tube and brought to a final volume of 10 ml with 80% (v/v) EtOH. A clear aliquot of 25 µL of supernatant was mixed with 250 µL glucose assay kit (Sigma G3293, Sigma-Aldrich, St. Louis, MO, USA) (a mixture of adenosine tri-phosphate, oxidized nicotinamide adenine dinucleotide, hexokinase, andglucose-6P dehydrogenase), and incubated for 15 min at 30 °C, and absorbance at 340 nm was read with a spectrophotometer (GENESYS 10 UV, Thermo Electron Corporation, Madison, WI, USA) to quantify glucose. Fructose was quantified by adding phosphoglucose isomerase (Sigma P9544) to the above aliquots, and determining the amount of glucose released. Finally, invertase (Sigma I4504) was added to obtain the combined concentration of glucose + fructose + sucrose. The pellet remaining after the above extraction was used to determine starch by enzymatic hydrolysis and subsequent quantification of the amount of glucose released [14,15,22]. Alpha-amylase (Sigma-Aldrich, A3403, Sigma-Aldrich, St. Louis, MO, USA) and amyloglucosidase (Sigma-Aldrich, A3042, Sigma-Aldrich, St. Louis, MO, USA) were used to hydrolyze the pellets. The supernatant from enzymatic hydrolysate was assayed for glucose as described above, and starch was calculated from the glucose concentration in the tissue residue.

Maize grain yield was determined each year by hand-harvesting and weighing ears from the central 9.12 m^2 of each plot at physiological maturity. Six randomly selected ears from each plot were shelled and moisture content and kernel percentage were determined. Grain yield per plot was estimated based on kernel percentage and expressed at 15.5% moisture content.

2.3. Experiment II

In 2015, three N levels were included as follows: (i) N0, no N fertilizer as control; (ii) N200, 200 kg N ha^{-1} split equally at emergence and V8 stages (because yield increase in response to the N150 treatments in Experiment I were modest, 200 kg N ha^{-1} was applied instead in 2015); and (iii) N300, 200 kg N applied as described for the N200 treatment, and an additional 100 kg N ha^{-1} applied one week before silking. Each plot consisted of eight 6.1 m long rows, and N was applied manually in the form of urea. Ear leaves from three plants were sampled every 4 h starting at 08.00 h at silking and 20 DAS. The middle (~10 cm) of the leaf blade was dissected to remove the midrib, and the leaf blade tissues from the two sides of the midrib were separated into two groups and immediately immersed into liquid N_2. One part was stored at −80 °C and destined for gene expression analysis, and the other one was lyophilized for carbohydrate analysis.

The freeze-dried ear leaves were ground using a coffee grinder and further processed with a Geno/Grinder 2010 (SPEX SamplePrep, Metuchen, NJ, USA). Sugars (glucose, fructose, and sucrose) were extracted and starch solubilized and converted to glucose as described for Experiment I, but quantified by HPLC (Shimadzu Corp., Kyoto, Japan) with a refractive index detector (Model RID-10A) as previously described [8].

Ear leaf samples from the collections at 12:00 at silking and 20 DAS were selected to investigate the expression pattern of genes involved in sucrose and starch metabolism as described previously [8]. Briefly, frozen leaves were ground in liquid N_2, and total RNA was extracted using a RNeasy Plant Mini kit (QIAGEN, Hilden, Germany). Reverse transcription was performed according to the manufacturer's instructions. Quantitative real-time PCR was conducted with *β-Tubulin* as a reference gene [23] on an ABI 7500 real-time PCR system with universal cycling conditions (95 °C for 2 min, 40 cycles of 95 °C for 15 s, and 60 °C for 1 min) following the manufacturer's instructions. Primers were designed as follows: *Tpt* (forward, 5′-GGA GAA AAG GAA AAA GAG CGC AT-3′; reverse, 5′-ACG ATG TTG TTG TTG TGT CCC-3′); *Agpsl1* (forward, 5′-TGG GAC AGT TAT ATG AAG CGG A-3′; reverse, 5′-TCC ATG ATC TCC AGC ACA CTG-3′); *Agpll1* (forward, 5′-TCC TAA ACC TTC TAA GAT GGC G-3′; reverse, 5′-AAA CTG AAC CTT GGA GGC TGT-3′); *Ss1* (forward, 5′-GCT ATT GGC TCC ATT GCT CC-3′; reverse, 5′-TTC AAA ACT GTC GTT CGG CT-3′); *Bmy3* (forward, 5′-CCC ATG AGG ACG ACC TGC CA-3′; reverse, 5′-TTT ATC ACC CGC CCG TTT ATT TTG-3′); *Mex1-like* (forward, 5′-GGT TGG ACA GCC ACA CTT CT-3′; reverse, 5′-TGC AGA ACC AGT GAA CCA CA-3′); *Sps1* (forward, 5′-CGT TCC TCA TCA AAG ACC CCC-3′; reverse, 5′-ACG GAA AGA TAC CTG AGT GCC T-3′); *Sus2* (forward, 5′-CCA ACC GCA GTA GTA ATG GC-3′; reverse, 5′-CGG CTT GCC AGC AAA GAA AT-3′); *Sweet13a* (forward, 5′-CTG GGC GTT TGC TTT CG-3′; reverse, 5′-ACT TGC TCT TGT AGA TGC GGT A-3′); *Sweet13b* (forward, 5′-TGC GTA CTG CGT AGT TCC AT-3′; reverse, 5′-GGA GAT GAC GTT GCC TAG GAG-3′); *Sweet13c* (forward, 5′-CAA GAG TTT GAG ACA GCA GAG G-3′; reverse, 5′-CCA GGA AGG TCA TGA AGG AG-3′); *Sut1* (forward, 5′-GGC ACA AGT GGT TTC CGT TC-3′; reverse, 5′-TTT GCC TTT GTG GGG AGG TT-3′); *Sus1* (forward, 5′-CGT ACA CCG AGT CGC ACA AGA G-3′; reverse, 5′-TCC ACC AGC CCA GTC AAG TTC T-3′); β-Tubulin (forward, 5′-CTA CCT CAC GGC ATC TGC TAT GT-3′; reverse, 5′-GTC ACA CAC ACT CGA CTT CAC G-3′). Relative expression values were calculated according to the Pfaffl method [24], and expressed as ratios relative to values of normal N supply (N$_{200}$) at time 00:00 h at silking as shown in [8].

2.4. Statistical Analysis

Analysis of variance was employed to examine N treatment effects and treatment means were compared based on least significant difference (LSD) at the 0.05 level of probability. The ANOVA of repeated measurements presented in Figure 2 and Pearson correlations presented in Figure 5 were performed in R. For gene expression in Figure 3, the effects of N treatment and stage (and their interaction) were analyzed with the MIXED procedure in SAS (SAS Institute Inc., Cary, NC, USA). The stage and N treatment factors were considered as fixed and replication as random factors.

3. Results

3.1. Dry Matter Accumulation in Developing Ear (or Kernels) and Final Grain Yield

To provide context for the NCS and gene expression data, grain yield data are presented in Figure 1 (adapted from [8] and [13]). Briefly, compared to the N0 and N150 treatments, the final grain yield of the N300 treatment was increased by 2.1–4.3 fold in 2010 and by 1.2–2.0 fold in 2011, respectively, and significant differences were detected between N0 and N150 in 2010 (Figure 1). Similarly, in 2015, 1.3–1.8 fold increase of grain yield in N200 and N300 treatments was observed in comparison with N0. A 19.3% lower yield was observed in N300 relative to N200 in 2015.

In agreement with the grain yield, N application resulted in greater ear dry matter accumulation from pre-silking to maturity (Table 1). Significant N treatment effects on ear dry weight were observed at −6, 0, and 8 DAS. Particularly during the fast grain filling phase (i.e., from 18 to 57 DAS), dry matter accumulation rate in developing ears was much greater in N300 treatment than N0 and N150 (1.76 vs. 0.57–0.98 g plant^{-1} day^{-1}) in 2011. The accumulation rate under 200 and 300 kg N ha^{-1} was also markedly higher than that in the N0 control (3.13–3.86 vs. 1.82 g plant^{-1} day^{-1}) from 20 to 51 DAS in 2015. Compared to normal N supply (N200), the N300 treatment had 17–18% lower ear dry weight at 20 and 51 DAS in 2015.

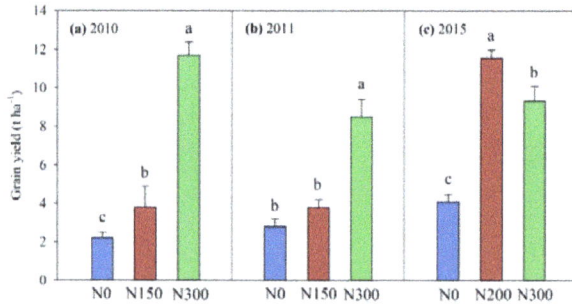

Figure 1. Maize grain yield in response to different nitrogen treatments in 2010 (**a**), 2011 (**b**), and 2015 (**c**) (adapted from [8] and [13]). N0, N150, N200 and N300 represent N treatments that received a total of either 0, 150, 200, or 300 kg N ha^{-1}. Nitrogen was applied as single application (N150) or as split applications (N200, N300). Vertical bars represent the standard error of four replications. Different letters above column indicate significant differences between nitrogen levels by LSD test ($p < 0.05$).

Table 1. Ear dry weight in response to N treatment in 2010, 2011, and 2015. N0, N150, N200, and N300 represent N treatments that received a total of either 0, 150, 200, or 300 kg N ha^{-1} in the form of single or split fertilizer applications.

Year	Treatment	Days after Silking				
		-10	-6	0	8	63
2010	N0	n.d.	0.1 ± 0.0 b	1.6 ± 0.4 b	4.6 ± 0.4 b	-	37.7 ± 3.9 b
	N150	0.1 ± 0.06 b	0.4 ± 0.2 ab	3.0 ± 1.6 b	8.9 ± 2.5 b	-	56.2 ± 8.2 b
	N300	0.21 ± 0.08 a	1.3 ± 0.4 a	11.6 ± 2.7 a	38.0 ± 4.4 a	-	175.4 ± 5.5 a
			-6	0	8	18	57
2011	N0	-	0.6 ± 0.0 b	1.1 ± 0.0 b	4.9 ± 0.9 c	7.9 ± 1.4 c	46.2 ± 6.0 b
	N150	-	0.9 ± 0.1 b	2.2 ± 0.4 b	13.4 ± 1.3 b	34.4 ± 2.5 b	56.7 ± 11.0 b
	N300	-	1.5 ± 0.2 a	5.1 ± 0.7 a	35.4 ± 2.4 a	84.2 ± 8.9 a	152.9 ± 6.3 a
				0		20	51
2015	N0	-	-	0.9 ± 0.1 b	-	13.4 ± 1.8 c	70.0 ± 6.2 c
	N200	-	-	2.2 ± 0.6 a	-	71.3 ± 2.9 a	191.0 ± 7.2 a
	N300	-	-	2.1 ± 0.5 a	-	59.2 ± 6.4 b	156.1 ± 13.0 b

N0, N150, N200 and N300 represent N treatments that received a total of either 0, 150, 200, or 300 kg N ha^{-1}. Nitrogen was applied as single application (N150) or as split applications (N200, N300). Values indicate mean $\pm$ standard error ($n = 4$); Different letters within column within each stage and year indicate significant differences between nitrogen level by LSD test ($p < 0.05$). Unit: g per ear.

3.2. Diurnal Changes of Carbohydrate in Ear Leaves

Pronounced diurnal pattern were observed for sucrose and starch concentrations in ear leaves at both silking and 20 DAS, while glucose and fructose concentrations changed to a lesser extent or did not change significantly during the 24-h periods (Figure 2). Sucrose levels reached the peak at 12:00 to 16:00 and then decreased, while starch accumulated over the course of the day to reach a maximum at 20:00, and was degraded during the night (Figure 2e–h).

Concentrations of all four non-structural carbohydrates were highly responsive to N availability ($p < 0.001$; Figure 2). Nitrogen deficiency (N0) caused lower leaf concentrations of glucose, fructose, and sucrose compared to N fertilized treatments, with particularly strong effects observed for glucose (0 and 20 DAS) and for fructose at 20 DAS. In contrasting to the soluble sugars, starch accumulation was much greater in the N-deficient leaves than in those from N200 and N300 treatments, implying preferential allocation of photoassimilates to starch than to sugars under N starvation.

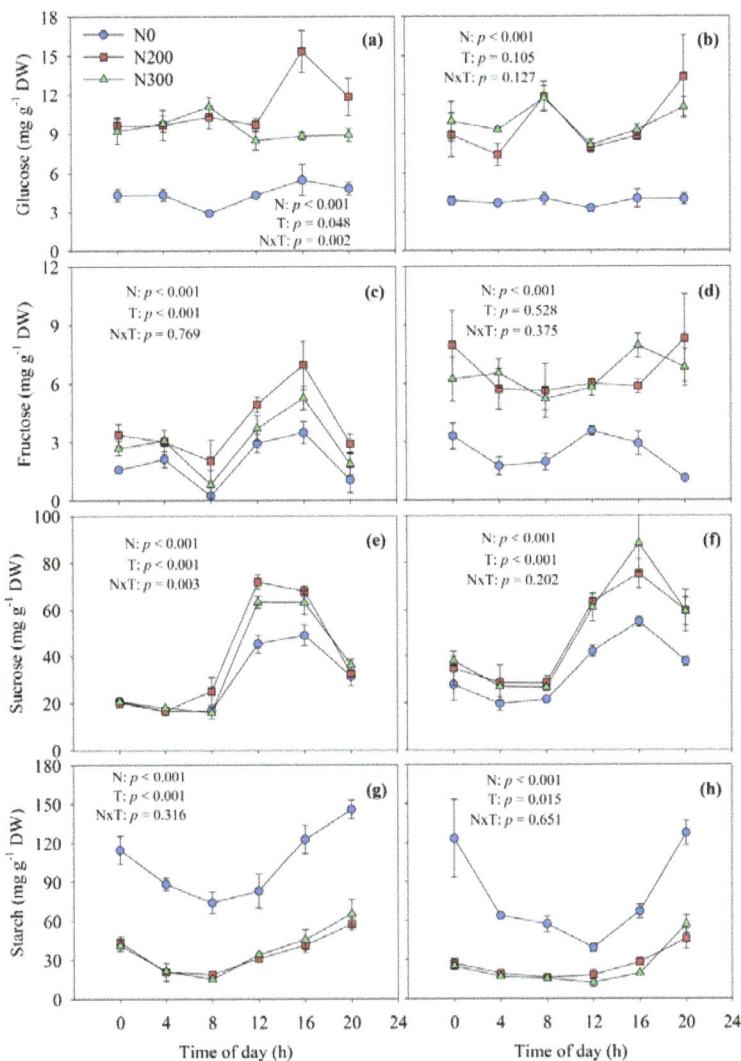

Figure 2. Diurnal dynamics of glucose, fructose, sucrose and starch concentrations in ear leaf at silking (**a,c,e,g**) and 20 days after silking (**b,d,f,h**) in 2015. Vertical bars represent the standard error of four replications. N0, N200, and N300 represent N treatments that received a total of either 0, 200, or 300 kg N ha^{-1} applied as split applications. N, N treatment; T, timepoint.

3.3. *Relative Transcript Abundance of Genes Involved in Sucrose and Starch Metabolism in the Ear Leaf*

Significant impacts of N availability on relative transcript abundance pattern were observed for all genes at silking or at 20 DAS or at both stages, except for *ZmSus2* (Figure 3). The transcript abundance of the phosphate/triose-phosphate translocator (*ZmTpt*) gene was lower in the N0 leaves than in the N200 and N300 leaves at both 0 and 20 DAS (Figure 3a), hinting that greater starch concentrations could in part be a result of reduced triose-phosphate export from the chloroplast. In agreement with the lower sucrose levels in N-deficient leaves, the relative transcript abundance of sucrose-phosphate synthase (*ZmSps1*) in N0 leaves was suppressed when compared to the N-fertilized treatments (Figure 3b). In contrast, the relative abundance of sucrose synthase (*ZmSus2*) transcripts was not significantly affected by N treatment (Figure 3c). Of the three *Sweet13* genes (responsible for sucrose export to apoplast), the expression levels of *Sweet13a* and *Sweet13b* were elevated in N-fertilized treatments at 20 DAS and relative transcript abundance of *Sweet13c* was higher in N300 than N0 at both silking and 20 DAS (Figure 3d–f). Relative transcript abundance of the sucrose transporter (*ZmSut1*) responsible for sucrose loading into the phloem increased with increasing N availability at silking, but the differences between N treatments were not different at 20 DAS (Figure 3g).

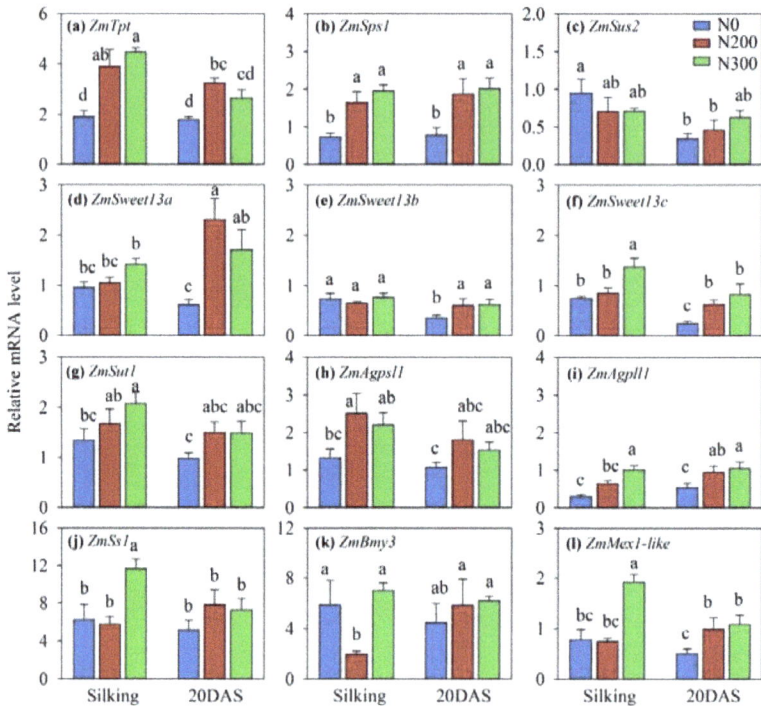

Figure 3. Relative transcript abundance of genes involved in the sucrose and starch metabolism in maize ear leaves sampled at 12:00 h at silking and 20 days after silking (DAS) in 2015. (**a**) *ZmTpt*, triose-phosphate translocator; (**b**) *ZmSps1*, sucrose-phosphate synthase1; (**c**) *ZmSus1*, sucrose synthase; (**d–f**) *ZmSweet13*, sugars will be eventually exported transporter 13; (**g**) *ZmSut1*, sucrose transporter1; (**h**) *ZmAgpsl1*, ADP glucose pyrophosphorylase small subunit leaf1; (**i**) *ZmAgpll1*, ADP glucose pyrophosphorylase large subunit leaf1; (**j**) *ZmSs1*, starch synthase1; (**k**) *ZmBmy3*, beta-amylase3; (**l**) *ZmMex1-like*, maltose exporter1-like; Vertical bars represent the standard error of four replicates. Different letters above column indicate significant differences between treatments across nitrogen level and stage by LSD test ($p < 0.05$). N0, N200, and N300 represent N treatments that received a total of either 0, 200, or 300 kg N ha^{-1} applied as split applications. Adapted from [8].

Of the genes encoding enzymes of starch biosynthesis, significant treatment effects were observed between N0 and N200 at silking for *ZmAgpsl1* (Figure 3h). Additionally, the relative transcript abundance of *ZmAgpll1* increased significantly in N-fertilized treatments at both silking and 20 DAS, while that of *ZmSs1* was significantly higher in the N300 compared to the N0 treatment at silking but not at 20 DAS (Figure 3i,j). Relative transcript level abundance of beta-amylase (*ZmBmy3*) exhibited an inconsistent response to N treatment at silking and did not differ at 20 DAS (Figure 3k). In contrast, the relative RNA levels of the maltose exporter gene (*ZmMex1-like*) were greater in the N300 than the N0 treatment at both silking and 20 DAS (Figure 3l).

3.4. NSC Dynamics in Developing Ears and Kernels

Over the course of ear development, the glucose, fructose, sucrose and starch concentration trends were consistent across N treatments and years (Figure 4). While glucose and fructose concentrations increased (Figure 4a–d), the concentrations of sucrose and starch decreased as the developing ear approached silking (Figure 4e–f). However, while glucose, fructose and sucrose concentrations were greater, starch concentrations were lower in kernels sampled 8 DAS than in mature kernels. Glucose and fructose concentrations in developing kernels decreased dramatically from 8 DAS to 18 DAS, and coincided with a steep increase in starch concentration during the same time period. Significantly negative relationships were observed between the hexoses and sucrose or starch (Figure 5b,c).

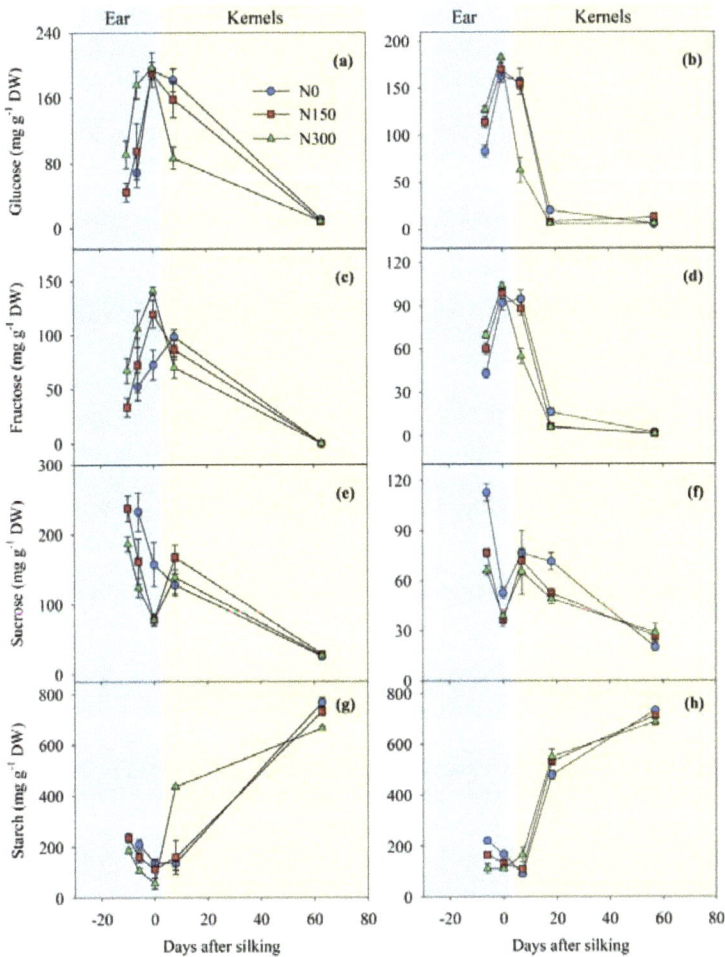

Figure 4. Temporal dynamics of glucose, fructose, sucrose and starch concentrations in the developing ear and kernels in 2010 (**a,c,e,g**) and 2011 (**b,d,f,h**). Day 0 indicates silking. Vertical bars represent the standard error of four replicates. Please note that carbohydrate concentrations were measured in the developing ears before and at silking, and in the kernels after silking. N0, N150, and N300 represent N treatments that received a total of either 0, 150, or 300 kg N ha^{-1}, with fertilizer in the N300 treatment applied as split application.

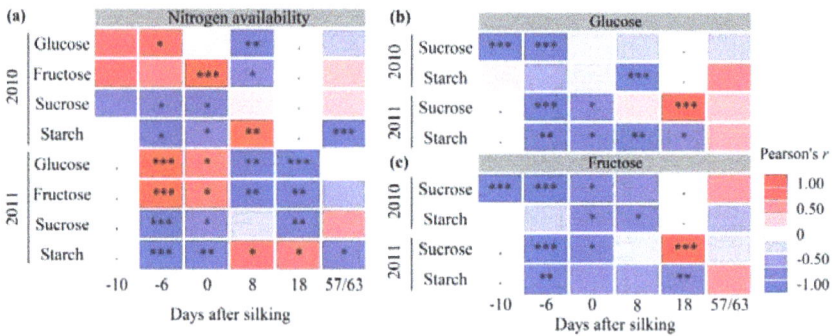

Figure 5. Pearson correlation coefficients between soil N availability and carbohydrate (**a**), between glucose and sucrose or starch (**b**), and between fructose and sucrose or starch (**c**) in developing ear in 2010 and 2011. Dots = not measured. *, $p < 0.05$; **, $p < 0.01$; ***, $p < 0.001$.

Prior to silking, the responses of glucose and fructose concentrations in developing ears to N supply were opposite to those of sucrose and starch. That is, while sucrose and starch concentrations in developing ears decreased with increasing N supply, glucose and fructose concentrations increased with additional N (all $p < 0.05$; Figure 4), which were consistent with the positive response to N supply (Figure 5a). However, in the growing kernels at 8 DAS, glucose and fructose concentrations in the N300 treatments were lower than in the N0 and N150 treatments (both $p < 0.05$). These results reveal a tendency towards an inverse relationship among N treatments and the concentrations of glucose and fructose when compared to pre-silking dynamics. Few differences in sucrose concentrations of the developing and mature kernels were found among N treatments in either year. In 2010, starch concentrations in kernels at 8 DAS were greater in the N300 treatment than the N0 and the N150 treatments, but at maturity, kernel starch concentration were lowest in N300, intermediate in N150, and highest in N0 ($p < 0.05$). In 2011, N fertilized treatments also had greater kernel starch concentrations in developing kernels ($p < 0.05$), and consistent with 2010, kernel starch concentration decreased with increasing N availability at the final harvest (N300 < N150 < N0; $p < 0.05$). In both years, kernel starch concentration at maturity was negatively correlated with N availability (Figure 5a).

4. Discussion

4.1. Nitrogen Application Increases Sugar Biosynthesis and Export from Leaves

Maize grain yield greatly depends on post-silking carbon assimilation in source leaves and allocation to developing ears [25–27]. It has been well documented that N starvation can dramatically suppress carbohydrate biosynthesis through leaf area development and photosynthesis [9,11,12]. At the same time, under N deficiency, leaves accumulate large amounts of carbohydrate as starch in the bundle sheath cells, at the expense of sugar export for growth [8,14,15,28,29]. Therefore, it is crucial to investigate the diurnal pattern of sugar and starch in leaves in response to N availability, particularly at the early and fast grain filling phase.

Of the four types of carbohydrate examined in this study, ear leaf glucose and fructose exhibited relatively weaker diurnal pattern, and largely much lower concentrations, than sucrose and starch at both silking and 20 DAS (Figure 2a–f), which was in agreement with a previous report [14]. The maximum sucrose concentration in ear leaves during the diurnal cycle occurred about 12:00 h to 16:00 h while starch accumulated almost linearly from morning until 20:00 h, regardless of N treatment (Figure 2e–h). The low levels and weak diurnal patterns in glucose and fructose are consistent with rapid assimilation of hexoses into sucrose and starch over the course of the day.

As expected, compared to N deficiency, N fertilization stimulated glucose, fructose, and sucrose biosynthesis as indicated by the greater ear leaf concentrations in N200 and N300 treatments (Figure 2).

Coupled with the lower starch concentrations in response to N fertilization, these results indicate a noticeable change in carbon allocation, including carbon flux into starch and sucrose, as a function of plant N status [8]. Triose-phosphate/phosphate translocators located at the inner envelope membrane of chloroplasts are key regulators of carbon flux from chloroplast to cytosol [30,31]. The increased transcript abundance of *ZmTpt* in N fertilized plants suggests is consistent with greater carbon movement to the cytosol for sucrose synthesis, and reduced carbon retention in the chloroplast for starch synthesis. In parallel, the expression of the gene encoding sucrose-phosphate synthase (*ZmSps1*) was up-regulated by N fertilization. Interestingly, unlike in other plants such as duckweed [28], the expression of some of the genes related to starch biosynthesis was upregulated in response to high N fertilization, which stands in contrast to the lower starch concentrations in N-sufficient leaves (Figure 2h–j). However, the transcript abundance of the maltose exporter (*ZmMex1-like*) was elevated by N fertilization, pointing toward greater carbon mobilization from starch and export from the chloroplast. As a storage carbohydrate, starch does not demonstrate regulatory activities, but it is a major integrator in the regulation of plant growth and generally is negatively correlated to plant biomass [32]. The lower starch levels and greater mobilization under sufficient N supply suggests an enhanced starch turnover, which increases the carbon availability for export and biomass production.

Sucrose export from leaves involves symplastic transport (plasmodesmata between mesophyll cells, bundle sheath cells, and vascular parenchyma cells), and phloem loading from the apoplast [33,34]. Previously, it has been shown that both the N-deficient and N-sufficient plants had conspicuous and visually normal plasmodesmata appearances between different types of cells in leaves, indicating an open symplastic route [8]. As to the sucrose phloem loading, the process mainly involves sugar transporters including SWEETs which are responsible for the sucrose efflux from vascular parenchyma cells to the apoplast [35,36], and SUTs for the sucrose uptake from apoplastic space [37]. Of the three *Sweet13* genes examined, expression of *Sweet13a* and *Sweet13c* was responsive to N availability, with greater transcript abundance observed under sufficient N supply (Figure 3), which may indicate greater SWEET protein abundance and enhanced sucrose efflux into apoplastic space. Compared to silking, transcription of *Sweet13* genes was more responsive to N availability at 20 DAS when fast grain filling occurs, suggesting a regulatory mechanism of *Sweet13* expression that accounts for carbon demand. In addition, *ZmSut1* transcript abundance increased in response to N fertilization (Figure 3g). Taken together, the results suggest that, despite reduced photosynthetic rates under N deficiency [14], sugar export from N-deficient ear leaves is impaired even during the grain filling phase.

4.2. Nitrogen Application Influences Carbohydrate Interconversions in Developing Ears

Concentrations of the four examined carbohydrates in ears pre-silking and kernels post-silking differed considerably in response to N applications (Figure 4). Opposing trends in glucose and fructose compared to sucrose and starch concentrations at pre-silking to silking suggest that sucrose unloaded from the phloem may be cleaved to glucose and fructose in the cob prior to transport to the kernels. After moving to the kernels, the monosaccharides appear to be synthesized (transiently) into sucrose (sucrose concentration increased shortly after silking, Figure 4) and into starch. These findings are supported by the Shannon hypothesis that the ^{14}C labeled sucrose is cleaved to fructose and glucose by acid invertase in the maternal pedicel and placento-chalazal tissue of kernels before being absorbed by the basal endosperm transfer cells in maize. Fructose and glucose are then used to form sucrose in the endosperm which in turn is used to synthesize starch in the amyloplasts [7,38–40]. Nitrogen fertilization may enhance the cleavage of sucrose to glucose and fructose in the cob prior to and at silking and the synthesis from glucose and fructose to sucrose shortly after silking (Figure 4), and thus increase kernel setting and filling. The distinct functions of these carbohydrates in processes leading to kernel set and early kernel growth, as well as a regulatory role of N status, is possibly mediated by the ratio of key amino acids such as asparagine and glutamine [5].

Shortly after silking, maize kernel glucose, fructose and sucrose concentrations exhibited a generally negative relationship with increasing N application (Figure 4a–f). However, starch accumulation in the developing kernels was positively correlated with N applications at 8 DAS and 18 DAS (Figure 4g,h). These results are consistent with studies by others that showed that increasing N supplies decreased the levels of C metabolites (sucrose and reducing sugars) in the developing kernels cultured in vitro at 20 days after pollination [16], and greater starch accumulation in developing kernels at 14 DAS in high compared to low N supply [41]. These interactions of carbon and N metabolism largely may be regulated by enzymes such as invertase, sucrose synthase and ADP-glucose pyrophosphorylase (AGPase) [16,42]. For instance, external N supply can increase AGPase activity, a limiting step in starch synthesis [16]. In any case, future studies are necessary to elucidate the molecular processes regulating carbohydrate movement, conversion and deposition in developing ears in response to various N regimes.

4.3. Source-Sink Carbon Partitioning and Grain Yield

Based on plant growth, sugar and starch dynamics, and ear development measured in this study and in the literature, we propose a conceptual model of the effect of N fertilization on source-sink carbon partitioning leading to greater grain yield (Figure 6). Although a critical value may exist (between 150 kg to 200 kg ha^{-1} applied N in this 3-yr study) beyond which grain yield tends to decrease, N fertilization generally increases N availability and stimulates shoot growth, especially leaf area development [9,11]. The enhanced soil N availability also results in higher leaf N content, driven by increased N assimilation [18,19]. In turn, this brings about greater leaf photosynthetic capacity in N fertilized maize [9,11,12], which enhances primary carbohydrate assimilation, i.e., triose-products. The increased carbohydrate assimilation offers precursors or carbon skeletons for amino acid biosynthesis, which in turn favors photosynthesis. On the other hand, greater triose production supplies the substrates for both sucrose and starch biosynthesis [30], while N fertilization increases the triose-phosphate/phosphate translocator activity, as well as *ZmSps1*, thus promoting more carbon flux to cytosol for sucrose synthesis, and lower carbon retention in the chloroplast as starch. The increased *ZmMex1-like* expression may further accelerate transient starch turnover or mobilization, and stimulate sucrose synthesis. Upregulation of *Sweet* and *Sut1* expression would lead greater abundance of SWEET and SUT1 proteins in turn mediating enhanced sucrose export from source leaves via phloem. At the early stages of ear development, especially the two weeks bracketing silking [6], N application enhances the cleavage of sucrose to glucose and fructose in the cob prior to and at silking and the synthesis from glucose and fructose to sucrose in the kernels shortly after silking. During the fast grain filling phase, generally from 18 DAS on [5], increased conversion of sugars to starch indicates greater filling rate in N-fertilized vs. no N treated maize kernels. In turn, the greater carbon demand imposes a feedback regulation stimulating carbon assimilation and export from leaves (Figure 6).

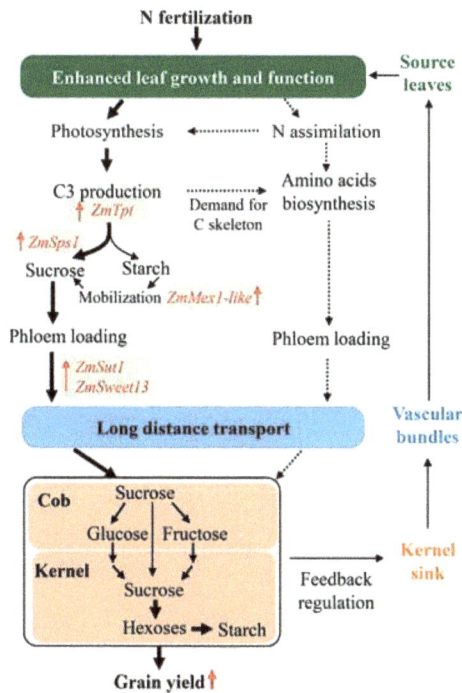

Figure 6. Schematic of a proposed metabolic pathway showing the source-sink carbon partitioning in determining the grain yield in maize under nitrogen fertilization.

5. Conclusions

In summary, greater leaf C assimilation and mobilization to the developing sink, and faster carbohydrate conversion in cobs and kernels as well as higher kernel starch accumulation around silking and during fast grain filling stages lead to the improvement of grain yield in N-fertilized maize plants. Further studies are needed to investigate the responses of key molecular players involved in sugar movement from maternal tissues to filial kernels to both low and high N availability, and their involvement in grain filling processes and yield formation.

Author Contributions: P.N., Y.P. and F.B.F. conceived and designed the experiments; P.N. and Y.P. performed the experiments, analyzed the data, and prepared the original draft. F.B.F. revised and edited the manuscript.

Funding: This research received no external funding.

Acknowledgments: The work in Northwest A&F University was supported by Natural Science Basic Research Plan in Shaanxi Province of China (No. 2018JQ3072), and the Fundamental Research Funds for the Central Universities (No. Z109021713).

References

1. Ort, D.R.; Long, S.P. Limits on yields in the Corn Belt. *Science* **2014**, *344*, 484–485. [CrossRef] [PubMed]
2. D Andrea, K.E.; Otegui, M.E.; Cirilo, A.G. Kernel number determination differs among maize hybrids in response to nitrogen. *Field Crop Res.* **2008**, *105*, 228–239. [CrossRef]
3. Echarte, L.; Andrade, F.H.; Vega, C.R.C.; Tollenaar, M. Kernel number determination in Argentinean maize hybrids released between 1965 and 1993. *Crop Sci.* **2004**, *44*, 1654–1661. [CrossRef]
4. Cantarero, M.G.; Cirilo, A.G.; Andrade, F.H. Night temperature at silking affects set in maize. *Crop Sci.* **1999**, *39*, 703–710. [CrossRef]

5. Seebauer, J.R.; Moose, S.P.; Fabbri, B.J.; Crossland, L.D.; Below, F.E. Amino acid metabolism in maize earshoots. Implications for assimilate preconditioning and nitrogen signaling. *Plant Physiol.* **2004**, *136*, 4326–4334. [CrossRef] [PubMed]

6. Hirel, B.; Le Gouis, J.; Ney, B.; Gallais, A. The challenge of improving nitrogen use efficiency in crop plants: towards a more central role for genetic variability and quantitative genetics within integrated approaches. *J. Exp. Bot.* **2007**, *58*, 2369–2387. [CrossRef] [PubMed]

7. Bihmidine, S.; Hunter, C.T.; Johns, C.E.; Koch, K.E.; Braun, D.M. Regulation of assimilate import into sink organs: update on molecular drivers of sink strength. *Front. Plant Sci.* **2013**, *4*, 177. [CrossRef]

8. Ning, P.; Yang, L.; Li, C.; Fritschi, F.B. Post-silking carbon partitioning under nitrogen deficiency revealed sink limitation of grain yield in maize. *J. Exp. Bot.* **2018**, *69*, 1707–1719. [CrossRef]

9. Ding, L.; Wang, K.J.; Jiang, G.M.; Biswas, D.K.; Xu, H.; Li, L.F.; Li, Y.H. Effects of nitrogen deficiency on photosynthetic traits of maize hybrids released in different years. *Ann. Bot.* **2005**, *96*, 925–930. [CrossRef]

10. Ciampitti, I.A.; Vyn, T.J. A comprehensive study of plant density consequences on nitrogen uptake dynamics of maize plants from vegetative to reproductive stages. *Field Crop Res.* **2011**, *121*, 2–18. [CrossRef]

11. Paponov, I.A.; Engels, C. Effect of nitrogen supply on leaf traits related to photosynthesis during grain filling in two maize genotypes with different N efficiency. *J. Plant Nutr. Soil Sci.* **2003**, *166*, 756–763. [CrossRef]

12. Uribelarrea, M.; Crafts-Brandner, S.J.; Below, F.E. Physiological N response of field-grown maize hybrids (*Zea mays* L.) with divergent yield potential and grain protein concentration. *Plant Soil* **2009**, *316*, 151–160. [CrossRef]

13. Peng, Y.; Zeng, X.; Houx, J.H.; Boardman, D.L.; Li, C.; Fritschi, F.B. Pre- and post-silking carbohydrate concentrations in maize ear-leaves and developing ears in response to nitrogen availability. *Crop Sci.* **2016**, *56*, 3218. [CrossRef]

14. Peng, Y.; Li, C.; Fritschi, F.B. Diurnal dynamics of maize leaf photosynthesis and carbohydrate concentrations in response to differential N availability. *Environ. Exp. Bot.* **2014**, *99*, 18–27. [CrossRef]

15. Peng, Y.; Li, C.; Fritschi, F.B. Apoplastic infusion of sucrose into stem internodes during female flowering does not increase grain yield in maize plants grown under nitrogen-limiting conditions. *Physiol. Plant.* **2013**, *148*, 470–480. [CrossRef]

16. Cazetta, J.O.; Seebauer, J.R.; Below, F.E. Sucrose and nitrogen supplies regulate growth of maize kernels. *Ann. Bot.* **1999**, *84*, 747–754. [CrossRef]

17. Ciampitti, I.A.; Murrell, S.T.; Camberato, J.J.; Tuinstra, M.; Xia, Y.; Friedemann, P.; Vyn, T.J. Physiological dynamics of maize nitrogen uptake and partitioning in response to plant density and nitrogen stress factors: II. Reproductive phase. *Crop Sci.* **2013**, *53*, 2588. [CrossRef]

18. Hirel, B.; Martin, A.; Tercé Laforgue, T.; Gonzalez Moro, M.B.A.; Estavillo, J.M. Physiology of maize I: A comprehensive and integrated view of nitrogen metabolism in a C4 plant. *Physiol. Plant.* **2005**, *124*, 167–177. [CrossRef]

19. Hirel, B.; Andrieu, B.; Valadier, M.H.; Renard, S.; Quilleré, I.; Chelle, M.; Pommel, B.; Fournier, C.; Drouet, J.L. Physiology of maize II: Identification of physiological markers representative of the nitrogen status of maize (*Zea mays*) leaves during grain filling. *Physiol. Plant.* **2005**, *124*, 178–188. [CrossRef]

20. Singletary, G.W.; Below, F.E. Nitrogen-induced changes in the growth and metabolism of developing maize kernels grown in vitro. *Plant Physiol.* **1990**, *92*, 160–167. [CrossRef]

21. Singletary, G.W.; Below, F.E. Growth and composition of maize kernels cultured in vitro with varying supplies of carbon and nitrogen. *Plant Physiol.* **1989**, *89*, 341–346. [CrossRef]

22. Zhao, D.; MacKown, C.T.; Starks, P.J.; Kindiger, B.K. Rapid analysis of nonstructural carbohydrate components in grass forage using microplate enzymatic assays. *Crop Sci.* **2010**, *50*, 1537. [CrossRef]

23. Lin, Y.; Zhang, C.; Lan, H.; Gao, S.; Liu, H.; Liu, J.; Cao, M.; Pan, G.; Rong, T.; Zhang, S. Validation of potential reference genes for qPCR in maize across abiotic stresses, hormone treatments, and tissue types. *PLoS ONE* **2014**, *9*, e95445. [CrossRef]

24. Pfaffl, M.W. A new mathematical model for relative quantification in real-time RT-PCR. *Nucleic Acids Res.* **2001**, *29*, e45. [CrossRef]

25. Braun, D.M.; Wang, L.; Ruan, Y.L. Understanding and manipulating sucrose phloem loading, unloading, metabolism, and signalling to enhance crop yield and food security. *J. Exp. Bot.* **2014**, *65*, 1713–1735. [CrossRef]

26. Sosso, D.; Luo, D.; Li, Q.; Sasse, J.; Yang, J.; Gendrot, G.; Suzuki, M.; Koch, K.E.; McCarty, D.R.; Chourey, P.S.; et al. Seed filling in domesticated maize and rice depends on SWEET-mediated hexose transport. *Nat. Genet.* **2015**, *47*, 1489–1493. [CrossRef]

27. Wang, L.; Lu, Q.; Wen, X.; Lu, C. Enhanced sucrose loading improves rice yield by increasing grain size. *Plant Physiol.* **2015**, 1170–2015. [CrossRef]

28. Zhao, Z.; Shi, H.; Wang, M.; Cui, L.; Zhao, H.; Zhao, Y. Effect of nitrogen and phosphorus deficiency on transcriptional regulation of genes encoding key enzymes of starch metabolism in duckweed (*Landoltia punctata*). *Plant Physiol. Biochem.* **2015**, *86*, 72–81. [CrossRef]

29. Schluter, U.; Colmsee, C.; Scholz, U.; Brautigam, A.; Weber, A.P.; Zellerhoff, N.; Bucher, M.; Fahnenstich, H.; Sonnewald, U. Adaptation of maize source leaf metabolism to stress related disturbances in carbon, nitrogen and phosphorus balance. *BMC Genom.* **2013**, *14*, 442. [CrossRef]

30. Zeeman, S.C.; Kossmann, J.; Smith, A.M. Starch: its metabolism, evolution, and biotechnological modification in plants. *Annu. Rev. Plant Biol.* **2010**, *61*, 209–234. [CrossRef]

31. Zeeman, S.C.; Smith, S.M.; Smith, A.M. The diurnal metabolism of leaf starch. *Biochem. J.* **2007**, *401*, 13–28. [CrossRef] [PubMed]

32. Sulpice, R.; Pyl, E.T.; Ishihara, H.; Trenkamp, S.; Steinfath, M.; Witucka-Wall, H.; Gibon, Y.; Usadel, B.R.; Poree, F.; Piques, M.C.O. Starch as a major integrator in the regulation of plant growth. *Proc. Natl. Acad. Sci. USA* **2009**, *106*, 10348–10353. [CrossRef] [PubMed]

33. Slewinski, T.L.; Braun, D.M. Current perspectives on the regulation of whole-plant carbohydrate partitioning. *Plant Sci.* **2010**, *178*, 341–349. [CrossRef]

34. Braun, D.M.; Slewinski, T.L. Genetic control of carbon partitioning in grasses: roles of *sucrose transporters* and *Tie-dyed* loci in phloem loading. *Plant Physiol.* **2009**, *149*, 71–81. [CrossRef] [PubMed]

35. Chen, L.; Qu, X.; Hou, B.; Sosso, D.; Osorio, S.; Fernie, A.R.; Frommer, W.B. Sucrose efflux mediated by SWEET proteins as a key step for phloem transport. *Science* **2012**, *335*, 207–211. [CrossRef] [PubMed]

36. Eom, J.; Chen, L.; Sosso, D.; Julius, B.T.; Lin, I.W.; Qu, X.; Braun, D.M.; Frommer, W.B. SWEETs, transporters for intracellular and intercellular sugar translocation. *Curr. Opin. Plant Biol.* **2015**, *25*, 53–62. [CrossRef]

37. Slewinski, T.L.; Meeley, R.; Braun, D.M. *Sucrose transporter1* functions in phloem loading in maize leaves. *J. Exp. Bot.* **2009**, *60*, 881–892. [CrossRef]

38. Felker, F.C.; Shannon, J.C. Movement of ^{14}C-labeled assimilates into kernels of *Zea mays* L: III. An anatomical examination and microautoradiographic study of assimilate transfer. *Plant Physiol.* **1980**, *65*, 864–870. [CrossRef]

39. Shannon, J.C. Movement of ^{14}C-labeled assimilates into kernels of *Zea mays* L: I. Pattern and rate of sugar movement. *Plant Physiol.* **1972**, *49*, 198–202. [CrossRef]

40. Shannon, J.C.; Dougherty, C.T. Movement of ^{14}C-labeled assimilates into kernels of *Zea mays* L: II. Invertase activity of the pedicel and placento-chalazal tissues. *Plant Physiol.* **1972**, *49*, 203–206. [CrossRef]

41. Canas, R.A.; Quillere, I.; Christ, A.; Hirel, B. Nitrogen metabolism in the developing ear of maize (*Zea mays*): analysis of two lines contrasting in their mode of nitrogen management. *New Phytol.* **2009**, *184*, 340–352. [CrossRef] [PubMed]

42. Nunes-Nesi, A.; Fernie, A.R.; Stitt, M. Metabolic and signaling aspects underpinning the regulation of plant carbon nitrogen interactions. *Mol. Plant* **2010**, *3*, 973–996. [CrossRef] [PubMed]

agronomy

MDPI

Article

Variability for Nitrogen Management in Genetically-Distant Maize (*Zea mays* L.) Lines: Impact of Post-Silking Nitrogen Limiting Conditions

Isabelle Quilleré [1], Céline Dargel-Graffin [1], Peter J. Lea [2] and Bertrand Hirel [1,*

[1] Institut Jean-Pierre Bourgin, INRA, Agro-ParisTech, Université de Paris-Saclay, 78000 Versailles, France;
     isabelle.quillere@inra.fr (I.Q.); celine.dargel-graffin@inra.fr (C.D.-G.)
[2] Lancaster Environment Centre, Lancaster University, Lancaster LA1 4YQ, UK; p.lea@lancaster.ac.uk
*   Correspondence: bertrand.hirel@inra.fr; Tel.: +33-1-30-83-30-89

Received: 25 October 2018; Accepted: 14 December 2018; Published: 19 December 2018

Abstract: The impact of nitrogen (N)-limiting conditions after silking on kernel yield (KY)-related traits and whole plant N management was investigated using fifteen maize lines representative of plant genetic diversity in Europe and America. A large level of genetic variability of these traits was observed in the different lines when post-silking fertilization of N was strongly reduced. Under such N-fertilization conditions, four different groups of lines were identified on the basis of KY and kernel N content. Although the pattern of N management, including N uptake and N use was variable in the four groups of lines, a number of them were able to maintain both a high yield and a high kernel N content by increasing shoot N remobilization. No obvious relationship between the genetic background of the lines and their mode of N management was found. When N was limiting after silking, N remobilization appeared to be a good predictive marker for identifying maize lines that were able to maintain a high yield and a high kernel N content irrespective of their female flowering date. The use of N remobilization as a trait to select maize genotypes adapted to low N input is discussed.

Keywords: genetic diversity; maize; nitrogen; remobilization; silking; uptake; ^{15}N-labeling

1. Introduction

The application of mineral nitrogen (N) fertilizers is one of the main agricultural practices used to maintain and restore soil fertility. It is able to stabilize or even increase yield for the majority of crop plants, including cereals such as maize. The applied mineral N is particularly soluble for easy uptake by plants, allowing the rapid assimilation of N during root and shoot vegetative growth [1], and thus ensuring the production of food for the constantly-growing world population [2].

Consequently, there has been, over the past 70 years, an almost five-fold increase in the total N applied to crops. In contrast, in harvestable material, such as grains used for human food and animal feed, the protein content has only increased by a factor of 3. This indicates that there was a 30% decrease in N-use efficiency (NUE), which can be defined as the yield obtained/unit of available N in the soil (supplied by the soil + N fertilizer) [3].

Plant NUE is the product of N-uptake efficiency (amount of N taken up/quantity of available N) and of the N-utilization efficiency (yield/absorbed N), [4]. There is large genetic variability of both N-uptake efficiency and N-utilization efficiency in many crops, notably in maize [5,6]. However, when examining the contribution of these two biological processes to the overall plant NUE, it has often been observed that the best performing maize varieties at high N fertilization input are not the best ones when the N fertilizer supply is lowered [7,8]. Such poor of performance under low N fertilization input is partly due to the occurrence of interactions between the genotype and the level of N present in the soil, but notably because most of the previous breeding strategies have been conducted under non-limiting N fertilization conditions. Therefore, the opportunities of selecting for high-yielding maize varieties when the N fertilization conditions are low have been missed, or not fully exploited [5].

Although it has been shown that maize productivity can be maintained under low-N input [9], high N fertilization rates have been, and are still, used in most high-yielding intensive agricultural maize production systems [10] and in breeding strategies [11,12]. However, under such high N fertilization inputs, over 50% and up to 75% of the mineral N applied to the field is not taken up by the plant and is lost by leaching into the soil [13]. Such a N loss leads to the eutrophication of freshwater and marine ecosystems [14–16] and to the emissions of N_2O (nitrous oxide), which has a global warming potential almost 300 times that of CO_2 [17]. The chemical synthesis of N fertilizers also increases crop production costs [18,19]. Altogether, both energy input for N fertilizer production and NUE are considered to be important indicators for the environmental impact of the production of most conventional food and energy crops such as maize [20].

It is estimated that an increase in agricultural production by at least 70%, will be necessary in order to feed the 9 billion people projected for the world population in 2050 [21]. Therefore, developing more sustainable agricultural practices based on fertilizer use rationalization and selecting or producing genetically-engineered genotypes exhibiting improved NUE [22,23] are possible ways of overcoming the detrimental impact of the overuse of N fertilizers [1,6].

In maize, 45–65% of the grain N, which acts a source of proteins for both humans and animals, is provided from pre-existing N in the stover before silking. Nitrogen translocation from the stover to the grain is very dependent upon the environmental conditions and/or the genotype [7]. The remaining 35–55% of the grain N originates from post-silking N uptake [7,24,25]. Therefore, to identify or select maize genotypes that exhibit improved N uptake before and after silking, it will be necessary to improve our understanding of the physiological and genetic determinants that govern these two biological processes. Such an improved understanding can be obtained using lines or hybrids for which their relative contribution is variable, and then proposing strategies to provide N when the plant needs it most.

In previous studies, remarkably large genetic variability was observed in the leaf metabolite content, leaf enzyme activities, leaf biomass-related components and KY of a core collection of maize lines [26], including races originating from different northern and southern countries of both America and Europe [27]. Therefore, in the present study, we have exploited the large genetic variability of these maize lines to examine the impact of N deprivation after silking on plant N uptake and plant productivity. Only a limited number of studies were undertaken to determine whether compensatory mechanisms such as N remobilization occur in genetically-distant maize lines when there is reduction in N availability during the grain filling period [7,24,25]. Whether these compensatory mechanisms could be used to select maize lines with improved NUE when there is a shortage of N after silking is discussed.

2. Materials and Methods

2.1. Plant Material

Seeds of the 15 maize inbred lines selected for the experiment were obtained from the core collection used for association genetics studies [27,28] of the Institut National de la Recherche Agronomique (INRA), Saint-Martin-de-Hinx, France. These inbred lines were classified into five main maize groups: Tropical (2 lines: EML1201 and P465P), European Flint (3 lines: Lo3, Lo32 and FV2), Northern Flint (3 lines: NYS302, C105 and ND36), Corn Belt Dent (6 lines: ND283, Mo17, FV252, SA24U, MBS847 and HP301) and Stiff Stalk (1 line: B73) [27]. This original classification was organized on the basis of genome sequence polymorphism of lines using Simple Sequence Repeat (SSR) microsatellite markers, and later on using Single Nucleotide Polymorphism (SNP) markers as previously described [27,29]. The group named Tropical contained lines from Argentina, Mexico and

Spain, whereas the Corn Belt Dent lines mostly originated from North America and contained two Popcorn lines (HP301 and SA24). Line FV252 also belonged to the Corn Belt Dent lines but originated from France. In the lines classified as European Flint, there were lines from France and Italy, whereas all the Northern Flint lines were from North America. The 15 maize lines were grown in a glasshouse at INRA, Versailles, France (N 48°48.133′, E 2°04.942′), until maturity without any additional light or heat from April 5th to September 19th, 2012. Seeds were first sown on coarse sand and after 2 weeks, when 2 to 3 leaves had emerged, 8 individual seedlings of each line of a similar height were transferred to pots for ^{15}N-labeling experiments and yield measurements. Each plant was transferred to a separate pot (diameter and height of 22 cm, volume 7l) containing clay loam soil and grown until maturity in the glasshouse. Clay loam soil was composed of a mixture of loam (washed fine silt with no minerals) and loam balls of ~0.5 cm diameter that ensured sufficient aeration of the roots and allowed the growth of the plant until maturity without lodging. Clay loam soil also allowed a constant flow of the provided nutrient solution.

All plants were watered four times a day with a complete nutrient solution containing 10 mM N (8 mM NO_3^- + 2 mM NH_4^+) [30]. The complete nutrient solution (N^+) contained 5 mM K^+, 3 mM Ca^{2+}, 0.4 mM Mg^{2+}, 1.1 mM $H_2PO_4^-$, 1 mM SO_4^{2-}, 1.1 mM Cl^- 21.5 μM Fe^{2+} (Sequestrene; Ciba-Geigy, Basel, Switzerland), 23 μM B^{3+}, 9 μM Mn^{2+}, 0.3 μM Mo^{2+}, 0.95 μM Cu^{2+} and 3.5 μM Zn^{2+}. The variation in the silking date between the 15 lines was approximately 5 weeks, starting from June 4th, 2012.

^{15}N-labeling was carried out on May 30th during vegetative growth, at the beginning of stem elongation (8 visible leaves, Biologishe Bundesanstalt Bunderssortenamt und Chemicshe Industrie (BBCH) 17) of the lines exhibiting the earliest development. Into each pot, 150 mL of a solution of 30 mM KNO_3^- containing 2% ^{15}N atom excess was applied to each pot in order to distribute on average 1.25 mg ^{15}N per plant. After silking, the plants were separated into 8 groups of 1 plant for each of the 15 lines. Four groups were watered with a complete nutrient solution (N^+) and the other four groups with a low N (N^-) solution. The eight groups were randomly distributed in the greenhouse to ensure homogenous plant growth irrespective of the line and the N treatment. The composition of the low N solution (N^-) was similar to that of the complete nutrient solution except that 8 mM NO_3^- + 2 mM NH_4^+, 1 mM SO_4^{2-} and 1.1 mM Cl^- were replaced by 0.3 mM NO_3^-, 2.2 mM SO_4^{2-} and 2.3 mM Cl^- respectively. Kernel yield (KY), kernel number (KN), thousand kernel weight (TKW) and kernel N content (%NK) were determined according to the methods described in [31] using individual plants from N^+ and in N^- in each of the 8 groups, making four replicates in total for each of the two N feeding conditions.

2.2. Determination of N Content and ^{15}N-Abundance

Plants were harvested when all the kernels were matured. Plant samples were separated in two different batches one corresponding to the shoots (leaves + stalk + sheaths + husk = SDW (Shoot Dry Weight)) and the other to the ear (cob + kernels). After the drying (70 °C in an oven) and weighing of shoots and kernels, the material was ground to obtain a homogeneous fine powder. A sub-sample of 1 mg was used to determine total N content and ^{15}N-abundance by an elemental automated analyser (Roboprep CN, SERCON Europa Scientific Ltd, Crewe, UK) coupled to an isotope ratio mass spectrometer (Tracermass, PDZ Europa Scientific Ltd, Crewe, UK) calibrated for measuring ^{15}N-natural abundance. As the amount of N present in the cobs is very low at plant maturity [32], it was not considered in calculating the plant N budget.

^{15}N abundance was calculated as atom per cent (A%), and defined as A% = 100 × (^{15}N) (^{15}N + ^{14}N), both in labeled plant samples and in unlabeled control plants. A% in the unlabeled control plants was close to 0.36634%, a value which corresponds to the natural abundance of atmospheric dinitrogen (N_2). ^{15}N enrichment (E%) of the plant samples was then defined as (A% sample − A% control). The amount of ^{15}N contained in the sample (Q) was calculated on a dry weight (DW) basis using the following formula: Q = DW × E% × N% with N% being the concentration of N in the sample. The amount of ^{15}N present in the kernels compared to that present in the whole plant (shoots + ear) was calculated using the following formula: Q kernels/ (Q kernels + Q shoots) and named ^{15}N-Harvest-Index ([15]NHI). The [15]NHI represents the proportion of ^{15}N that is absorbed at the vegetative stage and further remobilized to the kernels at the reproductive stage after silking. Roots were not considered to calculate the [15]NHI, since in previous studies [33], it was shown that under experimental conditions such as those employed in the present study the amount of ^{15}N present in the roots represents less than 5% of the total ^{15}N present in the whole plant.

2.3. Statistical Analysis

Statistical analysis of data was performed using the Student *t*-test functions of the XLStat-Pro 7.5 software, 2013 (Addinsoft, Bordeaux, France). Pearson correlation coefficients between yield-related traits, N accumulation before silking and the amount of N remobilized from the shoots to the kernels were calculated in order to identify their possible relationship. Yield, yield components, biomass accumulation and the N budget of the 15 lines grown in the two contrasting N feeding conditions, are presented as mean values for 4 plants with standard errors (SE) (SE = SD$/\sqrt{n}$), where SD is the standard deviation and *n* is the number of samples).

3. Results

3.1. Plant Agronomic Performances

Variations in KY in the glasshouse ranged from 27.3 g for line Lo32 to 105.8 g for line MBS847, with a mean of 67.3 g for all the lines. In previous field experiments [26], the variations in KY ranged from 32.5 g for line ND36 to 75.9 g for line MBS847, with a mean of 47 g for all the lines. The 15 lines grown in the glasshouse had a higher yield compared to those grown in the field, (40% higher) on average, explained by a higher KN and a higher TKW (20% on average). Correlation studies showed that between the glasshouse and the field experiment, there was a strong correlation of 0.8 (*p*-value < 0.0001) for KN and a good correlation of 0.63 (*p*-value < 0.01) for KY. In high-yielding lines (H) KY was >50 g and up to 120 g per plant and in low-yielding lines (L) KY was <50 g and down to 30 g per plant. Such a finding indicates that plant growth conditions were optimal in the glasshouse. Among the 15 lines examined, a large genetic variability was observed for the traits related to yield and plant N management. When a two-way ANOVA was performed, we found a significant genotypic effect (*p* < 0.0001) for all the measured traits. The variation in KY between the lowest and the highest yielding lines when the plants were grown either in N+ or in N− was up to 4-fold. When a comparison was made between N− or N+ conditions, two groups of lines could be identified on the basis of kernel production (Figure 1). One group did not show any significant reduction in KY in N−, whereas in the other group, a decrease in KY ranging from 20% in line FV252 to 80% in line B73 was observed (Table 1). Interestingly, following growth in N−, KY was not reduced in the two tropical lines EML1201 and P465P. For the Corn Belt, European Flint, Northern Flint lines and Stiff Stalk lines there was no clear relationship between their genetic characteristics and kernel production in N−. As shown in Table 1, the decrease in KY was due to a decrease in KN, whereas the reduction in N fertilization after silking did not have any marked impact on TKW. In agreement with a number of previous reports on maize kernel production and its genetic variability [34], our data showed that in N+ there was a positive correlation between KY and KN (*r* = 0.75, *p* < 0.001), whereas KY or KN and TKW were not correlated (*r* =0.42, *p* < 0.115 and *r* = 0.26, *p* < 0.359 respectively).

Following the reduction in post-silking N fertilization, four groups of lines were identified when kernel N content (%NK) was considered. The first group was represented by four lines (EML201, Lo3, NYS302 and ND283), which exhibited no change in KY and %NK. The second group (lines P465P, C105, Mo17, Lo32) exhibited on average a 10–15% decrease in kernel N and a concomitant decrease in KY. In the third group, there were six lines (FV252, ND36, SA24U, MBS847, HP301 and B73), which exhibited a decrease in KY and no change in the %NK. Only line FV2 was representative of group 4, in which there was a marked decrease of KY (71%) and an increase in % NK content of 19%.

There was no clear relationship between the four groups of lines defined in the present study on the basis on their agronomic performance in N^- and the five genetic groups they belong to (European Flint, Northern Flint, Corn Belt Dent, Tropical and Stiff Stalk; See Figure 1). Nevertheless, in the group of lines exhibiting a decrease in KY with no change in %NK in N^- (group 3 in Figure 1), there are four lines out of the six that belong the Corn Belt genetic group.

In most of the lines, total shoot biomass production at maturity (SDW) including stems, leaves, tassels and husks was unaffected following growth in N^- compared to N^+, except for lines NYS302, C105 and FV252, for which a 30% decrease was observed.

3.2. Plant Nitrogen Management

It can be seen that compared to the N^+-grown plants, most of the lines showed a significant reduction in the amount of Nt under N^- conditions, ranging from 15% in line Lo3 to 56% in line ND36.

We observed that there were no significant differences between the amount of total plant N (Nt) accumulated following growth in N^- (Table 1) and the Nt before siliking in N^+. This indicates that negligible amounts of N were later taken up by the plants under post-silking N-limiting conditions. There was an almost three-fold range of variation for Nt in N^- (Table 1).

The amount of N accumulated by the plants that was further translocated to the kernels via the remobilization process (^{15}N-Harvest index, ^{15}NHI) was estimated using the ^{15}N. In four lines (NYS302, P465P, C105 and ND36) an increase of up to 40% in the ^{15}N kernel content was observed in N^-. Such an increase was reduced to 20% in lines EML1201, Lo3 and FV252. In other lines such as ND283, Mo17, Lo32, SA24U, MBS847 and HP301, low N fertilization after silking did not have any significant impact on shoot N remobilization. Only in lines B73 and FV2 were lower amounts of ^{15}N (50%) translocated to the kernels, after the labeling period.

As shown in Figure 1, it was observed that neither the silking date nor KY had any direct relationship with the capacity of a line to take up or remobilize N. For example, only high yielding lines (H) such as EML1201 or low yielding lines (L) such as ND36 were able to remobilize more N under N limiting conditions after silking. We also observed that silking in line ND36 occurred much earlier compared to line EML1201.

Table 1. Agronomic performance and nitrogen management of the core collection of 15 maize inbred lines originating from Europe and America.

Group	Line	KY (g) ± SE			KN ± SE			%NK ± SE			SDW (g) ± SE			Nt (mg) ± SE			^{15}NHI ± SE		
		N$^+$	N$^-$	t-test	N$^+$	N$^-$	t-test	N$^+$	N$^-$	t-test	N$^+$	N$^-$	t-test	N$^+$	N$^-$	t-test	N$^+$	N$^-$	t-test
1	EML1201	92.6 ± 3.2	80.5 ± 8.5	ns	266 ± 8.4	216.5 ± 15.5	*	2.06 ± 0.12	1.94 ± 0.16	ns	136.2 ± 6.6	141.8 ± 14.3	ns	4.35 ± 0.19	3.07 ± 0.09	**	0.36 ± 0.01	0.47 ± 0.03	*
1	Lo3	86.1 ± 5.6	85.8 ± 6.2	ns	465 ± 12.9	469.3 ± 22.6	ns	1.86 ± 0.06	1.78 ± 0.03	ns	106.4 ± 12.8	91.4 ± 9.0	ns	3.09 ± 0.19	2.62 ± 0.16	(*)	0.46 ± 0.03	0.55 ± 0.02	*
1	NYS302	35.2 ± 4.0	29.8 ± 4.1	ns	238 ± 20.3	174.5 ± 11.5	ns	2.56 ± 0.13	2.36 ± 0.10	ns	133.3 ± 3.2	90.7 ± 7.9	**	3.69 ± 0.28	1.94 ± 0.13	*	0.21 ± 0.04	0.36 ± 0.04	(*)
1	ND283	34.1 ± 2.0	35.3 ± 1.5	ns	177.7 ± 10.5	197.8 ± 3.8	ns	2.14 ± 0.15	2.14 ± 0.20	ns	43.1 ± 1.5	48.9 ± 4.0	ns	1.19 ± 0.08	1.20 ± 0.12	ns	0.61 ± 0.03	0.67 ± 0.05	ns
2	P465P	56.7 ± 7.6	70.4 ± 1.3	ns	298.5 ± 60.1	311.7 ± 13.3	ns	2.33 ± 0.08	2.04 ± 0.10	(*)	155.4 ± 17.7	134.7 ± 14.3	ns	3.33 ± 0.22	2.44 ± 0.13	*	0.38 ± 0.08	0.61 ± 0.04	(*)
2	C105	52.2 ± 4.1	62.3 ± 8.3	ns	209 ± 14.0	227.3 ± 28.6	ns	2.17 ± 0.08	1.85 ± 0.07	ns	81.8 ± 3.0	61.7 ± 7.8	*	2.68 ± 0.11	1.87 ± 0.11	**	0.37 ± 0.01	0.62 ± 0.04	**
2	Mo17	94.4 ± 5.5	87.9 ± 15.7	ns	298.1 ± 21.3	295.8 ± 46.6	ns	2.31 ± 0.04	1.99 ± 0.13	(*)	155.3 ± 11.0	144.9 ± 7.5	ns	4.35 ± 0.23	3.31 ± 0.09	**	0.44 ± 0.03	0.47 ± 0.07	ns
2	Lo32	27.3 ± 0.4	24.8 ± 6.4	ns	223 ± 15.0	189 ± 47.8	ns	2.39 ± 0.04	2.15 ± 0.04	ns	126.9 ± 13.0	121.0 ± 18.1	ns	3.28 ± 0.49	2.27 ± 0.20	ns	0.17 ± 0.04	0.24 ± 0.08	ns
3	FV252	76.3 ± 3.5	59.1 ± 5.4	*	372.2 ± 20.1	295.4 ± 25.8	*	2.04 ± 0.13	1.93 ± 0.10	ns	99.9 ± 6.7	80.0 ± 6.6	(*)	3.07 ± 0.08	1.98 ± 0.15	***	0.48 ± 0.03	0.59 ± 0.04	*
3	ND36	37.2 ± 3.2	24.5 ± 1.5	(*)	245.7 ± 15.0	152.3 ± 3.8	(*)	2.58 ± 0.11	2.38 ± 0.27	ns	99.1 ± 16.9	62.0 ± 8.6	ns	3.03 ± 0.40	1.32 ± 0.19	(*)	0.30 ± 0.03	0.48 ± 0.02	ns
3	SA24U	122.5 ± 2.5	85.1 ± 8.6	**	793.3 ± 0.3	553.8 ± 81.3	**	1.97 ± 0.08	2.02 ± 0.04	ns	191.2 ± 1.5	126.3 ± 29.0	ns	4.73 ± 0.02	2.99 ± 0.34	ns	0.43 ± 0.01	0.60 ± 0.14	ns
3	MBS847	105.9 ± 8.7	59.9 ± 7.4	(*)	543 ± 32.3	311.4 ± 41.3	(*)	2.12 ± 0.08	2.15 ± 0.07	ns	119.3 ± 11.1	107.9 ± 12.3	ns	3.94 ± 0.31	2.56 ± 0.28	*	0.52 ± 0.04	0.49 ± 0.04	ns
3	HP301	49.3 ± 4.9	34.1 ± 3.6	(*)	468.9 ± 49.5	306.3 ± 39.1	(*)	2.32 ± 0.15	2.02 ± 0.17	ns	130.6 ± 2.1	160.2 ± 18.4	ns	3.12 ± 0.12	2.50 ± 0.10	*	0.32 ± 0.00	0.26 ± 0.04	ns
3	B73	85.4 ± 19.0	19.2 ± 3.0	(*)	391.5 ± 54.0	97.8 ± 4.3	*	2.29 ± 0.27	2.45 ± 0.22	ns	159.2 ± 12.7	156.5 ± 27.5	ns	4.38 ± 0.14	3.03 ± 0.72	(*)	0.38 ± 0.05	0.16 ± 0.03	*
4	FV2	54.4 ± 5.7	15.8 ± 2.6	**	258.6 ± 19.8	76.7 ± 10.3	***	2.04 ± 0.06	2.42 ± 0.16	*	75.5 ± 7.3	71.8 ± 7.0	ns	2.23 ± 0.20	1.55 ± 0.20	(*)	0.48 ± 0.04	0.24 ± 0.05	*

KY = Kernel Yield; KN = Kernel Number; %NK = Kernel nitrogen content; SDW = Shoot Dry Weight; Nt = Total Nitrogen in the whole plant (shoots + ear); ^{15}NHI = N-Harvest index. N$^+$ = Nitrogen non-limiting conditions. N$^-$ = Post-silking N-limiting conditions. Group = Four groups of lines identified on the basis of KY and on the %NK (see also Figure 1). Asterisks indicate a t-test with a p-value: (*) $p < 0.10$; * $p < 0.05$; ** $p < 0.01$; *** $p < 0.001$; ns, nonsignificant.

Line	KY	%N K	Group	NUp	NRem	Class	KYL N$^+$	KYL N$^-$	Sd
EML1201	=	=	1	-	+		H	H	M
Lo3	=	=	1	-	+	N-tolerant	H	H	M
NYS302	=	=	1	-	+		L	L	M
ND283	=	=	1	=	=		L	L	E
P465P	=	-	2	-	+		H	H	M
C105	=	-	2	-	+	N-semi	H	H	M
Mo17	=	-	2	-	=	tolerant	H	H	L
Lo32	=	-	2	=	=		L	L	M
FV252	-	=	3	-	+		H	H	E
ND36	-	=	3	-	+		L	L	E
SA24U	-	=	3	-	=	N-semi	H	H	L
MBS847	-	=	3	-	=	sensitive	H	H	M
HP301	-	=	3	-	=		L	L	L
B73	-	=	3	-	-		H	L	L
FV2	-	+	4	-	-	N-sensitive	H	L	E

Figure 1. Schematic representation of the distribution of the 15 maize inbred maize lines originating from Europe and America in relation to their agronomic performance and their mode of N management under post-silking N-limiting conditions. On the left of the panel (Line), colored boxes represent the five groups of maize lines obtained from different countries of Europe and America (Tropical: pale green; European Flint: Blue; Northern Flint: dark green; Corn Belt Dent: yellow and Stiff Stalk: orange). Kernel yield (KY), kernel number (KN) and kernel N content (%NK) were determined for the 15 maize lines grown in the glass house (Table 1). For the different measured traits including KY, %NK, N remobilization (NRem) and N uptake (NUp) the red background color indicates an increase (+), the green background color a decrease (-) and the blue background color no change (=) under N-limiting conditions after silking until plant maturity. Groups 1, 2, 3 and 4 represent the four groups of lines identified on the basis of KY and %NK. Class corresponds to a proposed ranking of the lines based on their N management response when post-silking N fertilization is reduced. On the right of the panel, the period for the silking date (Sd) is shown. E = early (from 4th to the 18th of June); M = medium (from the 19th to the 29th of June); L = late (from the 30th of June to the 6th of July), along with the level of kernel yield in N$^+$ (KYL N$^+$) and in N$^-$ (KYL N$^-$); H = High > 50 g and up to 120 g per plant; L= low < 50 g and down to 30g per plant. Detailed yield data are presented in Table 1.

4. Discussion

Studying the genetic and physiological basis of N uptake and N utilization efficiency in maize could help to improve our understanding of how these two processes contribute to the agronomic performance of maize. These studies could further improve N use efficiency in this crop, which is of major economic importance worldwide [35]. Genetic variability in maize still remains to be fully exploited, in particular when there is a shortage of N during key steps in the developmental cycle, such as the post-silking period. In this investigation, using a core collection of 15 maize inbred lines representative of the genetic diversity of both America and Europe, we examined the impact of N limiting conditions on N uptake and N remobilization in maize, following the silking process.

At the beginning of stem elongation, maize plants were labeled with $^{15}NO_3^-$ in order to estimate the proportion of N remobilized from the vegetative tissues by measuring the ^{15}N content in the kernels. In a previous study, it has been shown that ^{15}N-labeling at the beginning of rapid stem growth appears to be a useful tool for investigating the genetic variability of N remobilization using a large number of genotypes, as the proportion of N taken up after silking significantly contributed to the N budget of the whole plant [32]. Under agronomic conditions, residual N present in the soil could interfere with the estimation of the amount of N remobilized after silking [32]; therefore, plants were grown in large pots in a glasshouse and watered with a nutrient solution. Such semi-controlled growth conditions, allowed a good development of the roots both during the vegetative phase under N$^+$ conditions and following the post-silking N-limiting period. In agreement with this, we observed that in the glasshouse, KY was on average 40% higher compared to plants grown in the field. This is likely due to more favorable environmental conditions, including temperature and mineral nutrient availability, which have a greater effect on the lines of tropical origin.

Total plant N at maturity (shoots + kernels) was also measured in order to quantify the amount of N taken up by the plants and whether N limiting conditions after silking had any impact on the total plant N content at maturity. For most lines, N limiting conditions after silking had a negative impact on whole plant N uptake (30% decrease on average), except in lines ND283 and Lo32. Similar amounts of N were taken by these two lines⁻ until maturity and there was no marked impact on KY. Kernel yield, represents sink-strength and was much lower in ND283 and Lo32 compared to the other lines (Figure 1). Such a finding suggests that the two lines were able to take up enough N before the silking period, and were, in turn, more tolerant to a post-silking N-deficiency. It has been previously reported that inbred lines can be tolerant to N-deficiency as their KY remains practically unchanged under N limiting conditions, which indicates that they have a higher NUE [8]. In contrast, in a survey describing the adaptation of maize to low N environments, it was concluded that such an adaptation was due to the ability of modern hybrids to accumulate more N after silking while maintaining their productivity in terms of KY [36]. Such results confirm that compared to lines, hybrids are able to take up more N after silking [32]. Whether, the ability of lines ND 283 and Lo32 to take up enough N before silking is due to the root architecture [37] or to a higher efficiency of the inorganic N transport system [38] is currently under investigation.

An interesting result was the finding that N-limiting conditions during the post-silking period had, for most of the lines used in the present investigation, no impact on shoot biomass production. In maize hybrids, the occurrence of a linear relationship between post silking N uptake and the stover dry weight was previously reported. However, such a positive relationship was found only in old hybrids released before 1991 [36]. We did not observe such a genotypic-dependent control of dry matter accumulation and partitioning during the grain filling period. This is likely because lines produce, in general, less biomass compared to hybrids, and because, even if post-silking N uptake is reduced, there is still enough N to sustain shoot growth and development until maturity.

When examining plant performance in terms of KY and kernel N content, four main groups were identified. The first was represented by five lines belonging to the Tropical, European, Northern Flint and Corn Belt Dent genetic groups, in which no changes in KY and kernel N content were observed. The finding that both KY and kernel quality remained unchanged can be explained by the fact that more N was remobilized after silking in order to compensate for the shortage in N during the grain filling period. Therefore, we have classified these lines as being tolerant to a post-silking N stress (Figure 1).

In the second group, lines from four different genetic groups (Tropical, European, Northern Flint and Corn Belt Dent), were also present. Kernel Yield was not reduced, and three of them (P465P, C105, Mo17) were among the most highly productive, although there was a decrease in kernel N content. Two lines belonging to group 2 (P465P and C105) were characterized by an increase in N remobilization. However, in these two lines, such an increase was presumably not sufficient to compensate for the decrease in kernel N content resulting from the lack of N after silking. However, as KY was not reduced, the four lines belonging to group 2 were classified as semi-tolerant to a post-silking N stress.

In the third group of lines represented mostly by Corn Belt Dent lines (4 out of 6 lines in total), N-limiting conditions after silking led to a decrease in KY without any marked changes in the kernel N content. In two lines belonging to this group (FV252 a corn Belt Dent line and ND36 a Northern Flint line), N remobilization was higher, but apparently was not able to compensate for the decrease in KY. However, unlike two lines of group 2 (P465P and C105), the kernel N content remained similar to that of plants grown under non-limiting post-silking N conditions.

Group 4 was represented only by the Europen Flint line FV2 (Table 1). In this line, under N-conditions, there was a marked decrease in KY (more than 3-fold), which moved the ranking in terms of yield from a high yielding line to a low yielding line. Line FV2 was also unique in terms of response to post-silking N-limiting conditions, as both N uptake and N remobilization were lower, while kernel N content was higher. Such an increase in kernel N content can be explained by the strong sink strength limitation compared to most of the other lines (four-fold reduction in KN). The N concentration increased in the reproductive organs as the amount of available N in the shoots (although reduced in N^-) remained relatively high (Table 1). A similar pattern of N management strategy was observed for line B73, except that under our experimental conditions, the increase in kernel N content (%NK), although visible, was not significant. Such findings are in agreement with those of Rajcan and Tollenaar 1999 [39], who showed that the proportion of N in the maize grain can be variable, depending on post-silking N uptake and on the source: sink ratio.

In most of the studies aimed at investigating the effect of N application rates on KY and kernel N content, only a limited number of genotypes (mostly represented by hybrids) have been studied. In some cases, it has been observed that the kernel N content was higher when more N was remobilized from the shoots [40]. In other studies, it was proposed that enhancing post-silking N uptake, rather than N remobilization, was a possible way to increase kernel N accumulation [41]. When QTLs for post-silking N uptake were investigated, only low genetic variations for this trait were generally observed, and thus, a low number of chromosomal regions involved in the control of this trait were detected [42]. Such findings are consistent with the results obtained in the present study, as line FV2, which was used to produce the inbred line population for the detection of QTLs for post-silking N uptake, is one of the least efficient lines in terms of total plant N uptake either in N^+ or in N^- (Table 1, Figure 1).

Although it has been shown that N^- remobilization can be maximized if a large amount of N is accumulated before silking, there was no correlation between N accumulation before silking and the amount of N remobilized from the shoots to the kernels, under either N^+ or in N^- conditions ($r = 0.31$, $p = 0.26$ and $r = 0.16$, $p = 0.57$ respectively). This finding indicates that such a positive correlation between these traits, and one that is representative of N management, is not necessarily found when considering genetically-distant maize lines.

It can therefore be concluded that for metabolic traits [26] in maize lines exhibiting a large genetic variation, N management strategies are much more diverse compared to those found in genotypes originating from a closely-related genetic background and selected in specific areas of the world. Such a large genetic diversity can be exploited irrespective of plant female flowering date and of KY potential, as late-flowering, high yielding lines are able to maintain their agronomic performance even when there is a post-silking N deficiency stress.

5. Conclusions

Exploiting more extensive maize genetic diversity using lines of different geographical origin appears to be a promising way to select lines and then to produce hybrids that are able to maintain high agronomic performance, notably when less N is available during the post-silking period. Although such an exploitation of maize genetic resources could be limited by the fact that a number of these lines may be adapted to specific environmental conditions either in tropical or in temperate regions, it will help to improve our understanding of how these lines are able to maintain high yields under low N conditions. Such knowledge, combined with the benefit of modern genetic techniques, could be used for future breeding strategies, which up to now have generally been conducted under high N fertilization input [5], using genotypes originating from specific areas in the world.

Author Contributions: I.Q. and B.H. conceived and designed the experiments; I.Q. and C.D.-G. performed the experiments; I.Q. and B.H. analyzed the data; C.D.-G. contributed reagents/materials/analysis tools; B.H. and P.J.L. wrote the paper.

Acknowledgments: This study was funded by INRA (Institut National de la Recherche Agronomique), which also covers the cost of publishing in open access. We also thank Anne Marmagne and Michel Lebrusq for excellent technical assistance in ^{15}N analyses and in growing the plants, respectively.

Conflicts of Interest: The authors declare no conflict of interest.

References

1. Hirel, B.; Tétu, T.; Lea, P.J.; Dubois, F. Improving nitrogen use efficiency in crops for a sustainable agriculture. *Sustainability* **2011**, *3*, 1452–1485. [CrossRef]
2. Galloway, J.N.; Leach, A.M.; Bleeker, A.; Erisman, J.W. A chronology of human understanding of the nitrogen cycle. *Philos. Trans. R. Soc. Lond. B* **2013**, *368*. [CrossRef] [PubMed]
3. Lassaletta, L.; Billen, G.; Garnier, J.; Bouwman, L.; Velazquez, E.; Mueller, N.D.; Gerber, J.S. Nitrogen use in the global food system: Past trends and future trajectories of agronomic performance, pollution, trade, and dietary demand. *Environ. Res. Lett.* **2016**, *11*, 095007. [CrossRef]
4. Pathak, R.R.; Lochab, S.; Raghuram, N. Improving nitrogen-use efficiency. *Compr. Biotechnol.* **2011**, *4*, 209–218.
5. Haegele, J.W.; Cook, K.A.; Nichols, D.M.; Below, F.E. Changes in nitrogen use traits associated with genetic improvement for grain yield of maize hybrids released in different decades. *Crop Sci.* **2013**, *53*, 1256–1268. [CrossRef]
6. Han, M.; Okamoto, M.; Beatty, P.H.; Rothstein, S.J.; Good, A.J. The genetics of nitrogen use efficiency in crop plants. *Ann. Rev. Genet.* **2015**, *49*, 269–289. [CrossRef]
7. Gallais, A.; Coque, M. Genetic variation and selection for nitrogen use efficiency in maize: A synthesis. *Maydica* **2005**, *50*, 531–547.
8. Mastrodomenico, A.T.; Hendrix, C.C.; Below, F.E. Nitrogen use efficiency and the genetic variation of maize expired plant variety protection germplasm. *Agriculture* **2018**, *8*, 3. [CrossRef]
9. Hossard, L.; Archer, D.W.; Bertrand, M.; Colnenne-David, C.; Debraeke, P.; Ernfors, M.; Jeuffroy, M.H.; Munier-Jolain, N.; Nilsson, C.; Sanford, G.R.; et al. A meta-analysis of maize and wheat yields under low-input vs. conventional and organic systems. *Agron. J.* **2016**, *108*, 1155–1167. [CrossRef]
10. Meng, Q.; Cui, Z.; Yang, H.; Zhang, F.; Chen, X. Establishing high-yielding maize systems for sustainable intensification in China. *Adv. Agron.* **2018**, *148*, 85–109.
11. Bänziger, M.; Edmeades, G.O.; Beck, D.; Bellon, M. *Breeding for Drought and Nitrogen Stress Tolerance in Maize: From Theory to Practice*; CIMMYT: Mexico City, Mexico, 2000.
12. Chen, K.; Camberato, J.J.; Tuinstra, M.R.; Kumudini, S.; Tollenaar, M.; Vyn, T.J. Genetic improvement in density and nitrogen stress tolerance traits over 38 years of commercial maize hybrid release. *Field Crops Res.* **2016**, *196*, 438–451. [CrossRef]
13. Cameron, K.C.; Di, H.J.; Moir, J.L. Nitrogen losses from the soil/plant system: A review. *Ann. Appl. Biol.* **2013**, *162*, 145–173. [CrossRef]
14. Moss, B. Water pollution by agriculture. *Philos. Trans. R. Soc. Lond. B Biol. Sci.* **2008**, *363*, 659–666. [CrossRef] [PubMed]
15. Withers, P.J.A.; Neal, C.; Jarvie, H.P.; Doody, D.G. Agriculture and eutrophication: Where do we go from here? *Sustainability* **2014**, *6*, 5853–5875. [CrossRef]
16. Verzeaux, J.; Alahmad, A.; Habbib, H.; Nivelle, E.; Roger, D.; Lacoux, J.; Decocq, G.; Hirel, B.; Catterou, M.; Spicher, F.; et al. Cover crops prevent the deleterious effect of nitrogen fertilization on bacterial diversity by maintaining the carbon concentration of ploughed soil. *Geoderma* **2016**, *281*, 49–57. [CrossRef]
17. Smith, K.A. Changing views of nitrous oxide emissions from agricultural soil: Key controlling processes and assessment at different spatial scales. *Eur. J. Soil Sci.* **2017**, *68*, 137–155. [CrossRef]
18. Gellings, C.W.; Parmenter, K.E. Energy efficiency in fertilizers production and use. In *Efficient Use and Conservation of Energy*; Gellings, C.W., Ed.; Encyclopedia of Life Support Systems; UNESCO Publications: Oxford, UK, 2016; Volume II, pp. 123–136.
19. Hirel, B.; Lea, P.J. Genomics of nitrogen use efficiency in maize: From basic approaches to agronomic applications. In *The Zea Mays Genome, Compendium of Plant Genomes*; Bennetzen, J., Flint-Garcia, S., Hirsch, C., Tuberosa, R., Eds.; Springer: Cham, Switzerland, 2018; ISBN 978-3-319-97427-9.
20. Oita, A.; Malik, A.; Kanemoto, K.; Geschke, A.; Nishijima, S.; Lenzen, M. Substantial nitrogen pollution embedded in international trade. *Nat. Geosci.* **2016**, *9*, 111–115. [CrossRef]
21. McKenzie, F.C.; Williams, J. Sustainable food production: Constraints, challenges and choices by 2050. *Food Secur.* **2015**, *7*, 221–223. [CrossRef]

22. Ceccarelli, S. GM crops, organic agriculture and breeding for sustainability. *Sustainability* **2014**, *6*, 4273–4286. [CrossRef]

23. Swain, E.Y.; Rempelos, L.; Orr, C.H.; Hall, G.; Chapman, R.; Almadni, M.; Stockdale, EA.; Kidd, J.; Leifert, C.; Cooper, J.M. Optimizing nitrogen use efficiency in wheat and potatoes: Interaction between genotypes and agronomic practices. *Euphytica* **2014**, *199*, 119–136. [CrossRef]

24. Coque, M.; Gallais, A. Genetic variation among European maize varieties for nitrogen use efficiency under low and high nitrogen fertilization. *Maydica* **2007**, *52*, 383–397.

25. Ciampitti, A.; Vyn, T.J. Grain nitrogen source changes over time in maize: A review. *Crop Sci.* **2013**, *53*, 366–377. [CrossRef]

26. Cañas, R.A.; Yesbergenova-Cuny, Z.; Simons, M.; Chardon, F.; Armengaud, P.; Quilleré, I.; Cukier, C.; Gibon, G.; Limami, A.M.; Nicolas, S.; et al. Exploiting the genetic diversity of maize using a combined metabolomic, enzyme activity profiling, and metabolic modeling approach to link leaf physiology to kernel yield. *Plant Cell* **2017**, *29*, 919–943. [CrossRef] [PubMed]

27. Camus-Kulandaivelu, L.; Veyrieras, J.B.; Madur, D.; Combes, V.; Fourman, M.; Barraud, S.; Dubreuil, P.; Gouesnard, B.; Manicacci, D.; Charcosset, A. Maize adaptation to temperate climate: Relationship between population structure and polymorphism in the *Dwarf8* gene. *Genetics* **2006**, *172*, 2449–2469. [CrossRef] [PubMed]

28. Bouchet, S.; Servin, B.; Bertin, P.; Madur, D.; Combes, V.; Dumas, F.; Brunel, D.; Laborde, J.; Charcosset, A.; Nicolas, S. Adaptation of maize to temperate climates: Mid-density genome-wide association genetics and diversity patterns reveal key genomic regions, with a major contribution of the *Vgt2 (ZCN8)* locus. *PLoS ONE* **2013**, *8*, e71377. [CrossRef] [PubMed]

29. Yesbergenova-Cuny, Z.; Dinant, S.; Martin-Magniette, M.L.; Quillere, I.; Armengaud, P.; Monfalet, P.; Lea, P.J.; Hirel, B. Genetic variability of the phloem sap metabolite content of maize (*Zea mays* L.) during the kernel-filling period. *Plant Sci.* **2016**, *252*, 347–357. [CrossRef] [PubMed]

30. Coïc, Y.; Lesaint, C. Comment assurer une bonne nutrition en eau et en ions minéraux en horticulture. *Hortic. Française* **1971**, *8*, 11–14.

31. Martin, A.; Belastegui-Macadam, X.; Quilleré, I.; Floriot, M.; Valadier, M.H.; Pommel, B.; Andrieu, B.; Donnison, I.; Hirel, B. Nitrogen management and senescence in two maize hybrids differing in the persistence of leaf greenness. Agronomic, physiological and molecular aspects. *New Phytol.* **2005**, *167*, 483–492. [CrossRef] [PubMed]

32. Gallais, A.; Coque, M.; Quilléré, I.; Le Gouis, J.; Prioul, J.L.; Hirel, B. Estimating proportions of N remobilization and of post-silking N uptake allocated to maize kernels by ^{15}N labelling. *Crop Sci.* **2007**, *47*, 685–691. [CrossRef]

33. Cliquet, J.B.; Deléens, E.; Mariotti, A. C and N mobilization from stalk and leaves during kernel filling by ^{13}C and ^{15}N tracing in *Zea mays* L. *Plant Physiol.* **1990**, *94*, 1547–1553. [CrossRef]

34. Bertin, P.; Gallais, A. Genetic variation for nitrogen use efficiency in a set a recombinant maize inbred lines I. Agrophysiological results. *Maydica* **2000**, *45*, 53–66.

35. Hirel, B.; Gallais, A. Nitrogen use efficiency—Physiological, molecular and genetic investigations towards crop improvement. In *Advances in Maize (Essential Reviews in Experimental Biology)*; Prioul, J.L., Thévenot, C., Molnar, T., Eds.; Society for Experimental Biology: Cambridge, UK, 2011; Volume 3, pp. 285–310.

36. Mueller, S.M.; Vyn, T.J. Maize plant resilience to N stress and post-silking N capacity changes over time: A review. *Front. Plant Sci.* **2016**, *7*, 53. [CrossRef] [PubMed]

37. Mu, X.; Chen, F.; Wu, Q.; Chen, Q.; Wang, J.; Yuan, L.; MI, G. Genetic improvement or root growth increases maize yield via enhanced post-silking uptake. *Eur. J. Agron.* **2015**, *63*, 55–61. [CrossRef]

38. Garnett, T.; Plett, D.; Conn, V.; Conn, S.; Rabie, H.; Rafalski, J.A.; Dhugga, K.; Tester, M.A.; Kaiser, B. Variation for N uptake system in maize: Genotypic response to N supply. *Front. Plant Sci.* **2015**, *6*, 936. [CrossRef]

39. Rajcan, I.; Tollenaar, M. Source: Sink ratio and leaf senescence in maize: II. Nitrogen metabolism during grain filling. *Field Crop Res.* **1999**, *60*, 255–265. [CrossRef]

40. Chen, Y.; Xiao, C.; Wu, D.; Xia, T.; Chen, Q.; Chen, F.; Yuan, L.; Mi, G. Effects of nitrogen application rate on grain yield and grain nitrogen concentration in two maize hybrids with contrasting nitrogen remobilization efficiency. *Eur. J. Agron.* **2015**, *62*, 79–89. [CrossRef]

41. Yang, L.; Guo, S.; Chen, Q.; Chen, F.; Yuan, L.; Mi, G. Use of stable nitrogen isotope to reveal the source-sink regulation of nitrogen uptake and remobilization during grain filling phase in maize. *PLoS ONE* **2016**. [CrossRef] [PubMed]
42. Coque, M.; Martin, A.; Veyrieras, J.B.; Hirel, B.; Gallais, A. Genetic variation for N-remobilization and post-silking N uptake in a set of maize recombinant inbred lines. 3. QTL detection and coincidences. *Theor. Appl. Genet.* **2008**, *117*, 729–747. [CrossRef]

agronomy

MDPI

Review

Breeding Maize for Tolerance to Acidic Soils: A Review

Liliane Ngoune Tandzi [1,*], Charles Shelton Mutengwa [1], Eddy Léonard Mangaptche Ngonkeu [2] and Vernon Gracen [3]

[1] Department of Agronomy, University of Fort Hare, P. Bag X1314, Alice 5700, South Africa; cmutengwa@ufh.ac.za

[2] Institute of Agricultural Research for Development (IRAD), University of Yaounde I, P.O. Box 2123, Messa, Yaounde, Cameroon; ngonkeu@yahoo.fr

[3] College of Agriculture and Consumer Sciences, University of Ghana, PMB LG 30, Legon, Accra, Ghana; vg45@cornell.edu

* Correspondence: tnliliane@yahoo.fr or LTandziNgoune@ufh.ac.za; Tel.: +27-063-459-4323

Received: 31 March 2018; Accepted: 23 May 2018; Published: 29 May 2018

Abstract: Acidic soils hamper maize (*Zea mays* L.) production, causing yield losses of up to 69%. Low pH acidic soils can lead to aluminum (Al), manganese (Mn), or iron (Fe) toxicities. Genetic variability for tolerance to low soil pH exists among maize genotypes, which can be exploited in developing high-yielding acid-tolerant maize genotypes. In this paper, we review some of the most recent applications of conventional and molecular breeding approaches for improving maize yield under acidic soils. The gaps in breeding maize for tolerance to low soil pH are highlighted and an emphasis is placed on promoting the adoption of the numerous existing acid soil-tolerant genotypes. While progress has been made in breeding for tolerance to Al toxicity, little has been done on Mn and Fe toxicities. More research inputs are therefore required in: (1) developing screening methods for tolerance to manganese and iron toxicities; (2) elucidating the mechanisms of maize tolerance to Mn and Fe toxicities; and, (3) identifying the quantitative trait loci (QTL) responsible for Mn and Fe tolerance in maize cultivars. There is also a need to raise farmers' and other stakeholders' awareness of the problem of Al, Mn, and Fe soil toxicities to improve the adoption rate of the available acid-tolerant maize genotypes. Maize breeders should work more closely with farmers at the early stages of the release process of a new variety to facilitate its adoption level. Researchers are encouraged to strengthen their collaboration and exchange low soil pH-tolerant maize germplasm.

Keywords: maize; low soil pH; toxicity; breeding; tolerance

1. Introduction

Maize (*Zea mays* L.) is among the most widely grown crops in the world after rice (*Oriza sativa* L.) and wheat (*Triticum aestivum* L.). It forms the basis for food security in some of the world's poorest regions in Africa, Asia, and Latin America and is produced on nearly 100 million hectares in 125 developing countries [1]. One of the major abiotic constraints of maize production is the occurrence of acidic soils, caused by a low potential of hydrogen (pH). Considerable grain yield reductions of maize under low soil pH have been reported in numerous studies. Dewi-Hayati et al. [2] reported that grain yield reduction in acid soils varied from 2.8 to 71%, whereas Tandzi et al. [3] found maize yield reduction under acid soils to be up to 69%. The variation in yield reduction under low soil pH is based on the level of acidity in the soil, the agro-climatic conditions of the environment, and the genetic potential of maize genotypes. Improving grain yield under acidic soil conditions is a major objective of maize breeding programs in many regions of the world. An estimated 3950 million ha, or 30% of global arable land, is covered by acidic soils [4–6]. The largest amount of potentially arable acid soils exists in the humid tropical zones, and comprises about 60% of the acid soils of the world [5]. The poor fertility of acidic soils is due to a combination of mineral toxicities (Al, Mn, and Fe) and nutrient deficits caused by the leaching or decreased availability of phosphorus (P), calcium (Ca), magnesium (Mg), sodium (Na), and micronutrients such as molybdenum (Mo), zinc (Zn), and boron (B) [7].

The development of high-yielding maize cultivars has been the target of selection and breeding procedures in tropical and subtropical regions with acid soils. Grain yield is often the product of interactions between plant genotypes and the environment during the cropping cycle [8]. A high level of heterosis and good combining ability are prerequisites for developing good, economically viable maize hybrids [9]. Conventional breeding, based on testcross data, has been widely used to estimate heterosis between populations or inbred lines, and used to assign inbreds to heterotic groups [10–15]. Combining ability analyses assess the potential value of inbred lines and identify the nature of gene action controlling various quantitative characters. This information is essential for maize breeding focusing on developing hybrids, synthetics, and improved open pollinated cultivars [16] under low soil pH. The advent of molecular genetics has enabled the use of DNA markers to tag genomic regions associated with tolerance to low soil pH, and the subsequent utilization of marker-assisted selection (MAS) to enhance the efficiency of maize breeding programs [17]. Additionally, the identification of key physiological processes associated with yield improvement and the determination of gene-to-phenotype associations can potentially increase the efficiency of breeding for acid tolerance, either through traditional or molecular methods [18].

This paper reviews maize improvement for tolerance to acidic soils using conventional and molecular technologies, with a special focus on the experimentations that have improved the acid tolerance of some maize genotypes. It also reviews the genetic, physiological, and biochemical mechanisms by which plants tolerate low soil pH stress. The adoption of existing and improved acid-tolerant maize genotypes is also taken into account. Challenges faced in breeding for acid tolerance are identified and suggestions for overcoming them are provided. Areas with limited information and research attention are also identified. The intensification of research efforts to fill the identified gaps in information could improve on the efforts already made in the development of high-yielding and high-quality acid-tolerant maize cultivars.

2. Acid Soils

2.1. Distribution of Acidic Soils

Acidic soils occur mainly in two global belts: the northern belt, with a cold, humid temperate climate, and the southern tropical belt, with warm and humid conditions [5,6,19]. The global distribution of acid soils is as follows: 40.9% in the Americas, 26.4% in Asia, 16.7% in Africa, 9.9% in Europe, and 6.1% in Australia and New Zealand. About 67% of the acid-soil area is under forests, 18% under savannas and prairie vegetation, 4.5% under arable lands, and less than 1% under perennial tropical lands [5]. In Cameroon, acid soils cover 75% of arable land [20,21], whereas in Kenya they cover only 13% of the total land area [22]. In South Africa, 5 million ha of soils are severely acidified with an estimated 11 million ha being moderately acidic [23]. In KwaZulu Natal, 85% of soils have pH < 5 and about half of these have an acid saturation of >10% [24]. The distribution of low soil pH (pH < 5) in the world is presented in Figure 1.

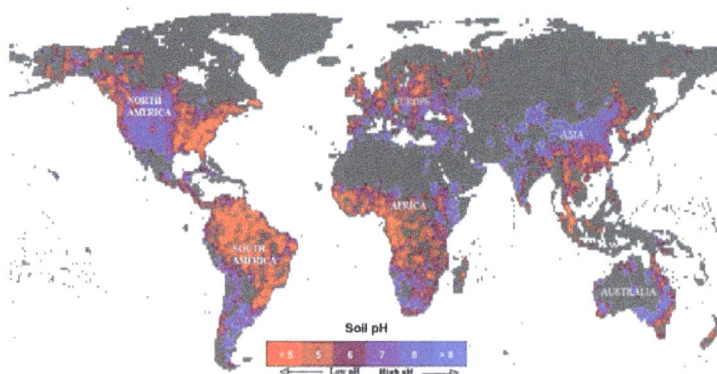

Figure 1. World distribution of soil pH (modified from [25]).

2.2. Acidification of Soils

Acidic soils are defined as soils with pH < 5.5 in the top layer [4–6]. The amount of hydrogen cation (H^+) activity in the soil solution determines the soil pH and is influenced by edaphic, climatic, and biological factors. High rainfall affects the rate of soil acidification when rainfall washes away bases (Ca^{2+}, Mg^{2+}, K^+, Na^+, and carbonate ion (CO_3^{-2})) from the soil. Hydrolysis results in a reduction in soil pH when a metal is dissolved in water, releasing protons. The hydrolytic displacement of base cations and the provision of additional acids from oxidation reactions are the main natural causes of soil acidification, which lead to base-deficient, aerated sands under strong leaching conditions such as high rainfall and drainage [26]. Poor agricultural practices (use of ammonium fertilizers and crop removal) also contribute to the acidification of the soil [27]. Soil acidification is intensified by the removal of cations through the harvesting of crops and by acid precipitation from polluted air [28,29].

2.3. Toxification of Acid Soils

Acid soil toxicity is caused by a combination of high solubility of toxic heavy metal elements (iron, copper, manganese, zinc, and aluminum), a lack of essential nutrients (phosphorus, magnesium, calcium, potassium, sodium), and low soil pH [30,31]. Low soil pH can therefore generate excesses of aluminum, iron, and manganese, which hamper crop production [32]. High Al and Fe oxides and hydroxide in low soil pH are responsible for P fixation, making it unavailable to plants [33–35]. All of these toxicities (Al, Mn, and Fe) should be considered when breeding for maize tolerance to low soil pH around the world.

2.3.1. Aluminum Toxicity

Al toxicity limits agricultural productivity by preventing crops to reach their yield potential [36–39]. Al toxicity (60 to 300 µg per liter of water in soil) can cause 25–80% yield losses in various crop plants [40]. Under Al toxicity, nitrogen (N), P, and potassium (K) uptake, which are essential nutrients responsible for the stimulation of root growth [41], become unavailable. Strong subsoil Al toxicity reduces plant rooting depth, increases susceptibility to drought, and decreases the use of subsoil nutrients [28]. Al toxicity effects result in root damage, which hamper nutrient uptake ability, resulting in nutrient deficiency in the plant [42,43]. Under Al toxicity, P deficiency leads to stunted plant growth, and thin and spindly stems with purpling leaves, which results in the reduction of grain yield [44,45]. The determination of the content of available Al (exchangeable and in the soil solution) is essential for an evaluation of the risk for plant production in acid soils. While most of the attention on acidic soils has been focused on Al toxicity, limited attention has been placed on Fe and Mn toxicities.

2.3.2. Iron Toxicity

Iron is the fourth most abundant mineral in the earth's crust after oxygen (O_2), silicon (Si), and Al. Fe toxicity is a disorder associated with large concentrations of reduced iron (Fe^{2+}) in the soil solution, which occurs in flooded soils [46]. The hydrolysis of Fe is more acidic than Al hydrolysis. Acidity resulting from Fe toxicity is normally buffered by Al hydrolysis reactions. However, once most of the soil Al ions have reacted, Fe hydrolysis takes over, leading to a profound decrease in soil pH [47].

In low soil pH, the anaerobic bacteria provide very high amounts of ferrous ion, which become toxic to plants. Acid soils that are poorly aerated or compacted can increase iron content to the point of toxicity. A high concentration of Fe^{2+} in the rhizosphere has antagonistic effects on the uptake of essentials nutrients (P, K, and Zn) by the plants, causing the accumulation of harmful organic acids or hydrogen sulphides, and consequently leading to plant yield reduction [48]. Yield reductions of 12 to 100% have been previously observed in rice growing in iron toxic soils [49–51], depending on the level of iron toxicity, genetic background of genotypes, and soil fertility status. High iron availability in the soils can also lead to direct or indirect toxicity in the plants [52]. High toxic levels of accumulated Fe in plants can damage lipids, proteins, and deoxyribonucleic acid (DNA). Direct effects of iron toxicity also include damage to cell structures leading to reduced plant growth and injury to foliage [53–56]. In tolerant genotypes, excess iron is known to precipitate on roots, forming an iron plaque that acts as a barrier against iron while ensuring the utilization of essential nutrients by those plants (a process known as indirect iron toxicity). In contrast, an imbalance of nutrients has been observed in susceptible plants growing in soils with toxic levels of iron [48,57,58]. The critical concentration of iron toxicity symptoms varies from 10 to 500 mg/L depending on the nutrient status in the plants and the presence of the reduction products in the environment [51].

2.3.3. Manganese Toxicity

Manganese (Mn) is an essential trace element throughout all stages of plant development, which becomes toxic when taken up in excessive quantities. Mn is deficient in maize plants when its level is less than 15 ppm; it is low when it is between 15 and 25 ppm; sufficient between 26 and 150 ppm; high between 151 and 200 ppm; and excessive or toxic when its concentration is higher than 200 ppm [59].

Mn toxicity is associated with Al and Fe hydrolysis, the primary reactions causing soil acidity. Soil acidification further enhances the solubility of Mn, and thus increases its bioavailability to toxic levels in natural and agricultural systems [60,61]. The effects of Mn toxicity are more pronounced in sensitive plants with a decrease in soil pH, which further increases the solubility of Mn [47,61]. The first symptoms of Mn toxicity appear on the oldest leaves of plants as chlorosis, which later progresses to necrosis [62]. In addition, plants exposed to excess Mn exhibit a very strong inhibition of chloroplast structure and functions, reduced photosynthetic and transpiration rates, and inhibition of carbon dioxide (CO_2) fixation as a result of stomatal closure [63,64]. To date, there is a very limited number of published reports on manganese toxicity in plants. Therefore, this area of study requires more investigations.

2.4. Management of Acidic Soils

A number of management practices are used to correct low soil pH. Liming is the most commonly recommended management practice [65–67]. Kisinyo [68] found that the application of both lime and P fertilizer are important for P and N fertilizer recovery efficiencies necessary for healthy maize growth under acid soils. However, the application of lime and/or fertilizer is not always affordable for small-scale farmers and is not environmentally friendly [12,69]. Additionally, liming affects the topsoil and does not remove acidity in the subsoil, where it poses a severe problem to developing roots [70,71]. Mwangi et al. [72] reported that farm yard manure is a better amendment for correcting soil pH because it has a strong buffering capacity that contains both soil acidity and alkalinity. However, the general recommendations are very high (10 tons per hectare) and the manure is not always available [73].

The addition of crop residues to soils can result in an increase in soil pH [74–76]. Hoyt and Turner [74] found an increase in soil pH of about 0.5 of a pH unit when lucerne meal was added to acid soil, but observed a decline of pH 20 days after incubation. It has been generally observed that the addition of residues causes an initial rise in soil pH, which is then followed by a decline in pH [75,77]. The use of acid-tolerant maize cultivars constitutes an efficient and permanent alternative to increase yields in acidic soils [78].

3. Mechanisms of Tolerance to Low Soil pH

The mechanisms of tolerance to low soil pH occur at physiological, biochemical, and molecular levels. However, some of the mechanisms are still poorly understood, and thus require more research. Knowledge of the mechanisms of tolerance of maize genotypes to Al, Fe, and Mn toxicities could facilitate the development of acid soil-tolerant genotypes.

The exclusion of Al in the root apoplast as well as intracellular tolerance to Al toxicity [79,80] have been suggested as important mechanisms for Al tolerance in maize [36–38]. Aluminum exclusion is related to the ability of Al-tolerant plants to excrete organic acids (predominantly citric acid and oxalate in maize) and phenolics from the root apex [37,81,82]. The secreted organic anions (OA) bind with Al to form a complex (Al–OA) which protects the root apex, thus allowing it to continue growing [38,80,81]. The exudation of other organic compounds may be important for the chelation of Al in the root apex [38], even though little is known about their mechanisms of action [79]. Other ligands released in the root apex include phenolic compounds, flavonoids, succinates, phosphates, uridine diphosphate-glucose, and polysaccharides in the form of mucilages [38,79–81,83,84]. Mucilage is a gelatinous material made of high molecular weight polysaccharides exuded from the most external layers of the root apex and is an important mechanism of resistance to metals [79,85].

Physiological, molecular, and biochemical studies have shown that the modification of the cell wall composition imparts resistance to Al toxicity to some genotypes [79,84–86]. The most important internal tolerance mechanisms are Al-binding proteins, the chelation of Al using organic acids (such as citric acid) and other organic ligands in the cytosol, the compartmentalization or sequestration of Al in the vacuoles, the evolution of Al-tolerant enzymes (such as Nicotinamide adenine dinucleotide phosphate—specific isocitrate dehydrogenase), elevated enzyme activity [37,86,87], and the Al-induced synthesis of callose (such as 1, 3-glucan) [78,88]. The phospholipid composition of the plasma membrane plays an important role in Al toxicity since it creates a negative charge on the surface of the membrane and increases the sensitivity to Al [79,89].

The mechanisms of tolerance to Mn toxicity are not clearly established [64]. Different plant species and genotypes respond differently for tolerance to Mn toxicity [64,90]. The tolerance of maize plants to Mn toxicity has been attributed to tolerance to high tissue concentrations of Mn [63]. Stoyanova et al. [64] evaluated four cultivars of maize under high and toxic concentrations and found that the most tolerant genotype, Kneja 434, expressed a stronger internal capacity of protection against the phytotoxicity of Mn and a higher potential of Mn detoxification. More research is needed to better understand the mechanisms of plant tolerance to Mn and Fe toxicities. Connolly and Guerinot [53] stated that cells store iron with specialized iron-storage proteins called ferritins, which play an important role in iron homeostasis, but the significance of this finding remains unclear.

4. Breeding Maize for Tolerance to Low Soil pH

Tolerance to mineral elements can be defined as the ability of a plant to grow better, produce dry matter, develop fewer deficiency symptoms when grown at low or toxic levels of the mineral element, and give better yield [91,92]. The breeding of maize for tolerance to low soil pH has largely used conventional and molecular approaches, as elucidated in the following sections.

4.1. Conventional Breeding Methods

Breeding programs place emphasis on the grain and forage production of maize plants for human consumption and animal feed. Selection is therefore directed toward the improvement of plant growth characteristics under stress conditions. Conventional breeding makes use of heterosis, heterotic patterns, and heterotic groups, and combinability to develop improved maize hybrids with tolerance to low soil pH. Screening for tolerance to low pH can be performed in the laboratory using hematoxylin and nutrient culture assays [93–96], in the greenhouse or glasshouse [97,98], and under field conditions in sites that are characterized as 'hot spots' for soil acidity [3,12,99–101]. However, the use of nutrient growth solution is a poor predictor for genetic selection under Al toxic soils in the field. Ouma et al. [94] found that the nutrient culture screening for Al toxicity can predict field selection under Al toxic soils with an accuracy of 24 to 35%, depending on the saturation level of Al in a particular soil and the level of available phosphorus. Under low soil pH, Al ions tend to form highly stable complexes with phosphorus [102]. Under these conditions, maize cultivars with high P use efficiency have a good acid soil tolerance capacity.

4.1.1. Heterosis, Heterotic Patterns, and Heterotic Groups for Maize Tolerance to Low Soil pH

Heterosis (hybrid vigor) is the phenomenon in which the progeny of crosses between inbred lines or pure-bred populations perform better than the expected average of the two parents (mid parent heterosis) for particular traits [103] or better than the best parent (high parent heterosis). There are three probable causes of heterosis and none involves additive variance. The dominance hypothesis involves the action and interaction of favorable dominant alleles. The over dominance hypothesis proposes that heterozygous loci are superior to homozygous loci. The epistasis hypothesis involves the interaction of alleles from different loci, which is what occurs with regulatory loci that can turn other genes on or off. Ceballos et al. [104] found in their study in Mexico that in acidic soils, dominance was the most important source of variation, accounting for about 55% of the total variation. Additive and epistatic variance components each accounted for 22.5% of the total genotypic variation in the same study. Heterosis has been efficiently used in the development of high-yielding maize hybrids tolerant to low soil pH.

Inbreds can be classified into specific heterotic groups or patterns depending on their similarity in combining ability and heterotic response when crossed with genotypes from other genetically distinct groups [103]. Twenty-eight maize open pollinated cultivars (OPVs) were crossed in all possible combinations and the 378 varietal hybrids derived from the crosses were evaluated in 10 environments with acid soils in Brazil. From that study, the cultivars were classified into four heterotic groups based on combining ability data. The consistency of the proposed heterotic groups was confirmed by comparing intra- and inter-group first generation progeny (F_1) values and mid parent heterosis [105]. More often, significant differences have been recorded among environments and line x tester interaction across environments, which have highlighted the different responses of genotypes in various environments [12,13,15,106]. Four distinct heterotic groups were identified under acid soil and across environments in Cameroon based on the specific combining ability and yield superiority of F_1 hybrids over the best hybrid check from inbred line x tester crosses conducted in 2012–2014 [15]. In India, six parental inbreds were classified into three heterotic groups based on their specific combining ability for yield under low soil pH. In this case, the superior heterotic pattern was flint × dent [107].

The standard public sector view of heterosis is based on statistical estimates and often presents a complex view of heterosis and heterotic groups that is different from most private sector programs. In most private sector programs, breeders develop heterotic groups by using elite inbred lines as testers. The general process is to cross two inbreds with different desirable traits but which both combine well with an elite tester. The F_1 of this cross is selfed to produce a source population from which new inbreds are developed. The new inbreds are selected by the evaluation of hybrids between them and the tester. Any tester hybrids that outperform the best check hybrid by 10% or more are advanced and the new inbred is classed into an anti-tester heterotic group [15]. In practice, commercial maize hybrids consist of crosses of two unrelated inbreds, each derived from different families or heterotic groups. The diversity increases on pooling germplasm from different heterotic groups. The application of these methods could be useful in developing high-yielding maize genotypes with tolerance to low soil pH. Several research studies [12,13,15,103–107] have been conducted during the past few years under low soil pH conditions using conventional methods (including combinability and heterotic grouping) with the goal of developing high-yielding and tolerant maize genotypes. Despite the efforts and progress made by researchers, the grain yield loss of maize due to low soil pH is still very high. This might be because low soil pH is associated with others stresses such as water stress and poor soil fertility, thus making the selection process more complex.

4.1.2. Combining Abilities and Heritability of Maize Genotypes for Tolerance to Low Soil pH

The concept of combining ability is important in hybrid development, where you compare the performance of lines in hybrid combinations. Combining ability or productivity in crosses is defined as the ability of a parent or an inbred line to transmit desirable traits to a hybrid. There are two types of combining ability: general (GCA) and specific (SCA). SCA is the interaction of genes of two parents involved in a cross with a specific inbred in relation to its contributions in crosses with an array of other inbreds. It relates to non-additive gene effects and depends on how genes from each inbred complement the other [108]. Meanwhile, GCA refers to the average performance of a given genotype or parent in a series of hybrid combinations.

The nature and magnitude of gene action is an important factor in developing an effective breeding program [16]. In Cameroon, the evaluation of 121 testcross hybrids under acid soil environments showed that three parental inbred lines (Cam Inb gp1 17, 4001, and 9450) produced hybrids with the best specific combining ability [15]. Pandey et al. [99] found that parents versus crosses mean squares of grain yield were highly significant, indicating the importance of heterosis and non-additive gene effects for grain yield under acidic soil environments. Under low soil pH, Parentoni et al. [109] and Chen et al. [110] also found the non-additive gene effects to be more important than the additive effects for tolerance to low P. Moreover, Ouma [96] detected both additive and non-additive gene effects under high and low P in Kenyan acid soils. Under low soil pH, the yield-related traits are generally controlled by additive and non-additive gene effects with the predominance of non-additive gene effects. The breeding value and gene effects of maize genotypes for tolerance to low soil pH are presented in Table 1.

Table 1. Gene effects for maize tolerance to low soil pH.

Type of Maize Genotype	Combining Abilities in Low Soil pH	References
Hybrids	Importance of both additive and non-additive gene effects for yield and yield components in acid soil environments.	[100,101,103,111]
F$_1$ progenies	Both additive and non-additive gene actions with the predominance of non-additive effects were observed under acid soils.	[11,12,15,99]
Maize populations	Epistasis accounts for the higher proportion of the total variability of the total sum of squares in acid soil locations.	[104]
Maize hybrids (single and top cross)	Tolerance to Al toxicity in soil acidity was controlled by additive as well as non-additive gene effects, with the preponderance of additive effects.	[111–117]
Single cross hybrids	In acid soil with manganese toxicity, the contribution of non-additive gene effects was greater than the additive effects of genes.	[116,117]
Testcross hybrids	At Nkolbisson in Cameroon, where the soil acidity contains Mn toxicity, the effect of additive genes was higher than the effect of non-additive genes.	[14]

Grain yield is the most important quantitative trait and it involves multiple genetic effects. In the development of high-yielding maize hybrids, it is important to know the gene actions and their relative contribution in the expression of yield-related traits. Under low soil pH, yield is controlled by additive and non-additive gene actions [100,101,114] with the predominance of non-additive gene effects [11,12,15,99,103]. It is difficult to select for yield directly so the yield components that are usually controlled by additive gene effects are evaluated. Several researchers have reported high heritability and high genetic variance for different yield components in maize [118,119]. Rafique et al. [118] reported high heritability (>80%) for all of the traits collected (plant height, ear height, ear length, ear diameter, grain yield, kernel per row) showing heritable variation among genotypes. Rafiq et al. [119] found high heritability estimates (>80%) for all of the traits indicating the preponderance of additive gene action. Bello et al. [120] reported high phenotypic and genotypic coefficient of variations as well as high heritability for the number of grains per ear, ear weight, and plant and ear heights, and found that these traits were under the control of additive gene effects. Therefore, effective selection could be possible for the improvement of these yield components under normal conditions. Khan et al. [121] reported moderate heritability estimates under normal conditions for plant height (38.5%), ear height (32.6%), days to 50% tasseling (48.4%), 100-kernels weight (42.3%), ear length (49.5%), ear diameter (40.0%), and cob diameter (28.5%). Opposingly, Tandzi et al. [3] observed low heritability for all of the agronomic traits (anthesis-silking interval (9%), plant height (27%), ear height (40%), ear per plant (14%), ear aspect (28%), plant aspect (17%), and yield (8%)) recorded under low soil pH.

4.1.3. Secondary Traits Associated with Tolerance to Acidic Soils

The identification of secondary traits is very useful in breeding maize for tolerance to low soil pH due to their correlations with yield. Since yield is mostly controlled by non-additive gene action, secondary traits could be used as indirect predictors of yield in acid soil environments. In addition, secondary traits can also be useful for the genotypic characterization of plants in response to low soil pH stress. Baligar et al. [122] reported that at 41% Al saturation, tolerant maize genotypes produced high shoot and root weight as well as high nutrient efficiency ratio (NER) for P, Ca Mg, and Fe. The greater grain yield of newer acid-tolerant Argentinean maize hybrids was mainly related to an increase in harvest index [123]. Increased harvest index was more associated with a greater increase in grain yield components including kernel number and/or kernel weight than with an increase in shoot biomass. The characteristic most frequently associated with genetic yield improvement in maize under stress conditions is delayed leaf senescence or "stay green" [124,125]. Different secondary traits associated with tolerance to low soil pH at different stages of maize development have been identified over the years (Table 2). Welcker [101] observed moderately strong to strong correlations of anthesis-silking interval, plant height, and ears per plant with yield under low soil pH conditions. However, Tandzi et al. [3] observed weak to moderately weak correlations of anthesis-silking interval and plant height with yield, suggesting the unreliability of these traits as predictors of performance under low soil pH conditions. Leaf area, photosynthetic rate, stress tolerance index, and stress susceptibility index showed strong correlations with yield (Table 2); these traits could be reliable predictors of yield due to their probable stability across environments. Some secondary traits are specific to certain mineral toxicities in the soil. For instance, chlorophyll content could be a useful secondary trait under Mn toxicity [64].

Table 2. Secondary traits associated with tolerance to low soil pH.

Secondary Traits *	References
Anthesis silking interval (−0.65 in 1999 and −0.66 in 2000), plant height (+0.65 in 1999, +0.71 in 2000), and ears per plant (+0.50 in 1998, +0.74 in 1999 and +0.74 in 2000) were strongly related to yield.	[101]
Leaf area (+0.75) and photosynthetic rate (+0.78) were highly and positively correlated with grain yield.	[101]
Seminal root length measured at leaf stage 4 appeared to be the most sensitive trait for tolerance to low pH under laboratory conditions.	[12,101]
Relative Net Root Growth (RNRG) was found to predict field performance under Al toxic soils by between 24% and 35%.	[96]
Plant height (0.36), ear height (0.28), and stress tolerance index (0.94) were highly and positively correlated with yield. Anthesis-silking interval (−0.13), plant aspect (−0.4), ear aspect (−0.47), and stress susceptibility index (−0.90) were negatively correlated with yield.	[3]

* Values in brackets refer to correlation coefficients between yield and secondary trait.

4.2. Application of Molecular Tools in Breeding for Maize Tolerance to Acidic Soils

Molecular approaches have led to significant achievements of higher tolerance and the identification of genes responsible for acidic soil tolerance, specifically Al tolerance [29]. Molecular markers have been used in maize breeding to map blocks of genes associated with economically important traits, often termed as quantitative trait loci (QTL) [126]. Molecular markers are modern diagnostic tools, which may therefore help breeders to develop cultivars with tolerance to Al, Fe, and Mn toxicities. Molecular markers could be used in marker-assisted selection (MAS) [127], once genes linked to each of these toxicities have been identified.

Members of two different families of transporters, *ALMT* (Al-activated malate transporter) and *MATE* (multidrug and toxic compound extrusion), have been identified to facilitate organic acid anion efflux in Al-tolerant plants [128–131]. Using maize association mapping population and three independent F_2 populations derived from the crosses B73 × CML247, B73 × CML333, and B73 × NC350, Krill et al. [131] identified six candidate genes, including *ALMT2*, in response to Al toxicity. However, only four of the candidate genes, *Zea mays AltSB like* (*ZmASL*), *Zea mays aluminum-activated malate transporter2* (*ALMT2*), *S-adenosyl-L-homocysteinase* (*SAHH*), and *Malic enzyme* (*ME*), were identified by both the association and linkage mapping studies. Froese and Carter [132] found that the positive allele of marker wmc331 linked to *ALMT1* was not widely present in the winter wheat germplasm used in their study but was present only in the most tolerant cultivar. In addition to Al tolerance, genes in the *ALMT* family are also known to influence other physiological processes such as guard cell regulation, fruit quality, and seed development [102].

The *MATE* gene has been cloned and identified in several crops such as *Arabidopsis* [55], barley [133], wheat [134], sorghum [129], rice [135], bean [136], poplar [137], soybean [138], and maize [130,131] (Table 3). The Al-tolerant allele for the *MATE1* locus was found to contain a tandem gene triplication and to have higher levels of gene expression compared to the sensitive allele with a single gene copy [131]. A gene-specific marker, Cit7, was developed based on sequence of the barley *HvMATE* gene to improve the marker-assisted selection of barley genotypes under Al toxicity [139]. The presence of three copies of *MATE1* in Al-tolerant maize lines, Al237, C1006, and IL677a, originating from regions of highly acidic soils in South America and Africa underscores the role of gene copy number variation in the adaptation of plants to acidic soils, and further suggests that genome structural changes may be a rapid evolutionary response to new environments [131].

Mattiello et al. [140] identified 44 candidate genes located within or near intervals of QTL for Al tolerance, with several functions such as cyclins (RNA binding protein, a protease inhibitor) and xyloglucan endo transglycosylase protein 8 precursor that may work together and contribute to maintaining root growth in acid soil with toxic levels of Al. The use of more powerful genomic technics such as the Hi-C sequencing method could provide new evidence for the existence of spatial Al-tolerant gene clusters [141].

To date, there are no reports on genes or QTL associated with tolerance to Fe and Mn toxicities. The identification of QTL linked to secondary traits correlated with yield performance under conditions of either Fe or Mn toxicity could further enhance the efficiency of maize breeding for tolerance to low soil pH. QTL associated with secondary traits such as days to silking, anthesis-silking interval, and stay green characteristic under stressed environments [142,143] could have the potential to be utilized as indirect molecular predictors of performance of plants exposed to Fe and Mn toxicities. Moreover, it is relevant to check whether any of the previously identified genes or QTL for Al tolerance have pleiotropic effects for tolerance to Fe and Mn toxicities. Also, there is need for the application of the recent advances in proteomic and metabolomics to provide a greater understanding of the mechanisms involved in the tolerance of plants to low soil pH.

Table 3. Genes associated with Al tolerance in different crops.

Gene	Crop	Reference
MATE (multidrug and toxic compound extrusion)	Maize *Arabidopsis* Sorghum Rice Bean Poplar Soybean	[129–131] [55] [129] [135] [136] [137] [138]
Zea mays AltSB like (ZmASL), *Zea mays aluminum-activated malate transporter2 (ALMT2),* *S-adenosyl-L-homocysteinase (SAHH),* *Malic enzyme (ME)*	Maize	[139]
HvMATE	Barley	[135,139]
ALMT (Al-activated Malate Transporter)	Wheat, rice, tobacco, barley	[128,132,134,144,145]
BnALMT1 and *BnALMT2* *GmALMT1* *ScALMT1*	Rape Soybean Rye	[146–148]
MsALMT1 *HlALMT1*	*Medicago sativa* *Holcus lanatus*	[149,150]

5. Successes in Breeding for Low Soil pH-Tolerant Maize Genotypes

The International Maize and Wheat Improvement Center (CIMMYT), in collaboration with several National Agricultural Research Systems (NARS) all over the world, have developed soil acidity-tolerant maize cultivars to increase maize productivity per unit area. Acid soil-tolerant maize populations have undergone recurrent selections to improve these populations for grain yield under both acid and normal soils [151].

Significant progress has been made in the development of tolerant maize genotypes to low soil pH since 1997. Case studies showing grain yield improvements of some maize genotypes under acidic soil conditions are summarized in Table 4. Means of yield, percentage yield reduction, and percentage yield increase of the genotypes (general, hybrids, single cross, best single cross, three-way cross, best three-way cross, OPV, and the best commercial check) are summarized in Table 4. The mean yield reduction percentage of genotypes varied from 36 to 51% in Latin America and Asia, and from 37 to 40% in Cameroon. These variations were influenced by several factors such as the level of acidity, the agro-climatic conditions, and the tolerance status of the genotypes to low soil pH. The average grain yield of acid soil-tolerant OPVs such as Sikuani was 3.2 t/ha when evaluated across 13 acid soil environments. On the other hand, top cross hybrids developed in 1995 from crosses of OPVs and inbred lines had an average grain yield of 3.84 t/ha when evaluated across six acid soil environments [151]. Reference [103] reported that an addition of 60 kg ha^{-1} of phosphorus did not produce a significant grain yield difference for the acid soil-tolerant cultivar ATP-SR-Y but significantly increased the grain yield of the susceptible cultivar, Tuxpeno Sequia, indicating that ATP-SR-Y is a P-efficient cultivar. In Cameroon, a maize yield increase of 51% was obtained with some varieties under low soil pH, while a general yield reduction of 37% was observed in other varieties [3,12]. In Kenya, some single cross hybrids (KML 036 × MUL 863, KML 036 × S39615-1, MUL 863 × MUL 1007, MUL 125 × POOLB 26-1, MUL 817 × MUL 125) expressing superior tolerance to Al toxicity were identified [98]. The most common standard acid-tolerant maize cultivar developed in CIMMYT is CIMCALI 97 Balopia SA$_4$, referred to as 97BASA$_4$ [95].

Table 4. Case studies showing grain yield improvements of some maize genotypes evaluated under selected acidic soil environments from 1998 to 2015.

Genotype	Latin America and Asia (1997–2000)			Cameroon (1996–2014)		
	Mean					
	Yield (t/ha)	% Yield Reduction	% Yield Increase	Yield (t/ha)	% of Yield Reduction	% Yield Increase
General	2.62	44.6	19.5	3.15	37.3	5.7
Hybrids	2.64	44.9	20.1	3.0	39.5	1
Single cross	2.56	46.7	17.6	3.38	39.5	12.1
Best single cross	2.86	50.8	26.2	6.5	39	51.2
Three-way cross	2.70	41.6	21.8	-	-	-
Best three-way cross	3.10	41.3	31.9	-	-	-
OPV	2.11	35.9	-	2.97	36.9	-
Best commercial check	-	-	-	3.17	40	6.3

General = overall mean of genotypes, % yield reduction = the proportion of the yield that is lost under low soil pH compared to the control environment, % yield increase = proportion of yield gain under low soil pH compared to the check. Sources: [3,12,99].

Even though the expected yield of maize genotypes that are tolerant to low soil pH is still low compared to normal conditions, several OPVs and hybrids tolerant to this stress have already been developed and made available to producers. In addition to a high level of expression of the Al tolerance gene *MATE1*, some Maize cultivars such as Cateto also contain high copy numbers of the gene [129]. Several inbred lines and hybrids have been developed from crosses involving the cultivar Cateto. Perentoni et al. [45] classified two maize hybrids (L3 × Cateto and 228-3 × L22) as being P-use efficient based on their mean yield under low and high P environments. These single cross hybrids could be highly tolerant to low soil pH since they are capable of efficiently using phosphorus in a stress environment. Collaborative research should be encouraged in the regions where low soil pH is a major problem to maize production, as this may lead to improvements in the commercialization of low soil pH-tolerant cultivars.

6. Adoption of Acid Soil-Tolerant Maize

The adoption of improved maize cultivars remains an important issue in developing countries. Some acid soil-tolerant maize cultivars have not been adopted by farmers even when available, because farmers' selection criteria are generally not considered in the breeding process [51]. About 80% of farmers recycle improved cultivars, including hybrids, contrary to the recommendations [152] in most African countries. Although the superiority of hybrids compared to OPVs has been reported, OPVs will continue to be more important than hybrids under acid soils over the coming years due to the limited resource-base of smallholder farmers [151]. For instance, several high-yielding, low soil pH-tolerant top crosses and hybrid maize genotypes have been identified in Cameroon but have not yet been released [3,12].

Numerous studies have provided insight into the selection criteria of farmers in different parts of Africa. Tolerance to acidic soils was never mentioned in these studies, despite the prevalence of the problem. In the bimodal humid forest area of Cameroon, smallholder farmers are willing to adopt high-yielding maize cultivars tolerant to poor soil fertility if they produce good quality grain and are soft and sweet tasting [68]. Regression analysis showed that household size, level of education, contact with extension agents, access to credit, and yield of improved maize cultivars were the factors that influence the adoption of improved maize cultivars in Nigeria [153]. Gender, age, farm size, income, and lack of access to extension influenced the low adoption rate of maize technologies in Western Kenya. Therefore, it has been recommended that policies should consider household structure, empower smallholder farmers economically, and improve access to extension services to enhance the adoption of improved maize in the country [154]. The provision of social infrastructure, especially access roads to market centers, and the extension of agricultural education to farmers would increase the spread of improved maize cultivars in Kenya [155]. The use of these approaches might also facilitate the adoption of acid-tolerant cultivars.

Farmers, in most cases, are not aware when breeders release new high-yielding cultivars with tolerance to low soil pH. More demonstration plots combined with farmer training and farmer field days could enhance the adoption rate of improved cultivars. Researchers should apply participatory selection approaches in maize breeding to share knowledge with farmers and other stakeholders who regularly interact with farmers. It is not clear whether the majority of researchers actually believe in involving farmers when evaluating new cultivars. Without famer involvement, practitioners end up issuing blanket recommendations that are not context-specific. Researchers should organize some workshops explaining practical aspects to be taken into account to deal with the problem of soil acidity. This could involve designing simpler ways in which farmers could measure soil pH, as well as the identification of indicator plants that grow in acidic soils. The adoption of acid-tolerant cultivars can be expected to improve once farmers have a good appreciation of the problem, as well as simple ways of detecting soil acidity in their fields. Farmers could also be encouraged to share seeds of some Al-tolerant cultivars to facilitate their widespread use.

7. Conclusions

Low soil pH is often combined with other stresses such as drought, low N, low P in the soil, and poor soil fertility. Several maize genotypes with different levels of tolerance to acidic soils have been developed and commercialized throughout the world, but the yield losses to soil acidity still remain high. Tailoring the crop to fit acidic or less fertile soils is more effective and more economically and environmentally friendly than changing the soil to fit the crops. There is a need to use combining ability and heterosis to efficiently develop high-yielding and more adapted acid-tolerant maize genotypes. The integrated use of molecular tools such as marker-assisted selection already applied in some areas is highly encouraged. In countries, mostly in Africa, where the application of molecular markers is non-existent or very limited due to lack of facilities, financial resources, and skilled personnel, the establishment of modern state-of-the-art laboratories and the training of human resources are critical. Researchers are encouraged to strengthen their collaboration through the sharing of data, findings, and germplasm exchange. The availability of maize germplasm with a broad genetic base for tolerance to low soil pH would increase the potential for the development of high-yielding cultivars with high levels of tolerance to low soil pH as well as toxicities of Al, Fe, and Mn. The mechanisms of Mn and Fe tolerance in maize are still not clearly established. More research should be devoted to maize tolerance to Mn and Fe toxicities. To raise the level of adoption of improved maize cultivars under acidic soils, farmers should be involved in the selection process through participatory breeding and selection approaches.

Author Contributions: L.N.T. had the idea of the review in this specific breeding area and wrote the first draft. C.S.M., E.L.M.N. and V.G. contributed to the overall preparation of the manuscript. C.S.M. and V.G. provided the technical guidance and editing support.

Funding: No funding was provided for this review paper.

Acknowledgments: We appreciate the Post-Doctoral Fellowship provided to the main author by the Govan Mbeki Research and Development Center (GMRDC) at the University of Fort Hare in South Africa.

Conflicts of Interest: The authors declare no conflict of interest.

Abbreviations

Al	aluminum
Mn	manganese
Fe	iron
pH	potential of hydrogen
QTL	quantitative trait locus
P	phosphorus

Ca	calcium
Mg	magnesium
Na	sodium
Mo	molybdenum
Zn	zinc
B	boron
DNA	deoxyribonucleic acid
ha	hectare
H+	hydrogen cation
Ca^{2+}	calcium cation
Mg^{2+}	magnesium cation
K+	potassium cation
Na+	sodium cation
$(CO_3)^{-2}$	carbonate ion
O_2	oxygen
Si	silicon
ppm	parts per million
CO_2	carbon dioxide
OA	organic anion
OPVs	open pollinated cultivars
SCA	specific combining ability
GCA	general combining ability
F_1	first generation progeny
ALMT	aluminum-activated malate transporter
MATE	multidrug and toxic compound extrusion
F_2	second generation progeny
CML	CIMMYT maize line
CIMMYT	International Maize and Wheat Improvement Center
NARS	National Agricultural Research Systems
t/ha	tons per hectare
ATP-SR-Y	acid tolerant population streak resistant yellow
ZmMATE	zea multidrug and toxic compound extrusion
HvMATE	Hordeum vulgare multidrug and toxic compound extrusion
AltSB	aluminum tolerant in sorghum
ALMT2	aluminum-activated malate transporter 2
SAHH	S-adenosyl-L-homocysteinase
ME	malic enzyme
RNA	ribonucleic acid
ALMT1	aluminum-activated malate transporter 1.

References

1. Prasanna, B.M. Maize in the developing world: Trends, challenges, and opportunities. In *Addressing Climate Change Effects and Meeting Maize Demand for Asia-B, Proceedings of the Extended Summary 11th Asia Maize Conference, Nanning China, 7–11 November 2011*; CIMMYT: Texcoco, Mexico, 2011.

2. Dewi-Hayati, P.K.; Sutoyo, A.; Syarif, A.; Prasetyo, T. Performance of maize single-cross hybrids evaluated on acidic soils. *Int. J. Adv. Sci. Eng. Inf. Technol.* **2014**, *4*, 30–33.

3. Tandzi, N.L.; Ngonkeu, E.L.M.; Youmbi, E.; Nartey, E.; Yeboah, M.; Gracen, V.; Ngeve, J.; Mafouasson, H. Agronomic performance of maze hybrids under acid and control soil conditions. *Int. J. Agron. Agric. Res.* **2015**, *6*, 275–291.

4. Wambeke, A.V. Formation, distribution and consequences of acid soils in agricultural development. In *Plant Adaptation to Mineral Stress in Problem Soils*; Wright, J.M., Ferrari, A.S., Eds.; Special Publication Cornell University Agricultural Experiment Station: Ithaca, NY, USA, 1976; 15p.

5. Von Uexkull, H.R.; Mutert, E. Global extent, development and economic impact of acid soils. *Plant Soil* **1995**, *171*, 1–15. [CrossRef]

6. Dalovic, I.G.; Jockovć, Đ.S.; Dugalić, G.J.; Bekavac, G.F.; Purar, B.; Šeremešić, S.I.; Jockovć, M.Đ. Soil acidity and mobile aluminum status in pseudogley soils in the čačak–kraljevo basin. *J. Serb. Chem. Soc.* **2012**, *77*, 833–843. [CrossRef]
7. Gupta, N.; Gaurav, S.S.; Kumar, A. Molecular basis of aluminium toxicity in plants: A review. *Am. J. Plant Sci.* **2013**, *4*, 21–37. [CrossRef]
8. Araus, L.J.; Serret, D.M.; Edmeades, O.G. Phenotyping maize for adaptation to drought. *Front. Physiol.* **2012**, *3*, 1–29. [CrossRef] [PubMed]
9. Izhar, T.; Chakraborty, M. Combining ability and heterosis for grain yield and its components in maize inbreds over environments (*Zea mays* L.). *Acad. J.* **2013**, *8*, 3276–3280.
10. Menkir, A.; Badu-Apraku, B.; The, C.; Adepou, A. Evaluation of heterotic patterns of IITA's lowland white maize inbred lines. *Maydica* **2003**, *48*, 161–170.
11. Welcker, C.; The, C.; Andreau, B.; De Leon, C.; Parentoni, S.N.; Bernal, J.; Felicite, J.; Zonkeng, C.; Salazar, F.; Narro, L.; Charcosset, A.; Horst, W.J. Heterosis and combining ability for maize adaptation to tropical acid soils: Implications for future breeding strategies. *Crop Sci.* **2005**, *45*, 2405–2413. [CrossRef]
12. The, C.; Mafouasson, H.; Calba, H.; Mbouemboue, P.; Zonkeng, C.; Tagne, A.; Worst, J.H. Identification de groupes hétérotiques pour la tolérance du maïs (*Zea mays* L.) aux sols acides des tropiques. *Cah. Agric.* **2006**, *15*, 337–346.
13. Badu-Apraku, B.; Oyekunle, M.; Akinwale, R.O.; Aderounmu, M. Combining ability and genetic diversity of extra-early white maize inbreds under stress and nonstress environments. *Crop Sci.* **2013**, *53*, 9–26. [CrossRef]
14. Qurban, A.; Arfan, A.; Muhammad, A.; Sajed, A.; Nazar, H.K.; Sher, M.; Hafiz, G.A.; Idrees, A.N.; Tayyab, H. Line × tester analysis for morpho-physiological traits of *Zea mays* L. Seedlings. *Int. J. Adv. Life Sci.* **2014**, *1*, 242–253.
15. Tandzi, N.L.; Gracen, V.; Ngonkeu, E.L.M.; Yeboah, M.; Nartey, E.; Mafouasson, A.H.; Woin, N. Analysis of combining ability and heterotic grouping of maize inbred lines under acid soil conditions, control soil and across environments. *Int. J. Curr. Res.* **2015**, *7*, 21553–21564.
16. Gowda, R.K.; Kage, U.; Lohithaswa, H.C.; Shekara, B.G.; Shobha, D. Combining ability studies in maize (*Zea mays* L.). *Mol. Plant Breed.* **2013**, *4*, 116–127.
17. Tang, G.-Q.; Li, X.-W. Optimal multiple trait selection for multiple linked quantitative trait loci. *Acta Genet. Sin.* **2006**, *33*, 220–229. [CrossRef]
18. Tollenaar, M.; Lee, E.A. Dissection of physiological processes underlying grain yield in maize by examining genetic improvement and heterosis. *Maydica* **2006**, *51*, 390–408.
19. Khan, A.A.; Mcneilly, T.; Azhar, F.M. Review: Stress tolerance in crop plants. *Int. J. Agric. Boil.* **2001**, *3*, 250–255. [CrossRef]
20. Bindzi, T. Les sols rouges du Cameroun. In *8e Réunion du Sous-Comité Ouest et Centre Africain de Correlation des Sols Pour la Mise en Valeur des Terres*; MINREST-FAO: Yaoundé, Cameroun, 1987.
21. Nwaga, D.; The, C.; Ambassa-Kiki, R.; Ngonkeu, M.E.L.; Tchiegang-Megueni, C. Selection of arbuscular mycorrhizal fungi for inoculating maize and sorghum grown in oxisol/ultisol and vertisol in Cameroon. In *Managing Nutrient Cycles to Sustain Soil Fertility in Sub-Saharan Africa*; Bationo, A., Ed.; CIAT: London, UK, 2004; p. 608.
22. Kisinyo, P.O.; Opala, P.A.; Gudu, S.O.; Othieno, C.O.; Okalebo, J.R.; Palapala, V.; Otinga, A.N. Recent advances towards understanding and managing Kenyan acid soils for improved crop production. *Afr. J. Agric. Res.* **2014**, *9*, 2397–2408. [CrossRef]
23. Venter, A.; Herselman, J.E.; VanderMerwe, G.M.E.; Steyn, C.; Beukes, D.J. Developing soil acidity maps for South Africa. In Proceedings of the 5th International Symposium on Soil Plant Interactions at Low pH, Bergville, South Africa, 12–16 March 2001; p. 21.
24. Roberts, V.G.; Smeda, Z. The distribution of soil fertility constraints in KwaZulu-Natal, South Africa. In Proceedings of the 5th International Symposium on Plant Soil Interactions at low pH, Bergville, South Africa, 12–16 March 2001; p. 12.
25. Cohen, J.; Slessarev, E. *Soil pHertility: The Current*; University of California: Santa Barbara, CA, USA, 2016; Available online: http://www.news.ucsb.edu/2016/017434/soil-fertility (accessed on 10 May 2018).
26. Fey, M.V. Acid soil degradation in South Africa: A threat to agricultural productivity. *FSSA J.* **2001**, 37–41.
27. Rowell, D.L. Soil acidity and alkalinity. In *Russell's Soil Conditions and Plant Growth*, 11th ed.; Wild, A., Ed.; Longman Scientific and Technical: London, UK, 1988; pp. 844–898.

28. Ulrich, B.; Mayer, R.; Khanna, P.K. Chemical changes due to acid precipitation in a loess-derived soil in central Europe. *Soil Sci.* **1980**, *130*, 193–199. [CrossRef]

29. Hede, A.R.; Skovmand, B.; López-Cesati, J. Acid soils and aluminum toxicity. In *Application of Physiology in Wheat Breeding Reynolds*; Ortiz-Monasterio, M.P., Mcnab, J.I., Eds.; CIMMYT: Texcoco, Mexico, 2001; pp. 172–182.

30. Foy, C.D.; Chaney, R.L.; White, M.C. The physiology of metal toxicity in plants. *Annu. Rev. Plant Physiol.* **1978**, *29*, 511–566. [CrossRef]

31. Bian, M.; Zhou, M.; Sun, D.; Li, C. Molecular approaches unravel the mechanism of acid soil tolerance in Plants. *Crop J.* **2013**, *2013*, 91–104. [CrossRef]

32. Zeigler, R.S.; Pandey, S.; Miles, J.; Gourley, L.M.; Sarkarung, S. Advances in the selection and breeding of acid-tolerant plants: Rice, maize, sorghum and tropical forages. In *Plant Soil Interactions at Low pH*; Kluwer Academic Publishers: Dordrecht, The Netherlands, 1995; pp. 391–406.

33. Sanchez, P.A.; Shepherd, K.D.; Soule, M.J.; Place, F.M.; Buresh, R.J.; Izac, A.-M.N.; Mokwunye, A.U.; Kwesiga, F.; Nderitu, C.G.; Woomer, P.L. Soil fertility management in Africa: An investment in natural resource. In *Replenishing Soil Fertility in Africa*; Buresh, R.J., Sanchez, P.A., Calhoun, F., Eds.; SSSA Special Publication: Madison, WI, USA, 1997; Volume 51, pp. 1–46.

34. Oburo, P.A. Effects of Soil Properties on Bioavailability of Aluminium and Phosphorus in Selected Kenyan and Brazilian Soils. Ph.D. Thesis, Perdue University, West Lafayette, IN, USA, 2008.

35. Kisinyo, P.O.; Othieno, C.O.; Gudu, S.O.; Okalebo, J.R.; Opala, P.A.; Maghanga, J.K.; Ng'etich, W.K.; Agalo, J.J.; Opile, R.W.; Kisinyo, J.A.; et al. Phosphorus sorption and lime requirements of maize growing acids soil of Kenya. *Sustain. Agric. Res.* **2013**, *2*, 116–123. [CrossRef]

36. Delhaize, E.; Ryan, P.R. Aluminum toxicity and tolerance in plants. *Plant Physiol.* **1995**, *107*, 315–321. [CrossRef] [PubMed]

37. Kochian, L.V.; Hoekenga, O.A.; Pineros, M.A. How do crop plants tolerate acid soils? Mechanisms of aluminum tolerance and phosphorous efficiency. *Annu. Rev. Plant Biol.* **2004**, *55*, 459–493. [PubMed]

38. Kochian, L.V.; Pineros, M.A.; Liu, J.; Magalhaes, J.V. Plant adaptation to acid soils: The molecular basis for crop aluminium resistance. *Annu. Rev. Plant Physiol.* **2015**, *66*, 571–598. [CrossRef] [PubMed]

39. Rao, I.M.; Miles, J.W.; Beede, S.E.; Horst, W.J. Root adaptations to soils with low fertility and aluminium toxicity. *Ann. Bot.* **2016**, *118*, 593–605. [CrossRef] [PubMed]

40. Barabasz, W.; Albińska, D.; Jaśkowska, M.; Lipiec, J. Ecotoxicology of Aluminium. *Pol. J. Environ. Stud.* **2002**, *11*, 199–203.

41. Osaki, M.T.; Watanabe, T.; Tadano, T. Beneficial effect of aluminium on growth of plants adapted to low pH soils. *Soil Sci. Plant Nutr.* **1997**, *43*, 551–563. [CrossRef]

42. Steiner, F.; Zoz, T.; Junior, A.S.P.; Castagnara, D.D.; Dranski, J.A.L. Effects of aluminium on plant growth and nutrient uptake in young physic nut plants. *Semin. Ciênc. Agrar.* **2012**, *33*, 1779–1788. [CrossRef]

43. Silva, S. Aluminium toxicity targets in plants. *J. Bot.* **2012**, *2012*. [CrossRef]

44. Parentoni, S.N.; Souza, J.R.; Alves, V.M.C.; Gama, E.E.G.; Coelho, A.M.; Oliveira, A.C.; Guimaraes, P.E.O.; Guimaraes, C.T.; Vasconcelos, M.J.V.; Pacheco, C.A.P.; et al. Inheritance and breeding strategies for phosphorus efficiency in tropical maize (*Zea mays* L.). *Maydica* **2010**, *55*, 1–15.

45. Ouma, E.; Ligeyo, D.; Matonyei, T.; Were, B.; Agalo, J.; Emily, T.; Onkware, A.; Gudu, S.; Kisinnyo, P.; Okalebo, J.; et al. Development of maize single cross hybrids for tolerance to low phosphorus. *Afr. J. Plant Sci.* **2012**, *6*, 394–402.

46. Becker, M.; Asch, F. Iron toxicity in rice-conditions and management concepts. *Plant Nutr. Soil Sci.* **2005**, *168*, 558–573. [CrossRef]

47. Havlin, J.L.; Beaton, S.L.; Nelson and Nelson, W.L. *Soil Fertility and Fertilizers: An Introduction to Nutrient Management*; Pearson Prentice Hall: Upper Saddle River, NJ, USA, 2005.

48. Fageria, N.K.; Santos, A.B.; Barbosa-Filho, M.P.; Guimaraes, C.M. Iron toxicity in lowland rice. *J. Plant Nutr.* **2008**, *31*, 1676–1697. [CrossRef]

49. Sikirou, M.; Saito, K.; Dramé, K.N.; Saidou, A.; Dieng, I.; Ahanchédé, A.; Venuprasad, R. Soil-based screening for iron toxicity tolerance in rice using pots. *Plant Prod. Sci.* **2016**, *19*, 489–496. [CrossRef]

50. Ikehashi, H.; Ponnamperuma, F.N. Varietal tolerance of rice for adverse soils. In *Soils and Rice*; International Rice Research Institute IRRI: Los Banos, Philippines, 1978; pp. 801–825.

51. Sahrawat, K.L. Iron toxicity in Wetland rice and role of other nutrients. *J. Plant Nutr.* **2004**, *27*, 1471–1504. [CrossRef]

52. Saaltink, R.M.; Dekker, S.C.; Eppinga, M.B.; Griffioen, J.; Wassen, M.J. Plant-specific effects of iron-toxicity in wetlands. *Plant Soil* **2017**, *416*, 83–96. [CrossRef]

53. Wheeler, B.D.; Al-Farraj, M.M.; Cook, R.E.D. Iron toxicity to plants in base-rich wetlands: Comparative effects on the distribution and growth of Epilobium hirsutum and Juncus subnodulosus Schrank. *New Phytol.* **1985**, *100*, 653–669. [CrossRef]

54. Laan, P.; Smolders, A.; Blom, C.W.P.M. The relative importance of anaerobiosis and high iron levels in the flood tolerance of Rumex species. *Plant Soil* **1991**, *136*, 153–161. [CrossRef]

55. Liu, J.; Magalhaes, J.V.; Shaff, J.; Kochian, L.V. Aluminum-activated citrate and malate transporters from the MATE and ALMT families function independently to confer Arabidopsis aluminium tolerance. *Plant J.* **2009**, *57*, 389–399. [CrossRef] [PubMed]

56. Ayeni, O.; Kambizi, L.; Fatoki, O.; Olatunji, O. Risk assessment of wetland under aluminium and iron toxicities: A review. *Aquat. Ecosyst. Heath Manag.* **2014**, *17*, 122–128. [CrossRef]

57. Snowden, R.E.D.; Wheeler, B.D. Chemical changes in selected wetland plant species with increasing Fe supply, with specific reference to root precipitates and Fe tolerance. *New Phytol.* **1995**, *131*, 503–520. [CrossRef]

58. Tripathi, R.D.; Tripathi, P.; Dwivedi, S.; Kumar, A.; Mishra, A.; Chauhan, P.S.; Norton, G.J.; Nautiyal, C.S. Roles for root iron plaque in sequestration and uptake of heavy metals and metalloids in aquatic and wetland plants. *Metallomics* **2014**, *6*, 1798–1800. [CrossRef] [PubMed]

59. Schulte, E.E.; Kelling, K.A. *Understanding Plant Nutrients/Soil and Applied Manganese*; University of Wisconsin System Board of Regents and University of Wisconsin-Extension, Cooperative Extension: Madison, WI, USA, 1999; 4p.

60. Foy, C.D. Plant adaptation to acid, aluminium-toxic soils. *Commun. Soil Sci. Plant Anal.* **1988**, *19*, 959–987. [CrossRef]

61. Hong, E.; Ketterings, Q.; Mcbride, M. Manganese Agronomy Fact Sheet Series. In *Field Crop Extension*; College of Agriculture and Life Sciences: Ithaca, NY, USA, 2010; Volume 49.

62. Snowball, K.; Robson, A.D. *Nutrient Deficiencies and Toxicities in Wheat: A Guide for Field Identification*; CIMMYT: Texcoco, Mexico, 1991; 82p.

63. Doncheva, S.; Poschenriederb, C.; Stoyanovaa, Z.; Georgievaa, K.; Velichkovac, M.; Barcelób, J. Silicon amelioration of manganese toxicity in Mn-sensitive and Mn-tolerant maize varieties. *Environ. Exp. Bot.* **2009**, *65*, 189–197. [CrossRef]

64. Stoyanova, Z.; Poschenrieder, C.; Tzvetkova, N.; Doncheva, S. Characterization of the tolerance to excess manganese in four maize varieties. *Soil Sci. Plant Nutr.* **2010**, *55*, 747–753. [CrossRef]

65. Novak, J.M.; Busscher, W.J.; Laird, D.L.; Ahmedna, M.; Watts, D.W.; Niandou, M.A.S. Impact of biochar amendment on fertility of Southeastern Coastal Plan soil. *Soil Sci.* **2009**, *174*, 105112. [CrossRef]

66. Yuan, J.H.; Xu, R.K. The amelioration effects of low temperature biochar generated from nine crop residues on an acidic Ultisol. *Soil Use Manag.* **2010**, *27*, 110–115. [CrossRef]

67. Goulding, K.W.T. Soil acidification and the importance of liming agricultural soils with particular reference to the United Kingdom. *Soil Use Manag.* **2016**, *32*, 390–399. [CrossRef] [PubMed]

68. Kisinyo, P.O. Effect of lime and phosphorus fertilizer on soil chemistry and maize seedlings performance on Kenyan acid soils. *Sky J. Agric. Res.* **2016**, *5*, 097–104.

69. Tandzi, N.L.; Ngonkeu, E.L.M.; Nartey, E.; Yeboah, M.; Ngeve, J.; Mafouasson, H.; Nso-Ngang, A.; Bassi, O.; Gracen, V. Farmers' adoption of improved maize varieties in the humid forest area of Cameroon. *Int. J. Sci. Eng. Appl. Sci.* **2015**, *1*, 17–28.

70. Toma, M.; Sumner, M.E.; Weeks, G.; Saigusa, M. Long-term effects of gypsum on crop yield and subsoil chemical properties. *Soil Sci. Soc. Am. J.* **1999**, *39*, 891–895. [CrossRef]

71. Sierra, J.; Ozier-Lafontaine, H.; Dufour, L.; Meunier, A.; Bonhomme, R.; Welcker, C. Nutrient and assimilate partitioning in two tropical maize cultivars in relation to their tolerance to soil acidity. *Field Crop. Res.* **2006**, *95*, 234–249. [CrossRef]

72. Mwangi, T.J.; Ngeny, J.M.; Wekesa, F.; Mulati, J. *Acidic Soil Amendment for Maize Production in Uasin Gishu District, North Rift Kenya*; Kenya Agricultural Research Institute, National Agricultural Research Centre: Kitale, Kenya, 2000.

73. Van Averbeke, W.; Yoganathan, S. *Using Kraal Manure as a Fertilizer*; Department of Agriculture and the Agricultural and Rural Development Research Institute, Fort Hare: Alice, South Africa, 2003.

74. Hoyt, P.B.; Turner, R.C. Effects of organic materials added to very acid soils on pH, aluminium, exchangeable NH4, and crop yield. *Soil Sci.* **1975**, *119*, 227–237. [CrossRef]

75. Hue, N.V. Correcting soil acidity of a highly weathered Ultisol with chicken manure and sewage sludge. *Commun. Soil Sci. Plant Anal.* **1992**, *23*, 241–264. [CrossRef]

76. Noble, AD.; Zenneck, I.; Randall, P.J. Leaf litter ash alkalinity and neutralisation of soil acidity. *Plant Soil* **1996**, *179*, 293–302. [CrossRef]

77. Wong, M.T.F.; Nortcliff, S.; Swift, R.S. Method for determining the acid ameliorating capacity of plant residue compost, urban compost, farmyard manure and peat applied to tropical soils. *Commun. Soil Sci. Plant Anal.* **1998**, *29*, 2927–2937. [CrossRef]

78. Horst, W.J.; Puschel, A.K.; Schmohl, N. Induction of callose formation is a sensitive marker for genotypic aluminium sensitivity in maize. *Plant Soil* **1997**, *192*, 23–30. [CrossRef]

79. Bojorquez-Quintal, E.; Escalante-Magana, C.; Eschevarria-Machado, L.; Martinez-Estévez, M. Aluminum, a friend or foe of higher plants in acid soils. *Front. Plant Sci.* **2017**, *8*, 1767. [CrossRef] [PubMed]

80. Yang, Y.; Wang, Q.L.; Guo, Z.H.; Zhao, Z. Rhizosphere pH difference regulated by plasma membrane H + ATPase is related to differential Al-tolerance of two wheat cultivars. *Plant Soil Environ.* **2011**, *57*, 201–206. [CrossRef]

81. Zheng, S.J.; Ma, J.F.; Matsumoto, H. High aluminium resistance in buckwheat-1, Al-induced specific secretion of oxalic acid from root tips. *Plant Physiol.* **1998**, *117*, 745–751. [CrossRef]

82. Pellet, D.M.; Papemik, L.A.; Kochian, L.V. Multiple aluminium-resistance mechanisms in wheat (roles of root apical phosphate and malate exudation). *Plant Physiol.* **1995**, *112*, 591–597. [CrossRef]

83. Huang, C.F.; Yamaji, N.; Mitani, N.; Yano, M.; Nagamura, Y.; Ma, J.F. A bacterial-type ABC transporter is involved in aluminium tolerance in rice. *Plant Cell* **2009**, *21*, 655–667. [CrossRef] [PubMed]

84. Levesque-Tremblay, G.; Pelloux, J.; Braybrook, S.A.; Müller, K. Tuning of pectin methyl esterification: Consequences for cell wall biomechanics and development. *Plant* **2015**, *242*, 791–811. [CrossRef] [PubMed]

85. Che, J.; Yamaji, N.; Shen, R.F.; Ma, J.F. An Al-inducible expansion gene, Os EXPA 10 is involved in root elongation of rice. *Plant J.* **2016**, *88*, 132–142. [CrossRef] [PubMed]

86. Zhang, H.; Ma, Y.; Horst, W.J.; Yang, Z.-B. Spatial-temporal analysis of polyethylene glycol-reduced aluminium accumulation and xyloglucan endotransglucosylase action in root tips of common bean (Phaseolus vulgaris). *Ann. Bot.* **2016**, *118*, 1–9. [CrossRef] [PubMed]

87. Ma, J.F. Role of organic acids in detoxification of aluminum in higher plants. *Plant Cell Physiol.* **2000**, *41*, 383–390. [CrossRef] [PubMed]

88. Singh, S.; Tripathi, D.K.; Singh, S.; Sharma, S.; Dubey, N.K.; Chauhan, D.K.; Vaculík, M. Toxicity of aluminium on various levels of plant cells and organism: A review. *Environ. Exp. Bot.* **2017**, *137*, 177–193. [CrossRef]

89. Wagatsuma, T.; Khan, M.S.; Watanabe, T.; Maejima, E.; Sekimoto, H.; Yokota, T.; Nakano, T.; Toyomasu, T.; Tawaraya, K.; Uemura, H.; et al. Higher sterol content regulated by CYP51 with concomitant lower phospholipid contents in membranes is a common strategy for for aluminium tolerance in several plant species. *J. Exp. Bot.* **2015**, *66*, 907–918. [CrossRef] [PubMed]

90. Horst, W.J.; Fesht-Christoffers, M.; Naumann, A.; Wissemeier, A.H.; Maier, P. Physiology of manganese toxicity and tolerance in (Vigna unguiculata L.) Walp. *J. Plant Nutr. Soil Sci.* **1999**, *162*, 263–274. [CrossRef]

91. Graham, M.H.; Haynes, R.J.; Meyer, J.M. Changes in soil chemistry and aggregates stability induced by fertilizer application, burning and trash relation on a long-term sugarcane experiment in South Africa. *Soil Sci.* **2002**, *53*, 589–598. [CrossRef]

92. Hacisalihoglu, G.; Kochian, L.V. How do some plants tolerate low levels of soil zinc? Mechanisms of zinc efficiency in crop plants. *New Phytol.* **2003**, *159*, 341–350.

93. Magnacava, R.; Gardner, C.O.; Clark, R.B. Evaluation of inbred maize lines for aluminium tolerance in nutrient solution. In *Genetic Aspects of Plant Mineral Nutrition*; Gabelman, H.W., Loughman, B.C., Eds.; Martinus Nijhoff/Dr Junk, W. Publishers: Hague, The Netherlands, 1986; pp. 89–199.

94. Delhaize, E.; Craig, S.; Beaton, C.D.; Bennet, R.J.; Jagadish, V.C.; Randall, P.J. Aluminium tolerance in wheat (Triticum aestivum L.): Uptake and distribution of aluminium in root apices. *Plant Physiol.* **1993**, *103*, 685–693. [CrossRef] [PubMed]

95. Gudu, S.; Maina, S.M.; Onkware, A.O.; Ombakho, G.; Ligeyo, D.O. Screening of Kenyan maize germplasm for tolerance to low pH and aluminium for use in acid soils of Kenya. In Proceedings of the Seventh Eastern and Southern Africa Regional Maize conference, Nairobi, Kenya, 11–15 February 2002.

96. Ouma, E.; Ligeyo, D.; Matonyei, T.; Agalo, J.; Were, B.; Too, E.; Onkware, A.; Gudu, S.; Kisinyo, P.; Nyangweso, P. Enhancing maize grain yield in acid soils of Western Kenya using aluminium tolerant germplasm. *J. Agric. Sci. Technol.* **2013**, *3*, 33–46.

97. Cancado, G.M.A.; Loguercio, L.L.; Martins, P.R.; Parentoni, S.N.; Paiva, E.; Borem, A.; Lopes, M.A. Haematoxylin staining as a phenotypic index for aluminium tolerance selection in tropical maize (*Zea mays* L.). *Theor. Appl. Genet.* **1999**, *99*, 747–754.

98. Giaveno, C.D.; Filho, J.B.D.M. Field comparison between selection method at the maize seedling stage in relation to aluminium tolerance. *Sci. Agric.* **2002**, *59*, 397–401. [CrossRef]

99. Pandey, S.; Ceballos, H.; Magnavaca, R.; Bahia Filho, A.F.C.; Duque-Vargas, J.; Vinasco, L.E. Genetics of tolerance to soil acidity in tropical maize. *Crop Sci.* **1994**, *34*, 1511–1514. [CrossRef]

100. Salazar, F.S.; Pandey, S.; Narro, L.; Perez, J.C.; Ceballos, H.; Parentoni, S.N.; Bahia Filho, A.F.C. Diallel analysis of acid-soil tolerant and intolerant tropical maize populations. *Crop Sci.* **1997**, *37*, 1457–1462. [CrossRef]

101. Welcker, C. *Fitting Maize into Cropping Systems on Acid Soils of the Tropics*; Final Progress Report 01/10/1996-30/09/2000; INRA Guadeloupe: Paris, France, 2000; pp. 89–108.

102. Sharma, T.; Dreyer, I.; Kochian, L.; Pineros, M.A. The ALTM family of organic acid transporters in plants and their involvement in detoxification and nutrient security. *Front. Plant Sci.* **2016**, *7*, 1–12. [CrossRef] [PubMed]

103. Kwena, O.P. Recurrent Selection for Gray Leaf Spot (GLS) and Phaeosphaeria Leaf Spot (PLS) Resistance in Four Maize Populations and Heterotic Classification of Maize Germplasm from Western Kenya. Ph.D. Thesis, University of University of Nairobi, Nairobi, Kenya, 2008.

104. Ceballos, H.; Pandey, S.; Narro, L.; Perez-Velazquez, J.C. Additive, dominant, and epistatic effects for maize grain yield in acid and non-acid soils. *Theor. Appl. Genet.* **1998**, *96*, 662–668. [CrossRef]

105. Parentoni, S.N.; Magalhães, J.V.; Pacheco, C.A.P.; Santos, M.X.; Abadie, T.; Gama, E.E.G.; Guimarães, E.O.; Meirelles, W.F.; Lopes, M.A.; Vasconcelos, M.J.V.; et al. Heterotic groups based on yield-specific combining ability data and phylogenetic relationship determined by RAPD markers for 28 tropical maize open pollinated varieties. *Euphytica* **2001**, *121*, 197–208. [CrossRef]

106. Ifie, B.E. Genetic Analysis of Striga Resistance and Low Soil Nitrogen Tolerance in Early Maturing Maize (*Zea mays* L.) Inbred Lines. Ph.D. Thesis, University of Ghana, Accra, Ghana, 2013.

107. Rajendran, A.; Muthiah, A.; Joel, J.; Shanmugasundaram, P.; Raju, D. Heterotic grouping and patterning of quality protein maize inbreds based on genetic and molecular marker studies. *Turk. J. Biol.* **2014**, *38*, 10–20. [CrossRef]

108. Sprague, G.F.; Tatum, L.A. General versus specific combining ability in single crosses. *J. Am. Soc. Agron.* **1942**, *34*, 923. [CrossRef]

109. Parentoni, S.N.; Souza, J.R.; Alves, V.M.C.; Gama, E.E.G.; Coelho, A.M.; Oliveira, A.C.; Guimaraes, P.E.O.; Guimaraes, C.T.; Vasconcelos, M.J.V.; Magalhães, J.V.; et al. Breeding maize for Al tolerance, P use efficiency and acid soil adaptation for the cerrado areas of Brazil: EMBRAPA's experience. In Proceedings of the 3rd International Symposium on Phosphorus Dynamics in the Soil-Plant Continuum, Uberlândia, Brazil, 14–19 May 2006; pp. 129–131.

110. Chen, J.; Xu, L.; Cai, Y.; Xu, J. Plant growth habit, root architecture traits and tolerance to low soil phosphorus in an Andean bean population. *Euphytica* **2009**, *165*, 257–258.

111. Borrero, J.C.; Pandey, S.; Ceballos, H.; Magnacava, R.; Bahia, A.F.C. Genetic variances for tolerance to soil acidity in tropical maize population. *Maydica* **1995**, *40*, 283–288.

112. Magnavaca, R.; Gardner, C.O.; Clark, R.B. Comparisons of maize populations for aluminium tolerance in nutrient solution. In *Genetic Aspects of Plant Mineral Nutrition*; Gabelman, H.W., Loughman, B.C., Eds.; Martinus Nijjhoff: Dordrecht, The Netherlands, 1987.

113. Lima, M.; Furlani, P.R.; Miranda Filho, J.B. Divergent selection for aluminium tolerance in a maize (*Zea mays* L.) population. *Maydica* **1992**, *37*, 123–132.

114. Duque-Vargas, J.; Pandey, S.; Granados, G.; Ceballos, H.; Knapp, E. Inheritance of tolerance to soil acidity in tropical maize. *Crop Sci.* **1994**, *34*, 50–54. [CrossRef]

115. The, C.; Tandzi, N.L.; Zonkeng, C.; Ngonkeu, E.L.M.; Meka, S.; Leon, C.; Horst, W.J. Contribution of introduced inbred lines to maize varietal improvement for acid soil tolerance. In *Demand-Driven Technologies for Sustainable Maize Production in West and Central Africa*; Badu-Apraku, B., Fakorede, M.A.B., Lum, A.F., Menkir, A., Ouedraogo, M., Eds.; International Institute of Tropical Agriculture (IITA): Cotonou, Bénin, 2007; pp. 53–63.

116. Tekeu, H.; Ngonkeu, E.L.M.; Tandzi, L.N.; Djocgoue, P.F.; Bell, J.M.; Mafouasson, H.A.; Boyomo, O.; Petmi, C.L.; Fokom, R. Evaluation of maize (*Zea mays* L.) accessions using line x tester analysis for aluminum and manganese tolerance. *Int. J. Biol. Chem. Sci.* **2015**, *9*, 2161–2173. [CrossRef]

117. Petmi, C.L.; Ngonkeu, E.L.M.; Tandzi, N.L.; Ambang, Z.; Boyomo, O.; Bell, J.M.; Tekeu, H.; Mafouasson, H.; Malaa, D.; Noé, W. Screening of maize (*Zea mays* L.) genotypes for adaptation on contrasted acid soils in the humid forest zone of Cameroon. *J. Exp. Agric. Int.* **2016**, *14*, 1–15. [CrossRef]

118. Rafique, M.; Hussain, A.; Mahmood, T.; Alvi, A.W.; Alvi, B. Heritability and interrelationships among grain yield and yield components in maize (*Zea mays* L.). *Int. J. Agric. Biol.* **2004**, *6*, 1113–1114.

119. Rafiq, C.M.; Rafique, M.; Hussain, A.; Altaf, M. Studies on heritability, correlation and path analysis in maize (*Zea mays* L.). *Agric. Res.* **2010**, *48*, 35–38.

120. Bello, O.B.; Ige, S.A.; Azeez, M.A.; Afolabi, M.S.; Abdulmaliq, S.Y.; Mahamood, J. Heritability and genetic advance for grain yield and its component characters in maize (*Zea mays* L.). *Int. J. Plant Res.* **2012**, *2*, 138145.

121. Khan, M.H.; Ahmad, M.; Hussain, M.; Mahmood-ul-Hassan Ali, Q. Heritability and trait association studies in maize F_1 hybrids. *Int. J. Biosci.* **2018**, *12*, 18–26.

122. Baligar, V.C.; Pitta, G.V.E.; Schaffert, R.E.; Bahia Filho, A.F.D.C.; Clark, R.B. Soil acidity effects on nutrient use efficiency in exotic maize genotypes. *Plant Soil* **1997**, *192*, 9–13. [CrossRef]

123. Echarte, L.; Nagore, L.; Di Matteo, J.; Di, J.; Robles, M.; Cambareri, M.; Maggiora, A.D. *Grain Yield Determination and Resource Use Efficiency in Maize Hybrids Released in Different Decades: Chapter 2*; IITECH/Agricultural Chemistry: London, UK, 2013; pp. 1–36.

124. Duvick, D.N. What is yield? In *Developing Drought and Low N-Tolerant Maize*; Edmeades, G.O., Banziger, B., Mickelson, H.R., Pena-Valdivia, C.B., Eds.; CIMMYT: El Batan, Mexico, 1997; pp. 332–335.

125. Nguyen, H.T.; Blum, A. *Physiology and Biotechnology Integration for Plant Breeding*; Marcel Dekker Inc.: New York, NY, USA, 2004; 648p.

126. Xu, Y.; Crouch, J.H. Genomics of tropical maize, a staple food and feed across the world. In *Genomics of Tropical Crop Plants*; Moore, P.H., Ming, R., Eds.; Springer: New York, NY, USA, 2008; pp. 333–370.

127. Masojć, P. The application of molecular markers in the process of selection. *Cell. Mol. Biol. Lett.* **2002**, *7*, 499–509. [PubMed]

128. Sasaki, T.; Yamamoto, Y.; Ezaki, B.; Katsuhara, M.; Ahn, S.; Ryan, P.R.; Delhaize, E.; Matsumoto, H. A wheat gene encoding an aluminium-activated malate transporter. *Plant J.* **2004**, *37*, 645–653. [CrossRef] [PubMed]

129. Magalhaes, J.V.; Liu, J.; Guimaraes, C.T.; Lana, U.G.; Alves, V.M. A gene in the multidrug and toxic compound extrusion (MATE) family confers aluminum tolerance in sorghum. *Nat. Genet.* **2007**, *39*, 1156–1161. [CrossRef] [PubMed]

130. Maron, L.G.; Pineros, M.A.; Guimaraes, C.T.; Magalhaes, J.V.; Pleiman, J.K.; Mao, C.; Shaff, J.; Belicuas, S.N.J.; Kochian, L.V. Two functionally distinct members of the MATE (multi-drug and toxic compound extrusion) family of transporters potentially underlie two major aluminum tolerance QTL in maize. *Plant J.* **2010**, *61*, 728–740. [CrossRef] [PubMed]

131. Maron, L.G.; Guimaraes, C.T.; Krist, M.; Albert, P.S.; Birchler, J.A.; Bradburry, P.J.; Buckler, E.S.; Coluccio, A.E.; Danilova, T.; Kudrna, D.; et al. Aluminum tolerance in maize is associated with higher MATE1 gene copy number. *Proc. Natl. Acad. Sci. USA* **2013**, *110*, 5241–5246. [CrossRef] [PubMed]

132. Froese, P.S.; Carter, A.H. Single nucleotide polymorphisms in the wheat genome associated with tolerance of acidic soils and aluminium toxicity. *Crop Sci.* **2016**, *56*, 16621677. [CrossRef]

133. Furukawa, J.; Yamaji, N.; Wang, H.; Mitani, N.; Murata, Y.; Sato, K.; Katsuhara, M.; Takeda, K.; Ma, J.F. An aluminium-activated citrate transporter in barley. *Plant Cell Physiol.* **2007**, *48*, 1081–1091. [CrossRef] [PubMed]

134. Garcia-Oliveira, A.; Martins-Lopes, P.; Tolrà, R.; Poschenrieder, C.; Tarquis, M.; Guedes-Pinto, H.; Benito, C. Molecular characterization of the citrate transporter gene TaMATE1 and expression analysis of upstream genes involved in organic acid transport under Al stress in bread wheat (Triticum aestivum). *Physiol. Plant.* **2014**, *152*, 441–452. [CrossRef] [PubMed]

135. Yokosho, K.; Yamaji, N.; Ma, J.F. An Al-inducible MATE gene is involved in external detoxification of Al in rice. *Plant J.* **2011**, *68*, 1061–1069. [CrossRef] [PubMed]

136. Yang, X.Y.; Yang, J.L.; Zhou, Y.; Pineros, M.A.; Kochian, L.V.; Li, G.X.; Zheng, S.J. A de novo synthesis citrate transporter, Vigna umbellate multidrug and toxic compound extrusion, implicated in Al activated citrate efflux in rice bean (Vigna umbellata) root apex. *Plant Cell Environ.* **2011**, *34*, 2138–2148. [CrossRef] [PubMed]

137. Grisel, N.; Zoller, S.; Künzli-Gontarczyk, M.; Lampart, T.; Münsterkötter, M.; Brunner, I.; Bovet, L.; Métraux, J.-P.; Sperisen, C. Transcriptome responses to aluminium stress in roots of aspen (Populus tremula). *BMC Plant Biol.* **2010**, *10*, 185. [CrossRef] [PubMed]

138. Liu, J.; Li, Y.; Wang, W.; Gai, J.; Li, Y. Genome-wide analysis of MATE transporters and expression patterns of a subgroup of MATE genes in response to aluminium toxicity in soybean. *BMC Genom.* **2016**, *17*, 223.

139. Bian, M.; Jin, X.; Broughton, S.; Zhang, X.-Q.; Zhou, G.; Zhou, M.; Zhang, G.; Li, C. A new allele of acid soil tolerance gene from a malting barley variety. *BMC Genet.* **2015**, *16*, 92. [CrossRef] [PubMed]

140. Mattiello, L.; Rodrigues da Silva, F.; Menossi, M. Linking microarray data to QTLs highlights new genes related to Al tolerance in maize. *Plant Sci.* **2012**, *191*, 8–15. [CrossRef] [PubMed]

141. Schulz, T.; Stoye, J.; Doerr, D. Graph teams: A method for discovering spatial gene clusters in Hi-C sequencing data. *BMC Genom.* **2018**, *19*, 60–95. [CrossRef] [PubMed]

142. Ribeiro, P.F.; Badu-Apraku, B.; Gracen, V.E.; Danquah, E.Y.; Garcia-Oliveira, A.L.; Asante, M.D.; Afriyie-Debrah, C.; Gedil, M. Identification of Quantitative Trait Loci for grain yield and other traits in tropical maize under high and low soil-nitrogen environments. *Crop Sci.* **2018**, *58*, 321–331. [CrossRef]

143. Shi, S.; Azam, F.I.; Li, H.; Li, B.; Jing, R. Mapping QTL for stay green and agronomic traits in wheat under diverse water regimes. *Euphytica* **2017**, 213–246. [CrossRef]

144. Delhaize, E.; Ryan, P.R.; Hebb, D.M.; Yamamoto, Y.; Sasaki, T.; Matsumoto, H. Engineering high-level aluminium tolerance in barley with the ALMT1 gene. *Proc. Natl. Acad. Sci. USA* **2004**, *101*, 15249–15254. [CrossRef] [PubMed]

145. Krill, A.M.; Kirst, M.; Kochian, L.V.; Buckler, E.S.; Hoekenga, O.A. Association and linkage analysis of aluminium tolerance genes in maize. *PLoS ONE* **2010**, *5*, 111. [CrossRef] [PubMed]

146. Hoekenga, O.A.; Maron, L.G.; Pineros, M.A.; Cançado, G.M.A.; Shaff, J.; Kobayashi, Y.; Ryan, P.R.; Dong, B.; Delhaize, E.; Sasaki, T. AtALMT1, which encodes a malate transporter, is identified as one of several genes critical for aluminium tolerance in Arabidopsis. *Proc. Natl. Acad. Sci. USA* **2006**, *103*, 9738–9743. [CrossRef] [PubMed]

147. Ligaba, A.; Katsuhara, M.; Ryan, P.R.; Shibasaka, M.; Matsumoto, H. The BnALMT1 and BnALMT2 genes from rape encode aluminium-activated malate transporters that enhance the aluminium resistance of plant cells. *Plant Physiol.* **2006**, *142*, 1294–1303. [CrossRef] [PubMed]

148. Ligaba, A.; Katsuhara, M.; Sakamoto, W.; Matsumoto, H. The BnALMT1 protein that is an aluminium-activated malate transporter is localized in the plasma membrane. *Plant Signal. Behav.* **2007**, *2*, 255–257. [CrossRef] [PubMed]

149. Chen, Q.; Wu, K.-H.; Wang, P.; Yi, J.; Li, K.-Z.; Yu, Y.-X.; Chen, L.M. Over expression of MsALMT1, from the aluminium-sensitive Medicago sativa enhances malate exudation and aluminium resistance in tobacco. *Plant Mol. Biol. Rep.* **2013**, *31*, 769–774. [CrossRef]

150. Chen, Z.C.; Yokosho, K.; Kashino, M.; Zhao, F.-J.; Yamaji, N.; Ma, J.F. Adaptation to acidic soil is achieved by increased numbers of cis-acting elements regulating ALMT1 expression in Holcus lanatus. *Plant J.* **2013**, *76*, 10–23. [PubMed]

151. Narro, L.; Pandey, S.; De León, C.; Salazar, F.; Arias, M.P. Implications of soil-acidity tolerant maize cultivars to increase production in developing countries. In *Plant Nutrient Acquisition*; Ae, N., Arihara, J., Okada, K., Srinivasan, A., Eds.; Springer: Tokyo, Japan, 2001; pp. 447–463.

152. Nkonya, E.; Xavery, P.; Akonaay, H.; Mwangi, W.; Anandajayasekeram, P.; Verkuijl, H.; Martella, D.; Moshi, A. *Adoption of Maize Production Technologies in Northern Tanzania*; The United Republic of Tanzania, and the Southern African Centre for Cooperation in Agricultural Research (SACCAR): Gaborone, Botswana; CIMMYT: Texcoco, Mexico, 1998; ISBN 970-648-003-X 53.

153. Kudi, T.M.; Bolaji, M.; Akinola, M.O.; Nasa'i, D.H. Analysis of adoption of improved maize varieties among farmers in Kwara state, Nigeria. *Int. J. Peace Dev. Stud.* **2011**, *1*, 8–12.

154. Nyangweso, P.M.; Amusala, G.; Gudu, S.; Onkware, A.; Ochuodho, J.; Ouma, E.; Kisinyo, P.; Mugavalai, V.; Okalebo, J.R.; Othieno, C.O.; et al. Drivers of awareness and adoption of maize and sorghum technologies in Western Kenya. In Proceedings of the 19th International Farm Management Congress, Warsaw, Poland, 21–26 July 2013.

155. Mwabu, G.; Mwangi, W.; Nyangito, H. Does adoption of improved maize varieties reduce poverty? Evidence from Kenya. In Proceedings of the International Association of Agricultural Economists Conference, Gold Coast, Australia, 12–18 August 2006.

agronomy

MDPI

Article

Genotype-by-Environment Interaction and Yield Stability of Maize Single Cross Hybrids Developed from Tropical Inbred Lines

Hortense Noëlle Apala Mafouasson [1], Vernon Gracen [2], Martin Agyei Yeboah [2], Godswill Ntsomboh-Ntsefong [1] , Liliane Ngoune Tandzi [1,3,*] and Charles Shelton Mutengwa [3]

[1] Institute of Agricultural Research for Development (IRAD), P.O. Box 2123 Yaounde, Cameroon; mafouasson@yahoo.fr (H.N.A.M.); ntsomboh@yahoo.fr (G.N.-N.)
[2] West Africa Centre for Crop Improvement (WACCI), College of Basic and Applied Sciences, University of Ghana, PMB 30 Legon, Accra, Ghana; vg45@cornell.edu (V.G.); mayeboah@yahoo.com (M.A.Y.)
[3] Department of Agronomy, University of Fort Hare, P. Bag X1314, Alice 5700, South Africa; cmutengwa@ufh.ac.za
* Correspondence: tnliliane@yahoo.fr; Tel.: +27-634-594-323

Received: 27 March 2018; Accepted: 25 April 2018; Published: 1 May 2018

Abstract: Nitrogen (N) is one of the most important nutrients required for high productivity of the maize plant. In most farmers' fields in Sub-Saharan Africa (SSA), there is low availability of N in the soil mainly due to continuous cultivation of the land, crop residues removal, little or no application of fertilizers and rapid leaching. There is a need to develop low N tolerant and adapted maize genotypes. Evaluation of maize genotypes under different nitrogen conditions would therefore be useful in identifying genotypes that combine stability with high yield potential for both stress and non-stress environment. Eighty maize hybrids were evaluated at Mbalmayo and Nkolbisson in Cameroon, during 2012 and 2013 minor and major cropping seasons across 11 environments under low and high N conditions. The objectives of the study were: (i) to determine the effect of genotype x environment interaction (G × E) on grain yield and yield stability of single cross maize hybrids across low N and optimum N environments and (ii) to identify genotypes to recommend for further use in the breeding program. Yield data of 80 hybrids were analyzed initially and the analysis of 20 best performing genotypes was further performed for a better visualization and interpretation of the results. Combined analysis of variance showed highly significant G × E effects for grain yield. The GGE biplot analysis divided the study area into three mega environments: one related to the major cropping season while the two others were related to the minor cropping season. The grain yield of the 20 highest yielding hybrids ranged from 4484.7 to 5198.3 kg ha^{-1}. Hybrid 1368 × 87036 was the highest yielding in the minor season while the most outstanding hybrid, TL-11-A-1642-5 × 87036 was the best for the major season. The latter hybrid showed the potential for production across environments and should therefore be further tested in multiple environments to confirm consistency of its high yield performance and stability, and to facilitate its release as a commercial hybrid. High yielding but not stable hybrids across environments could be recommended for the specific environments where they performed well.

Keywords: Maize; hybrids; genotype × environment; stability

1. Introduction

Maize is one of the most important cereals in Sub-Saharan Africa (SSA) and a staple food for an estimated 50% of the population. It is an important source of carbohydrate, protein, iron, vitamin B, and minerals [1] and accounts for about 15% of the caloric intake of the population [2]. In Cameroon, maize is the most consumed cereal, much more than sorghum, rice and wheat [3]. Maize is grown in all the five agro ecological zones of the country, namely: Sudano-Sahelian Zone, High Guinea Savanna, Humid Forest Zone and the Western Highlands with a mono-modal rainfall pattern and the Humid Forest Zone with a bimodal rainfall pattern [4]. These agro-ecological zones are within an altitude ranging from zero and 4095 m above sea level. The crop is grown both by small and large scale farmers [3,4] under a wide range of conditions such as different soil types, soil fertility levels, moisture levels, different temperatures and cultural practices.

Low soil nitrogen (low N) limits maize yield production in diverse arable land. It is also one of the most important environmental constraints contributing to yield instability of maize. Ajala et al. [5] found large genotypic and phenotypic variances for maize grain yield under low and high N environments with low heritability estimates but with yield gain of more than 25% in all environments. Therefore, selection for grain yield and correlated traits under low N may result in improving maize performance under low N soil environments. Because of the high genotype × environment interactions involved, stressed experiments often produce rankings that differ significantly from one experiment to another, making it difficult to identify the best germplasm [6]. In general, maize yields are considerably low under the smallholders farming systems of the tropics than other environments predominantly due to lack of well-adapted and improved cultivars and due to genotype by environment (G × E) interaction [7]. Most farmers, especially small scale farmers usually grow varieties based on many criteria, but they usually do not consider the suitability of the variety to the environment which is usually influenced by many biotic and abiotic stresses among which is low soil nitrogen. Consequently, this always results in low yields compared to yields obtained in research stations [3]. In the bimodal forest zone of Cameroon, the average maize yield ranged from 0.8 to 1 t ha^{-1} at farmer level [8]. These low yields obtained by farmers are probably due to poor and unstable environmental conditions. Environmental conditions can fluctuate as a result of drought, reduced soil fertility, pressure from insects and diseases [9]. It has also been reported [9] that environmental conditions can further be amplified by socio-economic constraints faced by small scale farmers that result in suboptimal input application. The farmers usually have limited access to technology and inputs, especially fertilizer, irrigation facilities and pesticides and have no means to modify or condition the production environment [9]. The authors found that including selection under high priority abiotic stresses, such as drought and low N, in a routine breeding program and with adequate weighting can significantly increase maize yields in a highly variable drought-prone environment and particularly at lower yield levels.

Large genotype by environment interactions (G × E) commonly occur under stress conditions; consequently a variety which performs well in one environment during one season or year may not perform well in a different period or in a different site within the same region [10]. This is because genotypes exhibit different levels of phenotypic expression under different environmental conditions resulting in crossover performances [11]. Genotype × environment interaction is also the result from differences in the sensitivities of genotypes to the conditions in the target environment [11]. Genetic × environment interactions (G × E) are of major importance in developing improved genotypes across different environments. When G × E interaction effects are non-significant, means of evaluated varieties across environments are adequate indicators of genotypic performance across the environments. In this situation, the varieties are said to be stable across the environments [12]. Significant G × E indicates that selections from one environment may often perform differently in another and the variety is not stable across the environments [12]. Therefore, information on G × E may help in determining a breeding strategy. When G × E exists, it is necessary to determine whether there are important crossovers, i.e., rank changes of the genotypes in different environments, such that different winners are picked up in different environments [13]. When there is no change in rank of genotypes over environments, there is non-crossover type of interaction effects, and genotypes with superior means can be recommended for all the environments [13]. Breeders can also use information on G × E to choose appropriate locations for selection [13].

Maize growers need cultivars that are reliable and consistent across a wide array of stress conditions and have high yield potential that may be expressed when conditions become more favorable [14]. Plant breeders should therefore develop cultivars capable of withstanding unpredictable environmental variations [14]. In addition, the varieties developed should be stable across environments in order to be widely accepted by farmers throughout a region [14–17]. It is, therefore, important for newly improved maize cultivars to be evaluated at many sites and for a number of years before release [18,19]. Unfortunately, in these multi-location trials, varietal selection is often inefficient due to G × E and relative rankings of varieties usually differ across environments [15,17,18,20,21]. As a result, it becomes difficult to demonstrate the superiority of any single variety. This can be done through the use of various statistical models [19,20]. These statistical analyses give information on adaptability and stability of varieties across target environments. It would then be possible to identify varieties that are appropriate for a specific environment and those with stable performance across environments.

Many stability analysis models exist: joint regression analysis proposed by Eberhart and Russell [22] to estimate the average performance of a genotype in different environments relative to the mean performance of all genotypes in the same environment; multivariate analysis among which are the additive main effects and multiplicative interaction (AMMI) and the genotype and genotype by environment interaction (GGE) biplot analysis [19,20]. The AMMI model [23–25] and GGE biplot analysis [26] are the most common statistical tools used for the analysis of multi-environment trials (METs) [18,19]. The AMMI model combines analysis of variance for the genotype and environment main effects with principal components analysis of the G × E interactions [27]. The AMMI method captures a large portion of the G × E interaction sum of squares [28]. The AMMI can also help in informing important decisions in breeding programs, such as which genotypes exhibit specific adaptation and in selecting the testing environments [28]. This is important for new breeding programs that have not yet optimized their genotype testing networks. The results of an AMMI analysis are often presented in a biplot, which displays both the genotype and environment values and their relationships using the singular vector technique [29]. The GGE integrates the genotypic main effect with the G × E interaction effect (Yan et al., 2000). A GGE biplot can help in grouping mega-environments [27]. It can also help to identify more representative environments for cultivar evaluation [30] and to compare and rank genotypes using mean yields and stability [31]. The objectives of the study were to determine the effect of G × E on grain yield and yield stability of maize single cross hybrids across low N stress and optimal environments.

2. Materials and Methods

2.1. Germplasm

Eighty single-cross hybrids were used in this study with four of them serving as hybrid checks. The 76 hybrids were selected from 117 F1 developed by crossing thirty nine tropical inbred lines with three testers using a line × tester mating scheme [4]. The selection of hybrids was based on seeds availability. The lines included inbred lines from IRAD Cameroon, IITA, CIMMYT and lines from other African maize breeding programs. Of the 39 inbred lines, six were tolerant to low N, four to drought, five to acid soils and four to aluminum toxicity. The testers are parental lines of high yielding hybrids used as checks in this study. The four checks comprised three hybrids (87036 × Exp1 24, 9071 × Exp1 24, 87036 × 9071) from crosses among the 3 testers and hybrid 88069 × Cam inb gp1 17 a promising yellow hybrid of the national breeding program. The hybrid 87036 × Exp1 24 is a high yielding hybrid released in Cameroon and adapted to the Humid Forest Zone of Cameroon. Exp1 24 × 9071 is also a high yielding hybrid, developed from a cross between tropical lowland × temperate converted lines. Genotypes names and codes are presented in Table 1.

Table 1. Names and codes of 80 single cross hybrids evaluated across 11 environments in 2012 and 2013.

Genotypes	Code	Genotypes	Code
CLYN246 × 87036	G1	88069 × 87036	G39
TL-11-A-1642-5 × 87036	G2	ATP S8 30 Y-3 × 87036	G40
CLWN201 × Exp1 24	G3	CML 254 × Exp1 24	G41
J16-1 × Exp1 24	G4	CLYN246 × 9071	G42
1368 × 87036	G5	CLWN201 × 9071	G43
CLQRCWQ26 × Exp1 24	G6	CML343 × Exp1 24	G44
TL-11-A-1642-5 × Exp1 24	G7	CLQRCWQ26 × 9071	G45
TZ-STR-133 × 87036	G8	ATP S9 30 Y-1 × Exp1 24	G47
CLWN201 × 87036	G9	ATP S6-20-Y-1 × 9071	G48
ATP S6-20-Y-1 × Exp1 24	G10	J16-1 × 9071	G49
CLA 18 × Exp1 24	G11	CML 358 × Exp1 24	G50
ATP S6-20-Y-1 × 87036	G12	Entrada 3 × 87036	G51
Cam inb gp1 17 × 87036	G13	CML494 × 87036	G52
J16-1 × 87036	G14	J18-1 × 87036	G53
4001STR × 87036	G15	CML 444 × 87036	G54
CML343 × 87036	G16	Cam inb gp1 17 × 9071	G55
CLA 18 × 87036	G17	ATP S5 31 Y-2 × 9071	G56
CML395 × Exp1 24	G18	V-481-73 × Exp1 24	G57
CML451 × 87036	G19	Cla 17 × Exp1 24	G58
CML343 × 9071	G20	5057 × Exp1 24	G59
88069 × 9071	G21	ATP S8 30 Y-3 × Exp1 24	G60
CLQRCWQ26 × 87036	G22	KU1414 × Exp1 24	G61
ATP S6 20 Y-2 × Exp1 24	G23	TZ-STR-133 × Exp1 24	G62
4001STR × 9071	G24	ATP S5 31 Y-2 × Exp1 24	G63
ATP S5 31 Y-2 × 87036	G25	ATP S6 20 Y-2 × 9071	G64
ATP S9 30 Y-1 × 87036	G26	Cla 17 × 87036	G66
1368 × Exp1 24	G27	ATP S8 30 Y-3 × 9071	G67
CML165 × 87036	G28	CML 254 × 87036	G68
CML 358 × 87036	G29	88094 × 87036	G69
KU1414 × 87036	G30	TZ-STR-133 × 9071	G70
Entrada 29 × Exp1 24	G31	CML451 × 9071	G71
CML494 × 9071	G32	CML 254 × 9071	G72
CML 444 × Exp1 24	G33	TZMI 102 × 87036	G73
88069 × Exp1 24	G34	TZMI 102 × Exp1 24	G74
Cam inb gp1 17 × Exp1 24	G35	Ku1409 × 9071	G75
CLYN246 × Exp1 24	G36	Entrada 3 × 9071	G76
1368 × 9071	G37	5012 × 87036	G77
Ku1409 × 87036	G38	Ku1409 × Exp1 24	G78
Checks			
87036 × Exp1 24	G46	87036 × 9071	G79
Exp1 24 × 9071	G65	88069 × Cam inb gp1 17	G80

2.2. Experimental Sites

The 80 hybrids were evaluated at two locations (Mbalmayo and Nkolbisson) of the Humid Forest Zone of Cameroon with a bimodal rainfall pattern. These locations are among the maize growing areas of the Humid Forest Zone of Cameroon where are located the principal experimental sites of the Institute of Agricultural Research of Cameroon. Nkolbisson is located at 11°36' E and 3°44' N, 5 km from the main capital city 'Yaoundé'. The altitude is 650 m above sea level (a.s.l.). The mean annual rainfall is 1560 mm with bimodal distribution. The average daily temperature is 23.5 °C. The soil is sandy clay [32]. Mbalmayo is located at 11°30' E and 3°31' N. The altitude is 641 m a.s.l. The mean annual rainfall varies from 1017 to 1990 mm with bimodal distribution. The mean monthly temperature varies from 25 °C to 22 °C. The soil is sandy clay [33]. Based on the results of soil analysis in 2012 and 2013, the soil in Mbalmayo had a pH of 5.97 which is moderately acidic, while at Nkolbisson, pH was 4.54 and the soil classified as strongly acid.

The main cropping system in Nkolbisson is maize/groundnut/cassava in sole cropping or mixed cropping while in Mbalmayo, other cultivated crops include banana, melon, plantain and vegetables [32]. The hybrids were evaluated in a total of 11 environments. Each environment was assigned a code and consisted of a combination of site × year × season × nitrogen level (Table 2). The soil management consisted of two nitrogen levels; Low N (20 kg ha^{-1}) and Optimum

N (100 kg ha^{-1}). The geographical coordinates, climatic conditions of the localities and the 11 environments are described in (Table 2).

Table 2. Description of the eleven environments used to evaluate the 80 hybrids.

Site	Latitude, Longitude and Altitude	Environments Code	Year	Season	Soil Management	Average Rain Fall
Mbalmayo	3°31' N, 11°30' E, 641 m a.s.l.	E1	2012	Minor	Low N	488.87 mm
		E2	2012	Minor	Optimum N	488.87 mm
		E5	2013	Major	Low N	583.46 mm
		E6	2013	Major	Optimum N	583.46 mm
		E9	2013	Minor	Low N	499.66 mm
		E10	2013	Minor	Optimum N	499.66 mm
Nkolbisson	3° 44 N, 11°36 E, 650 m a.s.l.	E3	2012	Minor	Low N	281 (October–November) *
		E4	2012	Minor	Optimum N	281 (October–November) *
		E7	2013	Major	Low N	936 mm
		E8	2013	Major	Optimum N	936 mm
		E11	2013	Minor	Optimun N	662 mm

a.s.l. = above sea level; Low N = low soil nitrogen; Rainfall data were collected at Mbalmayo by IITA and at Nkolbisson by the Rice Project PRODERiP; Major season: From March to June; Minor season: From September to November; * Data for the entire season in this environment were not available.

2.3. Site Preparation and Soil Analysis

The soil was depleted of available Nitrogen by high density maize cropping without fertilizer application, and complete removal of organic matter after harvest [6], in order to establish low N plots in Mbalmayo and Nkolbisson [4]. This was done at Mbalmayo thrice between 2010 and 2011 and at Nkolbisson for six growing seasons between 2008 and 2012. To ensure the low N status of the sites [32], composite soil samples were collected before each cropping season and analyzed at the soil laboratory of the International Institute of Tropical Agriculture (IITA) Cameroon [4]. Soils were air-dried and ground to pass through a 2 mm sieve. Soil pH in water was determined in a 1:2.5 (w/v) soil: water suspension. Organic C was determined by chromic acid digestion and spectrophotometric analysis [34]. Total N was determined from a wet acid digest [35] and analyzed by colorimetric analysis [36]. Exchangeable Ca, Mg, K, and Na were extracted using the Mehlich-3 procedure [37] and determined by atomic absorption spectrophotometry. Exchangeable Al extracted using 1N KCl [38] and analyzed using the pyrocatechol violet method described by Mosquera and Mombiela [39]. Available P was extracted by Bray-1 procedure and analyzed using the molybdate blue procedure described by Murphy and Riley [40]. P expressed in ppm or µg/g; Al, Ca, Mg, K, and Na reported as cmol(+)/kg or me/100 g. Organic C and Total N expressed as % particle size (three fractions) was determined by the hydrometer method. The results of soil analysis are presented in Table 3.

Table 3. Soil characteristics at Mbalmayo and Nkolbisson before the trials in 2012 and nitrogen level in 2013.

Chemical Characteristics	Mbalmayo		Nkolbisson	
	0–10 cm	10–20 cm	0–10 cm	10–20 cm
Exchangeable Ca^{2+} (cmol kg^{-1})	5.92	2.58	1.53	0.88
Exchangeable Mg^{2+} (cmol kg^{-1})	1.15	0.63	0.77	0.46
Exchangeable K$^+$ (cmol kg^{-1})	0.11	0.06	0.38	0.24
Cation Exchange Capacity (cmol kg^{-1})	nd	nd	10.55	9.37
Organic Carbon %	1.30	0.58	1.87	1.51
C/N	9.90	8.03	15.90	12.94
Bray Phosphorus (mg kg^{-1})	2.11	0.99	13.85	3.10
pH 1:1 (H$_2$O)	5.97	5.04	4.54	4.36
Total Nitrogen %				
In 2012	0.13	0.07	0.12	0.12
In 2013	0.11	0.15	0.06	0.11

nd = Not determined.

2.4. Experimental Design

The experiment was established in two replicates of single row experimental units with an 8 × 10 alpha lattice design using 80 maize hybrids [4]. Rows were 5 m long in Mbalmayo and 4 m long at Nkolbisson. Between row spacing was 0.75 m while spacing between hills within the same row was 0.5 m. At planting, each hill received three seeds which were later thinned to two plants for a final density of 53,330 plants per hectare. Split fertilization, weed and pest control were done on each plot as described in Mafouasson et al. [4].

2.5. Data Collection

Data for grain yield was obtained as follows: grains were harvested at maturity from each row. The total number of ears and ear weight was recorded for each plot. Five ears were then randomly selected from each plot and their grains were shelled. The "Dickey John" moisture tester was used to measure the percent grain moisture at harvest. Grain yield ((kg ha^{-1}) for every entry from the data of fresh ear weight per plot (adjusted to 15% grain moisture) was calculated using the following formula [4]:

$$\text{Grain yield} \left(\text{kg ha}^{-1}\right) = \frac{\text{Fresh ear weight (kg/plot)} \times (100 - \text{MC}) \times 0.8 \times 10,000}{(100 - 15) \times \text{Area harvested/plot}}$$

where:

MC = moisture content in grains at harvest (%)
0.8 = Shelling coefficient
10,000 m^2 corresponds to 1 hectare
15% = moisture content required in maize grain at storage

2.6. Statistical Analysis

Data obtained was subject to combined analysis of variance (ANOVA) with the PROC GLM procedure in SAS [41] using the RANDOM statement set to the TEST option. Environments were considered as random effects while genotypes were treated as fixed effects. Entry means were adjusted for block effects with reference to lattice design [42]. Each environment was defined as year × season × site × nitrogen treatment and the means were separated using Tukey's test at $p < 0.05$ [4].

The AMMI statistical analysis of yield data was performed with Breeding View in the Integrated Breeding Platform Breeding Management System version 2.1 [43].

GGE biplot analysis was performed using Genstat 15th edition in order to identify genotypes that were suitable for the different environments as well as genotypes stable across the various environments, and to identify the different mega-environments. It was difficult to present the eighty hybrids on the AMMI and GGE biplot. Therefore, for a better visualization and interpretation of AMMI and GGE biplot, the top 20 best performing hybrids across environments and four checks were used for this analysis.

3. Results

3.1. Analysis of Variance for Grain Yield across Environments

The results of the combined ANOVA across environments for the 80 hybrids showed that genotype main effect (G), environment main effect (E) and G × E were all highly significant ($p < 0.001$) for grain yield (Table 4). The test environments contributed 60.13% of the total variation in the sum of squares for grain yield, while G and G × E sources of variation accounted for 6.81% and 33.05% of the total variation, respectively. The ratio of genotype (G) effect over genotype + genotype × environment (G + G × E) was 0.17.

Table 4. Combined analysis of variance for grain yield of 80 hybrids across eleven environments.

Source	df	Sum of Squares	% Contribution to Sum of Squares	Mean Square	Pr > F
Env	10	2,447,399,522	60.13	244,739,952	<0.0001
Rep (Env)	11	127,598,961		11,599,906	<0.0001
Block (Env × Rep)	220	382,845,773		1,740,208	0.0001
Genotype	79	277,051,837	6.81	3,506,985	<0.0001
Env × Genotype	790	1,345,460,727	33.05	1,703,115	<0.0001
Error	649	767,427,967		1,182,478	
Corrected Total	1759	6,096,183,487			
CV	26.15				
R^2	0.87				

Env = Environment; Rep= Replication; CV = Coefficient of variation; Pr = probability.

3.2. Yield Performance of the 20 Best Performing Hybrids and Four Checks across Eleven Environments

The 20 best performing hybrids were selected from the 80 hybrids evaluated across environments based on their highest mean yields across the 11 environments. The four checks were added to the 20 hybrids. Yield performance data of these 24 hybrids across eleven environments is presented in Table 5. The overall mean across the 11 environments for the 20 selected hybrids ranged from 4484.7 kg ha^{-1} to 5198.3 kg ha^{-1}. The highest yielding hybrid across environments was TL-11-A-1642-5 × 87036 with a yield of 5198 kg ha^{-1}. All the 20 hybrids selected yielded higher than the four checks. The best check across environments was Exp1 24 × 9071 (3912.4 kg ha^{-1}) followed by 87036 × Exp1 24 (3908.9 kg ha^{-1}). The bold and underlined mean yields are for those hybrids that were the highest yielding in each environment. TL-11-A-1642-5 × 87036 was the highest yielding in two optimum environments E4 and E8 with 9531 kg ha^{-1} and 8874 kg ha^{-1}. TL-11-A-1642-5 × Exp1 24 was the best performing in E2 (optimum) and E3 (low N) with 6427 and 5402 kg ha^{-1} respectively. Entrada 29 × Exp1 24 was also the highest yielding in two environments, E9 (low N) and E11 (optimum). One of the hybrid checks (87036 × Exp1 24) was not the best in any environments but was among the five highest yielding hybrids in E6 (optimum), E9 (low N) and E10 (optimum) with grain yield of 4232 kg ha^{-1} and 6410 kg ha^{-1} respectively (Table 5).

Table 5. Mean grain yield (kg ha^{-1}) of 20 hybrids and four checks across 11 environments in Mbalmayo and Nkolbisson in 2012 and 2013.

Genotypes	E1 LO	E2 OP	E3 LO	E4 OP	E5 LO	E6 OP	E7 LO	E8 OP	E9 LO	E10 OP	E11 OP	Mean Across
1368 × 87036	**4790**	6118	2547	8253	3956	4953	2038	5876	3247	4419	3734	4546
TZ-STR-133 × 87036	4382	5665	1508	8269	2225	5507	3604	7397	1602	5129	4166	4499
CLQRCWQ26 × Exp1 24	4253	4817	1520	6324	5404	6317	2778	6816	2163	5570	5110	4639
CLWN201 × 87036	4093	4739	2799	4987	5293	**7657**	3318	6535	2545	3223	4779	4540
CLYN246 × 87036	3979	4782	1327	7809	5157	7437	3175	7017	3922	5880	4940	5040
TL-11-A-1642-5 × Exp1 24	3940	**6427**	**5401**	7311	2807	4755	2987	3773	3718	4828	4449	4583
CML343 × 9071	3911	4774	1234	5672	3810	6142	2910	6011	4257	5062	5672	4486
CML395 × Exp1 24	3896	4396	1024	6999	3749	6599	2989	5127	3595	5742	6342	4572
CLQRCWQ26 × 87036	3557	4088	2546	5711	2144	7155	3646	5188	4550	5081	5783	4485
CLA 18 × Exp1 24	3426	4745	2317	7627	4370	5485	2358	5906	2911	5785	5376	4567
J16-1 × Exp1 24	3290	4707	2531	6424	4315	6054	2782	5789	4664	**7270**	3724	4694
CML451 × 87036	3284	4478	1528	6569	2879	5796	**4916**	7004	2847	4799	6569	4590
ATP S5 31 Y-2 × 87036	3198	4407	1425	6711	4636	5877	2554	6812	3240	4604	7038	4571
ATP S6-20-Y-1 × 87036	3076	4951	2447	6167	3171	4555	4717	8184	2655	4973	4829	4518
Cam inb gp1 17 × 87036	3043	4332	761	6957	3809	6089	3923	5703	4604	5624	4663	4499
ATP S6 20 Y-2 × Exp1 24	3036	3336	3651	6138	3684	5448	4337	6321	3074	4527	7549	4621
Entrada 29 × Exp1 24	2989	3667	1536	4077	4092	5742	3085	5846	**4905**	5981	**9114**	4602
CLWN201 × Exp1 24	2889	4491	1209	6464	**5707**	7568	3212	6532	4059	5708	4380	4750
TL-11-A-1642-5 × 87036	2734	3303	1448	**9531**	4985	6231	3531	**8874**	3230	5006	8619	5198
ATP S6-20-Y-1 × Exp1 24	2554	3220	2376	7367	4641	6739	4697	7719	2558	3281	7458	4761
Checks												
87036 × Exp1 24	2887	3801	623	2765	3241	6826	3350	6045	4232	6410	2710	3909
87036 × 9071	2175	2511	699	6253	1903	3273	2375	4011	3986	5968	2067	3211
Exp1 24 × 9071	1791	3475	984	4559	2850	5635	2281	7950	3778	4692	5155	3912
88069 × Cam inb gp1 17	782	1280	1799	4049	1804	2754	2783	4600	1952	4424	5032	2823
Means	3066	4132	1661	5882	3719	5413	3130	5832	3208	4993	4738	

LO = Low N environment; OP = Optimum environment.

3.3. Additive Main Effect and Multiplicative Interaction (AMMI) Analysis of 24 Hybrids for Grain Yield

The results of AMMI biplot analysis of the 24 hybrids evaluated in 11 environments showed that environment accounted for 59.82% of the total variation in the sum of squares, while genotype and genotype by environment interaction accounted for 7.89% and 32.28 % of variation observed in grain yield respectively (Table 6).

Table 6. Analysis of variance for additive main effects and multiplicative interaction model for grain yield of 24 hybrids across 11 environments.

Source	df	Sum of Squares	Mean Squares	Contribution to Total Variation (%)	F Probability
Genotypes	23	67,848,890	2,949,952	7.891443	<0.001
Environments	10	514,356,619	51,435,662	59.82435	<0.001
Interactions (G × E)	230	277,572,490	1,206,837	32.2842	
IPCA 1	32	80,827,507	2,525,860	29.22	<0.001
IPCA 2	30	58,189,285	1,939,643	20.96	<0.001
Residuals	168	138,555,699	824,736		

df = degree of freedom; G × E = Genotypes × Environment; IPCA = Interaction Principal Component Axis.

In the AMMI biplot (Figure 1) the genotype and environment main effects for grain yield are on the *x*-axis while the IPCA1 (Interaction Principal Component Axis 1) scores are on the *y*-axis. The vertical line is the grand mean for grain yield and the horizontal line (y-ordinate) represents the IPCA1 value of zero.

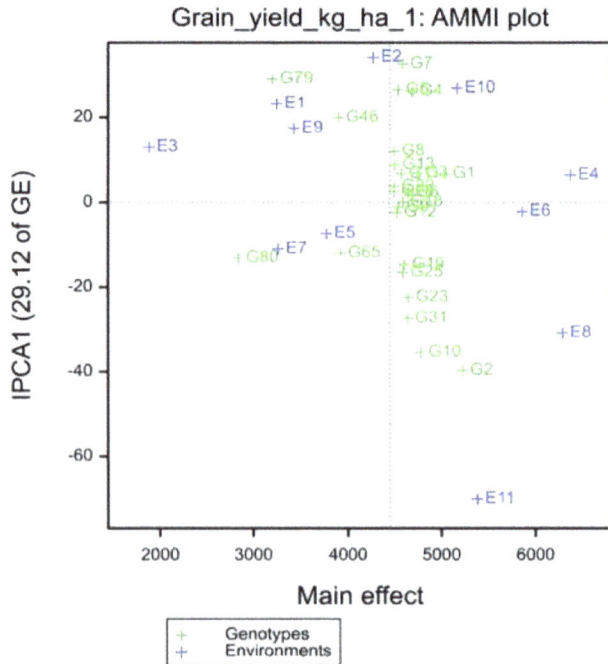

Figure 1. AMMI Biplot for grain of 24 maize hybrids showing genotypes and environments (E1-E11) plotted against their IPCA1 scores. (Codes for environments in Table 2).

In the AMMI biplot, the IPCA scores of a genotype are an indication of the stability of the genotype across environments. The more the IPCA score is close to zero, the more stable the genotype is across environments. The greater the IPCA scores, either positive or negative, the more specifically adapted a genotype is to certain environments. Accordingly, ATP S6-20-Y-1 × 87036 (G12) and CML395 × Exp1 24 (G18) had their IPCA1 close to zero and can be considered to have small interaction with the environments and to be the most stable hybrids (Figure 1). CML395 × Exp1 24 (G18) and ATP S6-20-Y-1 × 87036 (G12) had grain yield above the grand mean and CML395 × Exp1 24 (G18) was higher yielding than ATP S6-20-Y-1 × 87036 (G12) even though the difference was small. Among the 24 hybrids selected, 87036 × Exp1 24 (G46), Exp1 24 × 9071 (G65), 87036 × 9071(G79) and 88069 × Cam inb gp1 17 (G80) had grain yield response below the grand mean. The other 20 hybrids had grain yield above the grand mean. Among these 20 hybrids, TL-11-A-1642-5 × 87036 (G2) had the highest grain yield, followed by CLYN246 × 87036 (G1), ATP S6-20-Y-1 × Exp1 24 (G10) and CLWN201 × Exp1 24 (G3). TL-11-A-1642-5 × 87036 (G2) had a negative interaction with IPCA1. In contrast, TL-11-A-1642-5 × Exp1 24 (G7), 1368 × 87036 (G5) and J16-1 × Exp1 24 (G4) had yield above the grand mean with high positive IPCA1 scores. CLYN246 × 87036 (G1) and CLWN201 × Exp1 24 (G3) had comparable IPCA1 score and small interaction with environments. TL-11-A-1642-5 × 87036 (G2) was higher yielding than CLYN246 × 87036 (G1), but CLYN246 × 87036 (G1) was more stable than TL-11-A-1642-5 × 87036 (G2). ATP S6-20-Y-1 × Exp1 24 (G10), Entrada 29 × Exp1 24 (G31), ATP S6 20 Y-2 × Exp1 24 (G23), ATP S5 31 Y-2 × 87036 (G25) and CML451 × 87036 (G19) had grain yield above the grand mean and had negative interaction with IPCA1, and therefore negative interaction with the environments. Among the four low yielding hybrids, 88069 × Cam inb gp1 17 (G80) was the lowest yielding, followed by 87036 × 9071 (G79) which was the least stable among them.

G1 = CLYN246 × 87036; G2 = TL-11-A-1642-5 × 87036; G3 = CLWN201 × Exp1 24; G4 = J16-1 × Exp1 24; G5 = 1368 × 87036; G6 = CLQRCWQ26 × Exp1 24; G7 = TL-11-A-1642-5 × Exp1 24; G8 = TZ-STR-133 × 87036; G9 = CLWN201 × 87036; G10 = ATP S6-20-Y-1 × Exp1 24; G11 = CLA 18 × Exp1 24; G12 = ATP S6-20-Y-1 × 87036; G13 = Cam inb gp1 17 × 87036; G18 = CML395 × Exp1 24; G19 = CML451 × 87036; G20 = CML343 × 9071; G22 = CLQRCWQ26 × 87036; G23 = ATP S6 20 Y-2 × Exp1 24; G25 = ATP S5 31 Y-2 × 87036; G31 = Entrada 29 × Exp1 24; G46 = 87036 × Exp1 24; G65 = Exp1 24 × 9071; G79 = 87036 × 9071; G80 = 88069 × Cam inb gp1 17.

In AMMI biplot (Figure 2), environments are distributed from lower yielding in quadrant A (top left) and C (bottom left) to the higher yielding in quadrants B (top right) and D (bottom right).

This graph identified E1, E3, E5, E7 and E9 as low yielding environments. These were all low N environments in Mbalmayo and Nkolbisson in 2012 and 2013. Environments E4, E6, E8, E11 were identified as high yielding. These were optimum N plots in both locations in 2012 and 2013. The lowest yielding optimum environment was E2. The highest yielding environment was E4 (optimum N, minor season of 2012 at Nkolbisson) while the lowest was E3 (low N, minor season of 2012 at Nkolbisson).

The four highest yielding hybrids selected by AMMI for each environment are presented in Table 6. TL-11-A-1642-5 × 87036 (G2) appeared as the best hybrid in four (E5, E7, E8 and E11) out of 11environments. TL-11-A-1642-5 × 87036 (G2) was followed by TL-11-A-1642-5 × Exp1 24 (G7) which was the best in three environments (E1, E2 and E3) and 87036 × Exp1 24 (G46) was the highest yielding in two environments (E9 and E10). CLYN246 × 87036 (G1) appeared as the third in three environments and as fourth in three other environments. CLWN201 × Exp1 24 (G3) appeared as second, third and fourth in three different environments while ATP S6-20-Y-1 × Exp1 24 (G10) appeared as second in two environments and as third and fourth in two different environments.

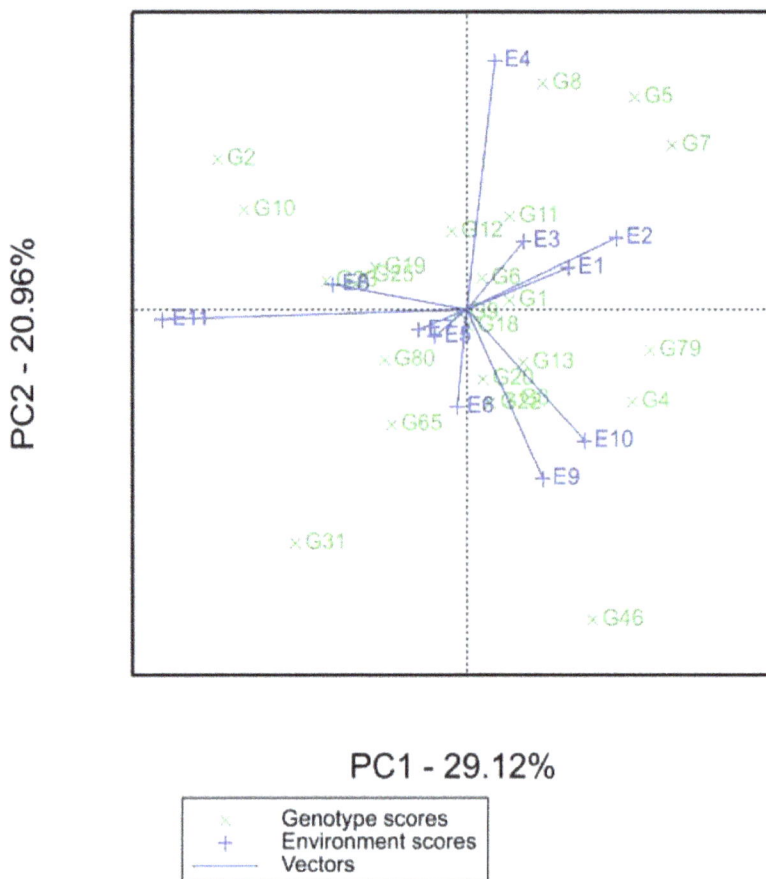

Figure 2. Biplot of the additive main effects and multiplicative interaction (AMMI) showing the relationship among 11 testing environments (E1–E11). (Codes of genotypes in Table 1 and environments in Table 2).

3.4. GGE Biplot Analysis of Best 20 Hybrids and Four Checks

The polygon view of the genotypes in the GGE biplot for 24 genotypes is presented in Figure 3. Primary (PC1) and secondary (PC2) scores were significant and explained 29.98% and 21.44% of the variation, respectively. Together they explained 51.42% of the genotype main effect and G × E interaction for the grain yield of maize hybrids evaluated in the 11 environments at Mbalmayo and Nkolbisson in 2012 and 2013.

The polygon view of a GGE biplot displayed the "which-won-where" pattern (Figure 3). The vertices of the polygon were the genotype markers located farthest away from the biplot origin in various directions, such that all genotype markers were contained within the resulting polygon. The biplot was divided into six sectors and three mega-environments and showed five vertex cultivars 1368 × 87036 (G5), TL-11-A-1642-5 × 87036 (G2), Entrada 29 × Exp1 24 (G31), 87036 × 9071 (G79) and 88069 × Cam inb gp1 17 (G80). The first mega-environment comprised E1, E2, E3 and E4 and had 1368 × 87036 as the highest yielding hybrid. These four environments were low N (E1and

E3) and optimum N (E2 and E4), minor season of 2012 at Mbalmayo and Nkolbisson. The second mega-environment consisted of E5, E6, E7, E8 and E11 and had TL-11-A-1642-5 × 87036 (G2) as the highest yielding hybrid. These environments were low N (E5 and E7) and optimum N (E8 and E11) of major season in 2013 at Mbalmayo and Nkolbisson plus E11 which is optimum N plot of minor season of 2013 at Nkolbisson. The third comprised E9 and E10 (low N, and optimum N plots of minor season in 2013 at Mbalmayo), with the highest yielding hybrid as 87036 × 9071 (G79). This mega-environment contained 87036 × Exp1 24 (G46). No environment fell within the sector with Entrada 29 × Exp1 24 (G31) and 88069 × Cam inb gp1 17 (G80), indicating that these hybrids were not the best in any of the mega-environments, or they were the poorest cultivars in some or all of the environments. Genotypes within the polygon were less responsive than the vertex genotypes.

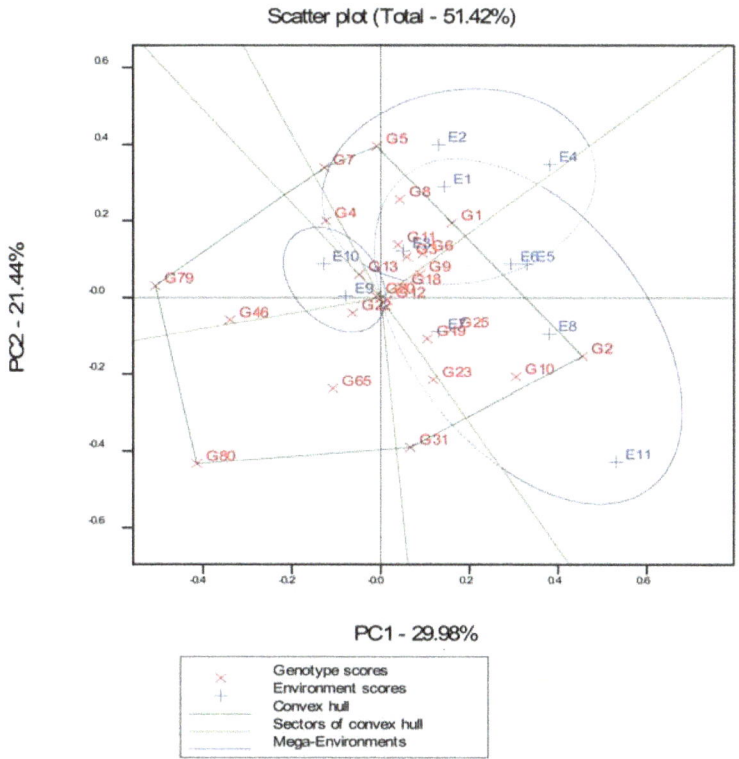

Figure 3. A "which won where" biplot based on grain yield of 24 single hybrids evaluated in 11 environments.

G1 = CLYN246 × 87036; G2 = TL-11-A-1642-5 × 87036; G3 = CLWN201 × Exp1 24; G4 = J16-1 × Exp1 24; G5 = 1368 × 87036; G6 = CLQRCWQ26 × Exp1 24; G7 = TL-11-A-1642-5 × Exp1 24; G8 = TZ-STR-133 × 87036; G9 = CLWN201 × 87036; G10 = ATP S6-20-Y-1 × Exp1 24; G11 = CLA 18 × Exp1 24; G12 = ATP S6-20-Y-1 × 87036; G13 = Cam inb gp1 17 × 87036; G18 = CML395 × Exp1 24; G19 = CML451 × 87036; G20 = CML343 × 9071; G22 = CLQRCWQ26 × 87036; G23 = ATP S6 20 Y-2 × Exp1 24; G25 = ATP S5 31 Y-2 × 87036; G31 = Entrada 29 × Exp1 24; G46 = 87036 × Exp1 24; G65 = Exp1 24 × 9071; G79 = 87036 × 9071; G80 = 88069 × Cam inb gp1 17.

Ranking of genotypes based on both mean grain yield and stability performance of the 20 best genotypes and four checks is presented in Figure 4 in order to identify the highest yielding and stable genotypes (Figure 4). Genotypes that are located at the center of the concentric circles are the ideal (highest yielding and stable). The GGE biplot identified CLYN246 × 87036 (G1) and TL-11-A-1642-5 × 87036 (G2) as superior since they were located close to the center of the concentric circles. Both were high yielding but TL-11-A-1642-5 × 87036 (G2) was the highest yielding and therefore the most desirable genotype. These hybrids were followed by CLQRCWQ26 × Exp1 24 (G6) and CLWN201 × 87036 (G9) (Figure 4).

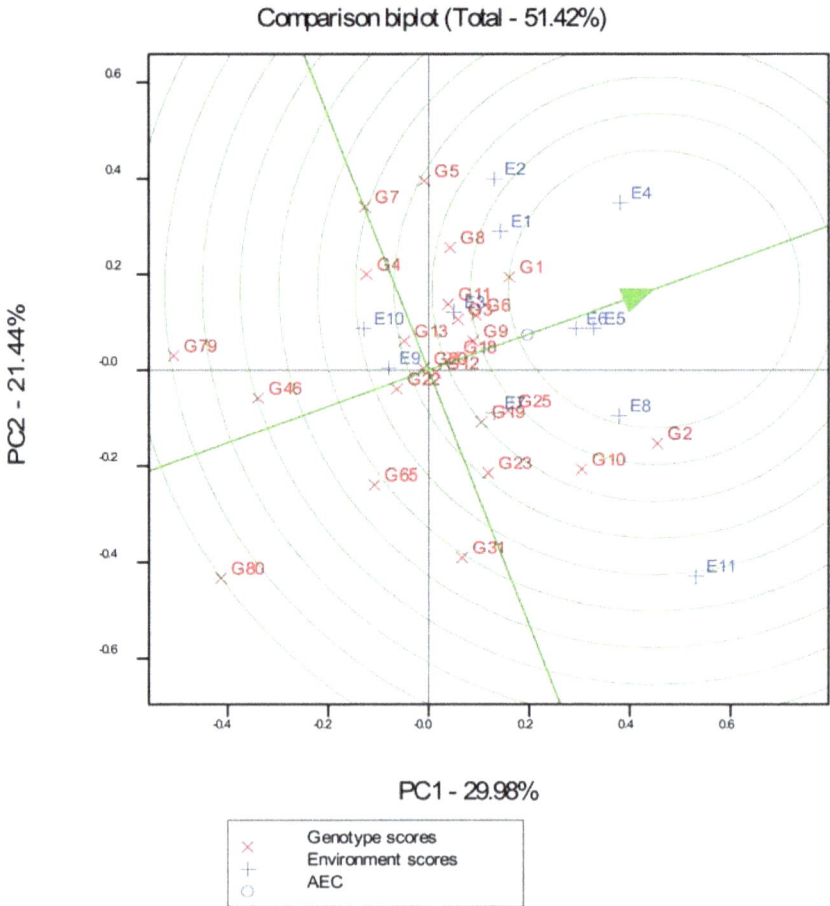

Figure 4. Comparison view of 24 hybrids with the ideal genotype based on average grain yield and stability for grain yield across 11 environments in 2012 and 2013.

G1 = CLYN246 × 87036; G2 = TL-11-A-1642-5 × 87036; G3 = CLWN201 × Exp1 24; G4 = J16-1 × Exp1 24; G5 = 1368 × 87036; G6 = CLQRCWQ26 × Exp1 24; G7 = TL-11-A-1642-5 × Exp1 24; G8 = TZ-STR-133 × 87036; G9 = CLWN201 × 87036; G10 = ATP S6-20-Y-1 × Exp1 24; G11 = CLA 18 × Exp1 24; G12 = ATP S6-20-Y-1 × 87036; G13 = Cam inb gp1 17 × 87036; G18 = CML395 × Exp1 24; G19 = CML451 × 87036; G20 = CML343 × 9071; G22 = CLQRCWQ26 × 87036; G23 = ATP S6 20 Y-2 × Exp1 24; G25 = ATP S5 31 Y-2 × 87036; G31 = Entrada 29 × Exp1 24; G46 = 87036 × Exp1 24; G65 = Exp1 24 × 9071; G79 = 87036 × 9071; G80 = 88069 × Cam inb gp1 17. The four checks were low yielding compared to the 20 hybrids selected. Hybrid 88069 × Cam inb gp1 17 (G80) was located far from the vertical axis at the left and far from the center of the concentric circle, therefore it was the most inferior hybrid in both mean grain yield and stability of performance.

4. Discussion

The greater variation contributed by environment than those from genotype and genotype × environment interaction indicated that the test environments were highly variable. This result is in agreement with Badu-Apraku et al. [18] who reported that contribution of test environments are much greater than from the other sources of variation in most multi-environmental trials. The highly significant G × E interaction for grain yield justified the use of AMMI and GGE biplots to decompose the G × E interactions and to determine the yield potential and stability of the evaluated single cross hybrids.

The results of the AMMI biplot analysis of the 24 hybrids evaluated in 11 environments also showed that environment effects accounted for 59.82% of the total variation in the sum of squares and was the highest value compared to the other components. The AMMI biplot revealed large variability among the 11 environments, but the yield range among the 24 hybrids was narrow. This is probably because the 20 hybrids were the best selected. ATP S6-20-Y-1 × 87036 and CML395 × Exp1 24 have IPCA1 scores near zero and therefore had small interaction with the environments. This small interaction with environments suggested that these hybrids are stable across environments [13].TL-11-A-1642-5 × 87036 was identified as the highest yielding hybrid. It was followed by CLYN246 × 87036, ATP S6-20-Y-1 × Exp1 24 and CLWN201 × Exp1 24. All these hybrids, except ATP S6-20-Y-1 × Exp1 24 are crosses between CIMMYT and IRAD lines. Acquaah [44] indicated that the development of adapted high yielding hybrids requires that the varieties used as parents are genetically divergent. The high yields obtained between CIMMYT and IRAD lines could therefore imply that they are genetically diverse. The negative interaction of TL-11-A-1642-5 × 87036 with the IPCA1 suggests that this hybrid was less sensitive to environmental changes and was likely to be adapted to unfavorable environments as indicated by Badu-Apraku et al. [18]. In contrast, TL-11-A-1642-5 × Exp1 24, 1368 × 87036 and J16-1 × Exp1 24 had large positive interaction with IPCA1 and might be more sensitive to environmental changes, and probably more adapted to favorable environments.

TL-11-A-1642-5 × 87036 was higher yielding than CLYN246 × 87036, but CLYN246 × 87036 was more stable than TL-11-A-1642-5 × 87036. Hybrids ATP S6-20-Y-1 × Exp1 24, Entrada 29 × Exp1 24, ATP S6 20 Y-2 × Exp1 24, ATP S5 31 Y-2 × 87036 and CML451 × 87036 had grain yield above the grand mean and negative interaction with the environments. Therefore, these hybrids were less sensitive to variation in the environments. They are also most likely to be adapted to unfavorable environments which in this study are low N environments.

AMMI biplot displayed the distribution of environments from low to high yielding in different quadrants of the graph. This graph placed all low N environments (E1, E3, E5, E7, E9) in the quadrants of lower yielding genotypes and showed the optimum environments (E4, E6, E8, E11) in quadrants of high yielding genotypes as expected.

The GGE biplot analysis of grain yield response and stability of 24 hybrids showed that PC1 explained 29.98% of total variation while PC2 explained 21.44% and together, the two axes accounted for 51.42%. This suggested that the biplot of PC1 and PC2 adequately approximated the environment centered data. The biplot for 24 hybrids was divided into six sectors and three mega-environments in which different cultivars should be selected and deployed to similar environments as suggested by Yan and Tinker [45]. According to Yan and Rajcan [46] a mega-environment is defined as the subset of locations that consistently share the best set of genotypes across years and the growing regions are relatively homogeneous with similar biotic and abiotic stresses and cropping system requirements.

In the polygon view, the vertex genotype in each sector represents the highest yielding genotype in the location that falls within that particular sector [13,26,45]. Accordingly, the biplot identified five vertex genotypes: 1368 × 87036, TL-11-A-1642-5 × 87036, Entrada 29 × Exp1 24, 87036 × 9071 and 88069 × Cam inb gp1 17.

Two out of the three mega-environments identified by the GGE biplot included both low and optimum N plots of the two locations, but they were related to different years and different growing seasons. The third mega-environment was related to one specific season of one specific year, but included two nitrogen treatments of one site. This could imply that the mega-environments constructed are based on growing seasons (minor or major) and not on different sites, or different nitrogen treatments. This suggests that seasons and years may have accounted more for significant environmental differences and to different genotypic responses to environments as indicated by Sibiya et al. [10]. It might probably be due to similar variation in rainfall amount and distribution as well as biotic stresses within seasons of each year which might have caused the 24 genotypes to have similar relative performance from one environment to another in the mega-environments. In the Bimodal Humid Forest Zone of Cameroon, there are two growing seasons, the major season and the minor season. During the minor season, the total rainfall was lower, the duration of the rainy period is usually shorter than in the major season. Moreover, during the minor season there is prevalence of many diseases such as fungal diseases (e.g., Maize leaf blight caused by Exserohilum turcicum) and maize stem borers among which the main species is Busseola fusca Fuller [47,48].

The results obtained suggest that highest yielding hybrids identified for each mega-environment should be proposed for environments similar to those of these mega-environments. Therefore, hybrids 1368 × 87036 could be proposed for the minor season and TL-11-A-1642-5 × 87036 for the major season. However, this should be done after further evaluation of hybrids in more environments including more locations, years and seasons as recommended by Yan and Tinker [45] who indicated the need for crossover interactions to be repeatable across the years so that target environments can be divided into mega-environments and genotypes be recommended based on METs (multi-environment trials). Yan and Tinker [13] indicated that an ideal genotype should be one that combines both high mean yield performance and high stability across environments; it should be on average environmental coordinate (AEC) on positive direction and have a vector length equal to the longest vector of the genotype as indicated by an arrow pointed to it. Accordingly, the GGE biplot identified TL-11-A-1642-5 × 87036 and CLYN246 × 87036 as closest to the ideal genotype. According to Badu-Apraku et al. [14], in the process of selecting for broad adaptation in maize production, an ideal genotype should have both high mean performance and high stability. TL-11-A-1642-5 × 87036 and CLYN246 × 87036 which were the highest yielding and the most stable hybrids across environments could therefore be selected for broad adaptation (production across environments). These hybrids were followed by CLQRCWQ26 × Exp1 24 and CLWN201 × Exp1 24. The top 20 hybrids performed better than the checks. The poor performance of the check 87036 × Exp1 24, a commercial hybrid, compared to the other hybrids might be due to the fact that it was developed many years ago and might not be adapted to changes (climatic, diseases) that might have occurred in the environments.

5. Conclusions

This study revealed that genotypes, environments and genotype × environment interaction were significant for grain yield. The genotypes therefore performed differently with respect to yield in each of the eleven test environments and their relative performance varied from one environment to another. AMMI analysis showed that environment effects accounted for a larger proportion of the total variation in the sum of squares for grain yield than genotype effects and genotype × environment effects. The AMMI biplot showed large variability among the environments but a narrow range for yields among hybrids. The GGE biplot classified the study area into three mega-environments. These mega-environments seemed to be related to the two growing seasons of the year (minor and major). High yielding hybrids were identified for each mega-environment and could be proposed for release for production in similar conditions. These hybrids are 1368 × 87036 for mega-environment 1, which is related to the minor season and TL-11-A-1642-5 × 87036 for Mega environment 2, which is

Agronomy **2018**, *8*, 62

related to the major season. The most outstanding hybrid was TL-11-A-1642-5 × 87036. This hybrid has the potential for production across environments and should therefore be tested further in multiple environments to confirm consistency of its high yield performance and stability to facilitate its release as a commercial hybrid. Hybrids which were selected as high yielding, but were not stable across environments could be recommended for the specific environments where they performed well. The results of this study should therefore be confirmed through further evaluation of hybrids at different locations of the Bimodal Humid Forest Zone during both minor and major seasons for several years.

Author Contributions: H.N.A.M. and V.G. conceived and designed the experiments; H.N.A.M. performed the experiments; H.N.A.M. and M.A.Y. analyzed the data; H.N.A.M., L.N.T., G.N.-N. and C.M. wrote the paper.

Acknowledgments: This study is part of the PhD work supported by Alliance for Green Revolution in Africa (AGRA) through the West Africa Centre for Crop Improvement (WACCI), University of Ghana. We are grateful to IRAD for the research facilities and to IITA and CIMMYT for supplying germplasm used in the study. We appreciate the post-doctoral fellowship awarded by Govan Mbeki Research and Development Centre (GMRDC) to Tandzi which facilitates the publication of the present paper.

Conflicts of Interest: The authors declare no conflict of interest.

References

1. Badu-Apraku, B.; Akinwale, R. Identification of early-maturing maize inbred lines based on multiple traits under drought and low N environments for hybrid development and population improvement. *Can. J. Plant Sci.* **2011**, *91*, 931–942. [CrossRef]
2. Ngo Nonga, F. *Durabilité des Activités Agricoles des Exploitations Familiales Agricoles à Base de Maïs du Grand Sud Cameroun, 2èmes Journées de Recherches en Sciences Sociales*; INRA SFER CIRAD: Lille, France, 2008; p. 20.
3. Derera, J. *Genetic Effects and Associations between Grain Yield Potential, Stress Tolerance and Yield Stability in Southern African Maize (Zea mays L.) Base Germplasm*; School of Biochemistry, Genetics, Microbiology and Plant Pathology, Faculty of Science and Agriculture, University of KwaZulu-Natal: Durban, South Africa, 2005; p. 175.
4. Mafouasson, A.H.N.; Kenga, R.; Gracen, V.; Yeboah, A.M.; Mahamane, N.L.; Tandzi, N.L.; Ntsomboh-Ntsefong, G. Combining Ability and Gene Action of Tropical Maize (*Zea mays* L.) Inbred Lines under Low and High Nitrogen Conditions. *J. Agric. Sci.* **2017**, *9*, 222–235.
5. Ajala, S.O.; Olaniyan, A.B.; Olayiwola, M.O.; Job, A.O. Yield improvement in maize for tolerance to low soil nitrogen. *Plant Breed.* **2018**, *137*, 118–126. [CrossRef]
6. Bänziger, M.; Edmeades, G.O.; Beck, D.; Bellon, M. *Breeding for Drought and Nitrogen Stress Tolerance in Maize: From Theory to Practice*; CIMMYT: Mexico, Mexico, 2000.
7. Abakemal, D.; Shimelis, H.; Derera, J. Genotype-by environment interaction and yield stability of quality protein maize hybrids developed from tropical-highland adapted inbred lines. *Euphytica* **2016**, *209*, 757–769. [CrossRef]
8. Citizens Association for the Defense of Collective Interests (ACDIC). *The Maize Crisis and the Misfortunes of Cameroon's Agriculture*; ACDIC: Yaounde, Cameroon, 2010; p. 46.
9. Bänziger, M.; Setimela, P.S.; Hodson, D.; Vivek, B. Breeding for improved drought tolerance in maize adapted to Southern Africa. "New directions for a diverse planet". In Proceedings of the 4th International Crop Science Congress, Brisbane, Australia, 26 September–1 October 2004; [CDROM]: International Crop Science Congress: Brisbane, Australia, 2004.
10. Sibiya, J.; Tongoona, P.; Derera, J.; Rij, N. Genetic analysis and genotype by environment (G X E) for grey leaf spot disease resistance in elite African maize (*Zea mays* L.) germplasm. *Euphytica* **2012**, *185*, 349–362. [CrossRef]
11. Falconer, D.S.; Mackay, T.F.C. *Introduction to Quantitative Genetics*; Longman: New York, NY, USA, 1996.
12. Miti, F. Breeding Investigations of Maize (*Zea mays* L.) Genotypes for Tolerance Tolow Nitrogen and Drought in Zambia. Ph.D. Thesis, University of Kwa-Zulu Natal, Pietermaritzburg, South Africa, 2007.
13. Yan, W.; Tinker, N.A. Biplot analysis of multi-environment trial data: Principles and applications. *Can. J. Plant Sci.* **2006**, *86*, 623–645. [CrossRef]

14. Kenga, R. *Combining Ability Estimates and Heterosis in Selected Tropical Sorghum (Sorghum bicolor (L.) Moench)*; Department of Plant Science, Faculty of Agriculture, Ahmadu Bello University Zaria: Zaria, Nigeria, 2001; p. 186.

15. Khalil, I.A.; Rahman, H.U.; Rehman, N.U.; Arif, M.; Khalil, I.H.; Iqbal, M.; Hidayatullah; Afridi, K.; Sajjad, M.; Ishaq, M. Evaluation of maize hybrids for grain yield stability in north-west of Pakistan. *Sarhad J. Agric.* **2011**, *27*, 213–218.

16. IITA. Maize 2014. Available online: http://www.iita.org/maize (accessed on 16 June 2014).

17. Etoundi, S.M.N.; Dia, K.B. Determinants of the adoption of improved varieties of Maize in Cameroon: Case of CMS 8704. Globalisation, Institutions and African Economic Development. In Proceedings of the African Economic Conference, Tunis, Tunisia, 12–14 November 2008; pp. 397–413.

18. Badu-Apraku, B.; Oyekunle, M.; Obeng-Antwi, K.; Osuman, A.S.; Ado, S.G.; Coulibay, N.; Yallou, C.G.; Abdulai, M.; Boakyewaa, G.A.; Didjeira, A. Performance of extra-early maize cultivars based on GGE biplot and AMMI analysis. *J. Agric. Sci.* **2012**, *150*, 473–483. [CrossRef]

19. Ndhlela, T. *Improvement Strategies for Yield Potential, Disease Resistance and Drought Tolerance of Zimbabwean Maize Inbred Lines*; Department of Plant Sciences (Plant Breeding), University of the Free State: Bloemfontein, South Africa, 2012; 295p.

20. Adu, G.B.; Akromah, R.; Abdulai, M.S.; Obeng-Antwi, K.; Kena, A.W.; Tengan, K.M.L.; Alidu, H. Assessment of Genotype by Environment interactions and Grain Yield Performance of Extra-Early Maize (*Zea mays* L.) Hybrids. *J. Biol. Agric. Healthc.* **2013**, *3*, 7–15.

21. Badu-Apraku, B.; Akinwale, R.O.; Menkir, A.; Obeng-Antwi, K.; Osuman, A.S.; Coulibaly, N.; Onyibe, J.E.; Yallou, G.C.; Abdullai, M.S.; Didjera, A. Use of GGE Biplot for Targeting Early Maturing Maize Cultivars to Mega-environments in West Africa. *Afr. Crop Sci. J.* **2011**, *19*, 79–96. [CrossRef]

22. Eberhart, S.A.; Russell, W.A. Stability parameters for comparing varieties. *Crop Sci.* **1966**, *6*, 36–40. [CrossRef]

23. Gauch, H.G. Model selection and validation for yield trials with interaction. *Biometrics* **1988**, *44*, 705–715. [CrossRef]

24. Gauch, H.G.; Zobel, R.W. Identifying mega-environments and targeting genotypes. *Crop Sci.* **1997**, *37*, 311–326. [CrossRef]

25. Zobel, R.W.; Wright, M.J.; Gauch, H.G. Statistical analysis of a yield trial. *Agron. J.* **1988**, *80*, 388–393. [CrossRef]

26. Yan, W.; Hunt, L.A.; Sheng, Q.; Szlavnics, Z. Cultivar evaluation and mega-environment investigation based on the GGE biplot. *Crop Sci.* **2000**, *40*, 597–605. [CrossRef]

27. Gauch, H.G.; Zobel, R.W. AMMI analyses of yield trials. In *Genotype by Environment Interaction*; Kang, M.S., Gauch, H.G., Eds.; CRC Press: Boca Raton, FL, USA, 1996; pp. 85–122.

28. Ebdon, J.S.; Gauch, H.G. Additive main effects and multiplicative interaction analysis of national turfgrass performance trials. *Crop Sci.* **2002**, *42*, 497–506. [CrossRef]

29. Gauch, H.G. Statistical analysis of yield trials by AMMI and GGE. *Crop Sci.* **2006**, *46*, 1488–1500. [CrossRef]

30. Cooper, M.; Stucker, R.E.; DeLacy, I.H.; Harch, B.D. Wheat breeding nurseries, target environments, and indirect selection for grain yield. *Crop Sci.* **1997**, *37*, 1168–1176. [CrossRef]

31. Yan, W. GGEbiplot-a windows application for graphical analysis of multi-environment trial data and other types of two-way data. *Agron. J.* **2001**, *93*, 1111–1118. [CrossRef]

32. The, C.; Ngonkeu, M.L.; Zonkeng, C.; Apala, H.M. *Evaluation and Selection of Maize (Zea mays L.) Genotypes Tolerant to Low N Soil. Optimizing Productivity of Food Crop Genotypes in Low Nutrient Soils*; IAEA-TECDOC-1721: Vienna, Austria, 2013; pp. 251–264.

33. Tchienkoua, M. *Soil and Land-Use Survey of the Northern Section of the Mbalmayo Forest Reserve (Southern Cameroon). A Study for the Selection of the IITA Humid Forest Station Site*; Ressource and Crop Management Research Monography N° 24; International Institute of Tropical Agriculture: Ibadan, Nigeria, 1996; p. 65.

34. Heanes, D.L. Determination of organic C in soils by an improved chromic acid digestion and spectrophotometric procedure. *Commun. Soil Sci. Plant Anal.* **1984**, *15*, 1191–1213. [CrossRef]

35. Buondonno, A.; Rashad, A.A.; Coppola, E. Comparing tests for soil fertility. II. The hydrogen peroxide/sulfuric acid treatment as an alternative to the copper/selenium catalyzed digestion process for routine determination of soil nitrogen-Kjeldahl. *Commun. Soil Sci. Plant Anal.* **1995**, *26*, 1607–1619. [CrossRef]

36. Anderson, J.M.; Ingram, J.S.I. *Tropical Soil Biology and Fertility: A Handbook of Methods*, 2nd ed; CAB International, The Cambrian News: Aberstwyth, UK, 1993; p. 221.

37. Mehlich, M. Mehlich 3 soil text extractant: A modification of the Mehlich 2 extractant. *Commun. Soil Sci. Plant Anal.* **1984**, *15*, 1409–1416. [CrossRef]

38. Barnhisel, R.; Bertsch, P. Aluminum. In *Methods of Soil Analysis, Part 2, Chemical and Microbiological Properties*; Page, A., Miller, R., Keeney, D., Eds.; American Society of Agronomy: Madison, WI, USA, 1982; pp. 275–300.

39. Mosquera, A.; Mombiela, F. Comparison of three methods for determination of soil Al in an unbuffered salt-extract. *Commun. Soil Sci. Plant Anal.* **1986**, *17*, 30–35. [CrossRef]

40. Murphy, J.; Riley, J.P. A modified single solution method for determination of phosphate in natural waters. *Anal. Chim. Acta* **1962**, *27*, 31–36. [CrossRef]

41. *Statistical Analysis System (SAS)*, version SAS/STAT 9.2; SAS Institute Inc.: Cary, NC, USA, 2008.

42. Cochran, W.G.; Cox, G.M. *Experimental Designs*; John Wiley & Sons: Hoboken, NJ, USA, 1960.

43. The IBP Breeding Management System Version 2.1. The Intergated Breeding Platform. May 2014. Available online: https://www.integratedbreeding.net/breeding-management-system (accessed on 13 May 2014).

44. Acquaah, G. *Principles of Plant Genetics and Breeding*; Wiley-Blackwell Publishers: Hoboken, NJ, USA, 2007.

45. Yan, W.; Tinker, N.A. An intergrated biplot analysis system for displaying interpreting, exploring genotyped by environments interactions. *Crop Sci.* **2005**, *45*, 1004–1016. [CrossRef]

46. Yan, W.; Rajcan, I. Biplot evaluation of test sites and trait relations of soybean in Ontario. *Crop Sci.* **2002**, *42*, 11–20. [CrossRef] [PubMed]

47. Aroga, R.; Ambassa-kiki, R.; The, C.; Enyong, L.; Ajala, S.O. On farm evaluation of performance of selecetd improved maize varieties in the forest zone of Central Cameroon. In Proceedings of the Seventh Eastern and Southern Africa Regional Maize Conference, Nairobi, Kenya, 5–11 February 2002; pp. 432–437.

48. Aroga, R.; Coderre, D. Abondance et diversité des foreurs de tiges et grains dans une biculture maïs-arachide au centre du Cameroun. *Afr. Crop Sci. J.* **2000**, *8*, 365–374.

agronomy

MDPI

Article

Combining Ability of Sixteen USA Maize Inbred Lines and Their Outbreeding Prospects in China

Ji-ying Sun [1,*,†] , Ju-lin Gao [1,*,†], Xiao-fang Yu [1,†], Jian Liu [2,†], Zhi-jun Su [1], Ye Feng [3] and Dong Wang [3]

1   College of Agronomy, Inner Mongolia Agricultural University, No.275, XinJian East Street,
     Hohhot 010019, China; yuxiaofang75@163.com (X.-f.Y.); barrysu@126.com (Z.-j.S.)
2   Vocational and Technical College, Inner Mongolia Agricultural University, Baotou 014109, China;
     silentliujian@163.com
3   Tongliao Academy of Agricultural Sciences, Qianjiadian, Tongliao 028015, China;
     fengye810914@126.com (Y.F.); nkywd@126.com (D.W.)
*   Correspondence: jiying-sun@imau.edu.cn (J.-y.S.); julin-gao@imau.edu.cn (J.-l.G.);
     Tel.: +86-139-4713-0409 (J.-y.S. & J.-l.G.)
†   These authors contributed equally to this work.

Received: 29 October 2018; Accepted: 23 November 2018; Published: 27 November 2018

Abstract: In China, there is an increasing need for greater genetic diversity in maize (*Zea mays* L.) germplasm and hybrids appropriate for mechanical harvesting. In order to test and distinguish American maize inbred lines with exceptional combining ability, four Chinese maize inbred lines (Chang7-2, Zheng 58, four-144 and four-287) were used to judge the combining ability and heterosis of 16 USA inbred lines by a NCII genetic mating method. The results showed that among the American inbred lines, 6M502A, LH208, NL001, LH212Ht, PHW51, FBLA and LH181 expressed good GCA for yield characteristics; while RS710, PHP76, FBLA, and PHJ89 showed excellent GCA for machine harvesting characteristics. Five hybrids (NL001 × Chang7-2, LH212Ht × Chang7-2, FBLA × four-144, LH181 × four-287, PHK93 × four-287) had better SCA values for yield characteristics, at 1.69, 1.07, 1.48, 1.84 and 1.05, respectively; while NL001 × Chang 7-2, 6M502A × Chang7-2, LH212Ht × Chang7-2, LH181 × four-287, PHW51 × Chang7-2 had better TCA values for yield characteristics, at 3.03, 2.80, 2.41, 2.19 and 1.91, respectively; NL001 × Chang7-2, 6M502A × Chang7-2, LH212Ht × Chang7-2, LH181 × four-287, PHW51 × Chang7-2 showed excellent Control Heterosis values, with 21.48%, 19.64%, 15.93%, 14.05% and 11.60% increases, respectively, compared to the check and potential for future utilization in Inner Mongolian corn production.

Keywords: USA inbred lines; combining ability; machine harvesting characteristics; yield characteristics; control heterosis

1. Introduction

The genetic diversity of maize germplasm in China is decreasing, due to fewer inbreds being used to produce modern, high-yielding maize hybrids. At the same time, the change of corn planting patterns in China has greatly increased the demand for full mechanization, which requires the improvement of maize varieties to be suitable for the machine harvesting of the grain. However, the lack of maize germplasm suitable for mechanical harvesting that has high combining ability, strong disease and pest resistance, and wide adaptability has become a bottleneck for maize breeding development in China [1–4]. North American germplasm plays an important role in China's corn yield potential, and their genetic contribution to Chinese corn has been increasing [5,6]. Using American maize germplasm is an effective way to improve the diversity of Chinese maize germplasm and screened favorable allele donors due to its clear genealogical origin and abundant genetic variation [7].

Previous studies have shown that the growth period, the silking stage, the ear height, the plant height, and the kernel moisture concentration at the R6 stage (physiological maturity) could be used to determine a maize germplasm suitable for mechanical harvesting [8]. General combining ability (GCA) is determined by the additive effects of genes, and can distinguish the genetic component of an inbred line and reflect its potential for utilization. Specific combining ability (SCA) is determined by the non-additive effect of genes, which is easily affected by the environment and cannot be stably inherited. It is used as a reference when sifting through hybrid combinations. The total combining ability (TCA) effect value is determined by the parental inbred GCA and SCA, it can be used as an index to evaluate combined hybrid performance. Control heterosis (CH) is considered to be the yield-increasing index for corn varieties in the national standard of China, the best hybrid combination can be selected by analyzing the control heterosis. The research used 16 inbreds from the expired Plant Variety Protection Act (ex-PVP) germplasm adapted to the USA Corn Belt and germplasm currently used in Chinese production as the basic materials. The present study aimed to determine the germplasm most suitable for mechanical harvesting, and with the most favorable agronomic traits and yield-related traits, as well as the combining abilities of the lines, by analyzing their GCA, SCA, TCA and CH, so as to clarify the breeding potential of USA germplasm in the Inner Mongolian Maize production area, and provide a reference for its utilization.

2. Materials and Methods

2.1. Germplasm and Experimental Sites

Sixteen diverse maize expired Plant Variety Protection Act (ex-PVP) inbred germplasms adapted to the USA Corn Belt were acquired from the North Central Regional Plant Introduction Station (http://www.ars-grin.gov/npgs, verified 24 August 2016), through the Maize Industrial Technology System Construction of Modern Agriculture of China by international communication.

The classification and pedigree sources of the sixteen USA inbreds and the four China inbreds are shown in Table 1. The China heterotic group A is similar to the USA heterotic group of stiff stalk synthetic (SS), while the China heterotic group B is similar to the USA heterotic group of non-stiff stalk synthetic (NSSS).

Table 1. Genealogical origin of USA maize inbred lines and China testers.

Number	Germplasm	Heterotic Group	Genealogical Origin
1	RS 710	NSS	1202 × 1250
2	LH191	SS	LH132 × Pioneer 3184
3	LH192	SS	LHE137 × LHE136
4	PHN34	SS	SC359 × PH157 specifically (SC359/PH157)X#4221
5	PHP76	NSS	(G50/PHEJ8)X812X
6	PHW51	SS	(PHDF2/PHG41)RXB333X
7	FBLA	SS	(B14////Mt42).A656(B14//Mt42)
8	6F629	NSS	88051B/4608H
9	6M502A	NSS	MAWU.4913
10	NL001	SS	1089HT × A634HT/B73
11	LH181	NSS	LH58 XL H122
12	LH208	SS	LH74 × CB59G
13	LH212Ht	NSS	LH123Ht × (LH123Ht X LH24)
14	Lp215D	NSS	Mo17 × Lp216D
15	PHJ89	NSS	PHT77 × PHG47
16	PHK93	NSS	PHB72 × PHT60 specifically (PHB72/PHT60)6K41K111K211
17	Zheng58	A	Ye 478improved line
18	Chang7-2	B	V59 × Huangzaosi

Table 1. *Cont.*

Number	Germplasm	Heterotic Group	Genealogical Origin
19	four-144	A	VMA724 improved line
20	four-287	B	four-444 × 255

A is generally suitable to be the female parent with high yield, and A is similar to SS, B is generally suitable to be the male parent with more pollen, and B is similar to NSS.

The trials were conducted in 2015 at the two main Maize production areas of Inner Mongolia in China—Hohhot and Tongliao. The weather condition of 2015 and basis soil fertilizer are as below (Tables 2 and 3).

Table 2. Weather condition of 2015 in Hohhot and Tongliao.

Experimental Sites	Latitude	Longitude	Solar Radiation	Average Temperature	Precipitation
			h per Year	°C	mm per Year
Hohhot	40°33′ N	110°31′ E	1780.5	17.8	275.4
Tongliao	43°42′ N	122°32′ E	1224.7	20.5	433.2

Table 3. Basis fertilizer of soil in Hohhot and Tongliao.

Experimental Sites	Organic Matter	Available N	Available P	Available K	Soil Type
	g/kg	mg/kg	mg/kg	mg/kg	
Hohhot	18.9	44.8	16.2	120.4	Sandy loam
Tongliao	20.4	55.6	18.2	167.9	Meadow chernozemic soil

2.2. Experimental Design

Hybrids were produced using the sixteen USA maize inbred lines as female parents and the four China test species as male parents. In a NC-II genetic mating design, 64 hybrid combinations were produced at Hainan province Ledong county experimental base (18°45′5.38″ N, 109°10′10.22″ E) in the winter of 2014.

In 2015, the 64 hybrid combinations and one control hybrid (Zhengdan 958) were planted at Hohhot and Tongliao. An α-lattice block design was used with five replications, 0.6 m row spacing, 0.25 m plant spacing, 40 plants per plot, with a density of 66,670 plants/ha. Two row plots. The rate of NPK fertilizer applied was N: 200 kg/ha, P_2O_5: 105 kg/ha and K_2O: 62 kg/ha. Phosphate fertilizer and potash fertilizer were applied as basal fertilizer once before planting and nitrogen fertilizer was applied by 30% (60 kg/ha) at V6 stage (six leaves with collars visible) and 70% (140 kg/ha) at V12 stage (twelve leaves with collars visible), respectively. Irrigation and other management measures during the whole growth period were similar to local farmer practices.

2.3. Measurements and Production Indicators

The days from field emergence to 50% silking and to maturity were recorded for each plot. During plant maturation, 10 plants were randomly selected, and their total height and ear height were measured.

Plant stand counts were tallied to confirm plant populations at the R6 plant growth stage, and ear stand counts were tallied to confirm ear number per ha. The two rows of each plot were manually harvested for determination of grain yield at physiological maturity, corn ears were tallied and weighed, the grain was removed manually to analyze for moisture content using seed moisture meter (PM-8188-A, KETT ELECTRIC LABORATORY, Tokyo, Japan), 300 randomly selected kernels were weighed to estimate average individual kernel weight. According to the average weight of the ear, select 10 ears of each plot to assess the number of rows per ear and grain number per row. The kernel weight and the yield were presented at 14% moisture content.

2.4. Data Statistical Analysis

Variance analysis of all the traits collected including the General and Specific Combining Abilities was performed by GLM of SAS software version [9], linear model was as followed [10]:

$$Y_{ijk} = \mu + m_i + f_j + (m \times f)_{ij} + e_{ijk} \tag{1}$$

where Y_{ijk} is the k observational value of the progeny of parents i and j, μ is the universal mean, m_i is i-th paternity effect, f_j is j-th maternal effect, $(m \times f)_{ij}$ is the interaction effect, e_{ijk} is the error term.

$$TCA_{ij} = G_i + G_j + S_{ij} \tag{2}$$

where TCA_{ij} is Total Combining Ability of the progeny of parents i and j, G_i is General Combining Ability of parent i, G_j is General Combining Ability of parent j, S_{ij} is Special Combining Ability of the progeny of parents i and j.

$$CH_{ij} = (YF_{ij} - YC)/YC \times 100\% \tag{3}$$

where CH_{ij} is the Control Heterosis of the progeny of parents i and j, YF_{ij} is the average yield of an individual hybrid combination by parents i and j, YC is the average yield of the control Zhengdan 958.

3. Results

3.1. Field Characteristic and Adaptability of the Trial Inbreds

Throughout the 2014 trial of field adaption, the sixteen USA ex-PVP inbred germplasms showed excellent adaption characteristics to the weather and soil condition in the two main production areas of Maize in Inner Mongolia—Hohhot (Table 4) and Tongliao (Table 5).

From Table 4, it can be seen that the days to silking of the Sixteen USA inbreds was 63–82 days, respectively, after the emergence, the days to maturity of the sixteen inbreds were 111–126 days, respectively; the plant height of the sixteen inbreds was 120–260 cm, respectively; the ear height was 39–100 cm, respectively; ASI (anthesis–silking interval) was −1 to −3, respectively, and the overall merit of adaptability was 4–8, respectively. From Table 5, we can see that the days to silking of the sixteen USA inbreds was 51–67 days, respectively, after emergence; the days to maturity of the sixteen inbreds was 112–123 days, respectively; the plant height was 140–246 cm, respectively; the ear height was 34–100 cm, respectively; the ASI was 0 to −3, respectively; and the overall merit of adaptability was 4 to 8, respectively. The field characteristics of the sixteen USA inbreds in Hohhot and Tongliao were suitable for acting as a hybrid parent together with the four China inbreds, and based on the adaptability of the sixteen inbreds to the conditions in Hohhot and Tongliao, the inbreds can adapt to grow in the Inner Mongolian maize production areas, so this study select the sixteen USA ex-PVP inbred germplasms and the 4 China test lines as trial materials.

3.2. Phenotypic and Grain Yield Traits

The machine-harvest characteristics of the maize hybrids in the different locations varied greatly. At Hohhot, the days to maturity between hybrids differed by 13, while at Tongliao hybrids matured over 33 days. The days to silking at Hohhot varied by 18 days, but that at Tongliao by 12 days. Plants were shorter, but with less variability at Hohhot, ranging from 159.5 to 278.8 cm. Meanwhile, at Tongliao, plant height ranged from 202.0 to 324.0 cm. Ear height of the hybrids at Hohhot was 47.5–116.0 cm, and that at Tongliao was 71.0–158.0 cm. At harvest time, the grain moisture content varied from 18.6% to 38.6% at Hohhot, and from 24.2% to 35.3% at Tongliao (Table 6).

A basic statistical analysis of the hybrid yield-based indicators of ear row number, kernel grains per row, 100-kernel weight and grain yield carried out at Hohhot and Tongliao (Table 7) showed that the extremes, average, standard deviation and variable coefficient of the measured traits at the different locations varied greatly. There was a greater range in kernel grains per row at Hohhot—from 29.5 to 46.5—compared to 32.1 to 42.6 at Tongliao. At Hohhot, ear row number varied from 11.6 to 18.0, with slightly more at Tongliao, from 12.8 to 18.4.

Table 4. Field characteristic and adaptability of the sixteen USA inbreds and four China inbreds in Hohhot. Values are the average $\pm$ 1 standard error.

Number	Inbred	Days to Silking Day	Days to Maturity Day	ASI Day	Plant Height cm	Ear Height cm	Seeding Potential	Disease Resistance (R6)	Anti-Lodging	Ear Evaluation	Overall Merit
1	RS710	72 ± 0.8	118 ± 0.7	−2 ± 0.8	182 ± 4.1	55 ± 2.6	1 [†]	2 [‡]	1 [§]	3 [¶]	7 [€]
2	LH191	78 ± 0.5	126 ± 1.4	−2 ± 0.8	260 ± 8.4	100 ± 5.0	1	2	1	3	7
3	LH192	79 ± 0.9	125 ± 0.7	−3 ± 0.7	200 ± 12.9	78 ± 2.5	1	2	1	3	7
4	PHN34	76 ± 0.4	124 ± 0.0	−2 ± 0.9	180 ± 3.7	79 ± 4.0	1	1	1	2	5
5	PHP76	72 ± 0.4	117 ± 0.0	−3 ± 0.5	165 ± 3.6	60 ± 1.5	1	1	1	3	6
6	PHW51	74 ± 1.1	122 ± 0.4	−2 ± 1.1	198 ± 13.2	85 ± 3.5	1	2	1	1	5
7	FBLA	74 ± 0.4	119 ± 0.0	−1 ± 0.9	181 ± 7.0	79 ± 2.5	1	1	1	1	4
8	6F629	75 ± 0.9	119 ± 0.9	−1 ± 0.4	202 ± 6.9	87 ± 2.2	1	3	1	3	8
9	6M502A	76 ± 0.5	118 ± 0.0	−3 ± 0.8	243 ± 6.5	100 ± 3.2	1	1	1	1	4
10	NL001	75 ± 0.8	111 ± 0.0	−3 ± 0.8	210 ± 5.8	80 ± 2.2	1	1	1	1	4
11	LH181	77 ± 0.5	122 ± 0.4	−2 ± 0.5	228 ± 5.7	75 ± 2.8	1	1	2	1	5
12	LH208	82 ± 0.0	118 ± 0.9	−2 ± 0.5	134 ± 6.8	48 ± 2.6	1	1	1	1	4
13	LH212Ht	71 ± 0.4	122 ± 0.0	−2 ± 0.4	195 ± 2.6	70 ± 1.4	1	1	1	1	4
14	Lp215D	81 ± 0.5	116 ± 0.4	−2 ± 0.5	190 ± 8.2	70 ± 3.2	1	1	1	1	4
15	PHJ89	63 ± 0.4	119 ± 0.9	−2 ± 0.4	120 ± 4.1	39 ± 2.2	1	2	1	1	5
16	PHK93	79 ± 0.5	123 ± 0.0	−2 ± 0.8	185 ± 7.6	50 ± 1.5	1	2	1	1	5
17	Zheng58	74 ± 0.5	126 ± 0.9	−2 ± 0.5	175 ± 4.5	80 ± 4.5	1	1	2	1	5
18	Chang7-2	73 ± 0.5	123 ± 0.0	0 ± 0.8	145 ± 8.2	38 ± 2.3	2	1	1	1	5
19	Four-144	69 ± 1.1	119 ± 1.1	−2 ± 1.5	178 ± 3.8	55 ± 1.9	1	1	2	1	5
20	Four-287	68 ± 0.5	120 ± 0.5	0 ± 0.5	156 ± 4.0	46 ± 1.1	1	2	1	1	5

Sowing date: 2 May 2014. Emergence date: 15 May 2014; [†] means the estimate of the seeding potential, with 1–5 representing the seeding potential from the best to the worst; [‡] means the estimate of the disease resistance in R6 growth stage, with 1–5 representing the disease resistance from the best to the worst; [§] means the estimate of the Anti-lodging, with 1–5 representing the Anti-lodging characteristic from the best to the worst; [¶] means the ear evaluation, with 1–5 representing from the best ear characteristic and yield potential to the worst; [€] means the sum of seeding potential, disease resistance, Anti-lodging, and ear evaluation; the lower the amount, the better the overall merit.

Table 5. Field characteristics and adaptability of the sixteen USA inbreds and four China inbreds in Tongliao. Values are the average ± 1 standard error.

Number	Inbred	Days to Silking Day	Days to Maturity Day	ASI Day	Plant Height cm	Ear Height cm	Seeding Potential	Disease Resistance (R6)	Anti-Lodging	Ear Evaluation	Overall Merit
1	RS710	51 ± 0.5	112 ± 0.5	−1 ± 0.0	140 ± 3.9	34 ± 1.0	1[†]	3[‡]	1[§]	3[¶]	8[ᶜ]
2	LH191	67 ± 0.5	123 ± 0.9	−1 ± 0.0	172 ± 4.1	56 ± 1.5	1	2	1	2	6
3	LH192	66 ± 0.5	120 ± 0.0	−2 ± 0.5	175 ± 3.6	58 ± 1.5	1	3	1	2	7
4	PHN34	63 ± 1.1	122 ± 0.0	−1 ± 0.5	210 ± 7.6	82 ± 2.0	1	2	1	2	6
5	PHP76	55 ± 0.5	116 ± 1.1	0 ± 0.9	171 ± 3.3	55 ± 1.5	1	2	1	3	7
6	PHW51	63 ± 0.5	122 ± 0.5	0 ± 0.5	205 ± 4.1	60 ± 3.5	1	2	1	2	6
7	FBLA	58 ± 1.0	118 ± 0.5	−3 ± 1.3	181 ± 1.9	45 ± 1.0	1	2	1	1	5
8	6F629	58 ± 0.4	115 ± 1.1	0 ± 0.7	206 ± 4.7	78 ± 0.6	1	1	1	3	6
9	6M502A	63 ± 0.5	117 ± 0.0	−2 ± 0.8	207 ± 2.6	73 ± 1.5	1	1	1	1	4
10	NL001	62 ± 0.4	115 ± 0.4	−2 ± 0.4	160 ± 5.8	63 ± 0.6	1	1	1	1	4
11	LH181	61 ± 0.4	118 ± 0.5	−1 ± 0.7	246 ± 7.9	72 ± 4.5	1	1	1	1	4
12	LH208	65 ± 0.8	120 ± 0.4	−2 ± 0.5	225 ± 3.2	75 ± 2.0	1	1	1	1	4
13	LH212Ht	64 ± 0.5	122 ± 0.5	−2 ± 0.8	245 ± 4.6	100 ± 6.5	1	1	1	2	5
14	Lp215D	56 ± 0.8	118 ± 0.4	0 ± 0.8	212 ± 2.6	60 ± 0.6	1	2	1	1	5
15	PHJ89	53 ± 0.9	115 ± 1.1	−1 ± 1.3	210 ± 3.2	63 ± 1.0	1	2	1	2	6
16	PHK93	64 ± 0.5	115 ± 1.1	−1 ± 0.4	205 ± 4.0	55 ± 1.5	1	1	1	2	5
17	Zheng58	66 ± 0.5	120 ± 0.9	−2 ± 0.4	189 ± 6.1	81 ± 0.6	1	1	2	1	5
18	Chang7-2	65 ± 0.5	121 ± 0.5	−1 ± 0.5	154 ± 3.2	47 ± 2.0	1	1	1	1	4
19	Four-144	57 ± 0.5	116 ± 0.0	1 ± 0.9	170 ± 2.1	52 ± 0.6	1	1	2	2	5
20	Four-287	58 ± 0.4	115 ± 0.4	1 ± 0.4	152 ± 2.9	43 ± 1.5	1	2	1	1	5

Sowing date: 1 May 2014. Emergence date: 14 May 2014; [†] means the estimate of the seeding potential, with 1–5 representing the seeding potential from the best to the worst; [‡] means the estimate of the disease resistance in R6 growth stage, with 1–5 representing the disease resistance from the best to the worst; [§] means the estimate of the Anti-lodging, with 1–5 representing the Anti-lodging characteristic from the best to the worst; [¶] means the ear evaluation, with 1–5 representing from the best ear characteristic and yield potential to the worst; [ᶜ] means the sum of seeding potential, disease resistance, Anti-lodging, and ear evaluation; the lower the amount, the better the overall merit.

Agronomy **2018**, *8*, 281

Table 6. Summary of machine-harvesting-related trait measurements for 64 maize hybrids grown in Hohhot and Tongliao in 2015.

| | Hohhot | | | | | Tongliao | | | | |
	Days to Maturity	Days to Silking	Plant Height	Ear Height	Grain Moisture	Days to Silking	Days to Maturity	Plant Height	Ear Height	Grain Moisture
	Day	Day	cm	cm	%	Day	Day	cm	cm	%
minimum	136.0	66.0	159.5	47.5	18.6	55.0	105.0	202.0	71.0	24.2
maximum	149.0	84.0	278.8	116.0	38.6	67.0	138.0	324.0	158.0	35.3
SD	0.65	0.97	7.25	3.05	0.80	0.83	0.57	11.10	4.86	0.96
average	144.7	77.1	225.0	75.4	29.9	61.4	121.7	280.0	110.0	28.6
CV	0.45	1.26	3.22	4.05	2.69	1.34	0.47	3.96	4.42	3.37

Table 7. Summary of yield-related trait measurements for 64 maize hybrids grown in Hohhot and Tongliao in 2015.

| | Hohhot | | | | Tongliao | | | |
	Kernels per Row	Rows per Ear	100-Kernel Weight	Grain Yield	Kernels per Row	Rows per Ear	100-Kernel Weight	Grain Yield
			g	t/ha			g	t/ha
minimum	29.5	11.6	23.6	6.5	32.1	12.8	26.7	7.7
maximum	46.5	18.0	39.6	15.3	42.6	18.4	41.0	16.8
SD	0.37	0.27	0.75	0.65	0.56	0.29	0.86	0.88
average	39.3	14.3	30.7	11.0	38.7	15.9	33.5	11.4
CV	0.95	1.93	2.46	5.97	1.43	1.83	2.56	7.74

At Hohhot, 100-kernel weight of the hybrids was lighter, and varied from 23.6 g to 39.6 g, compared to at Tongliao, which was 26.7 g to 41.0 g. There was a large variation in grain yield of the hybrids, 6.5 t/ha–15.7 t/ha and 7.7 t/ha–16.8 t/ha at Hohhot and Tongliao, respectively.

3.3. Analysis of Variance of Main Characteristics

Table 8 shows that the variances of five traits relating to the suitability of harvesting (the days to maturity, days to silking, plant height, ear height and moisture content at harvest) in paternal tester's heterosis, general combining ability (GCA) and specific combining ability (SCA) of maternal lines were all significant or highly significant. The environmental and gene interaction effects of all traits were highly significant.

Table 8. Analysis of variance for the machine-harvesting characteristics of the hybrids derived from USA maize inbred lines crossed with China inbred testers grown in two locations in 2015.

Variation Source	DF	Days to Maturity	Days to Silking	Plant Height	Ear Height	Grain Moisture at Harvest
Environment	1	50,669.1 **	23,531.3 **	290,554.0 **	115,065.8 **	155.3 **
Line	15	166.2 **	134.8 **	3133.8 **	801.0 **	49.4 **
Tester	3	467.4 **	158.4 **	22,005.8 **	15,824.8 **	118.3 **
Line × Tester	45	56.6 **	12.3 **	390.0 **	350.8 **	22.3 **
Line × Environment	15	69.6 **	30.1 **	516.3 **	167.3 **	29.3 **
Tester × Environment	3	316.5 **	2.7 *	2255.1 **	344.4 **	19.6 **
Line × Tester × Environment	45	55.3 **	8.6 **	504.9 **	363.0 **	30.2 **

* and ** in the column represents significance at the 0.05 and 0.01 probability level, respectively.

The four grain yield characteristics (kernel grains per row, the ear row number, 100-kernel weight and grain yield) expressed significant differences for line, tester, environment, and between the interaction (Table 9).

Table 9. Analysis of variance for the yield characteristics of the hybrids derived from USA maize inbred lines crossed with China inbred testers grown in two locations in 2015.

Variation Source	DF	Kernel Number per Row	Kernel Row Number	100-Kernel Weight	Grain Yield
Environment	1	29.4 **	248.7 **	755.0 **	16.0 **
Line	15	60.8 **	4.7 **	47.8 **	14.5 **
Tester	3	187.4 **	41.6 **	279.4 **	36.6 **
Line × Tester	45	12.6 **	1.7 **	18.9 **	3.7 **
Line × Environment	15	16.2 **	2.8 **	17.3 **	4.4 **
Tester × Environment	3	32.2 **	1.6 **	14.6 **	2.7 **
Line × Tester × Environment	45	6.7 **	0.7 **	18.0 **	2.6 **

** in the column represents significance at the 0.01 probability level, respectively.

3.4. General Combining Ability (GCA) Effect of USA Inbred Lines Suitable for Machine-Harvest Indexes and Grain Yield Characters

The GCA effect of both days to maturity and days to silking were significantly negative for RS710, PHP76, FBLA, 6F629, NL001, Lp215D and PHJ89, indicating that hybrids derived from these inbred lines had faster development, with shorter days to silking and to maturity (Table 10). Hybrids made from RS710, PHP76, FBLA, or PHJ89 resulted in shorter plants with lower ear heights, displaying lower GCA effect values. Additionally, the negative GCA effect values of grain water content at harvest of RS710, PHP76, FBLA, 6F629, LH208 and PHJ89, indicates that hybrids derived from these inbred lines had a faster grain dehydration rate (Table 10).

Table 10. The general combining ability of USA inbred lines for machine-harvest and yield characteristics.

Line	Days to Maturity	Days to Silking	Plant Height	Ear Height	Grain Moisture at Harvest	Kernels per Row	Kernel Rows	100-Kernel Weight	Grain Yield
RS710	−4.96 **	−5.47 **	−34.33 **	−13.96 **	−2.91 **	−4.85 **	−0.04	−1.47 **	−1.97 **
LH191	3.04 **	2.57 **	−3.72 **	−0.85	0.10	−0.77 **	0.18 **	1.20 **	−0.31 **
LH192	4.41 **	2.73 **	4.93 **	1.57 *	1.89 **	0.60 **	0.76 **	−0.66 **	−0.49 **
PHN34	3.50 **	2.07 **	13.54 **	11.26 **	2.06 **	0.67 **	−0.09 *	0.11	−0.28 *
PHP76	−1.88 **	−3.35 **	−13.09 **	−2.13 **	−0.86 **	0.59 **	−0.04	−2.07 **	−0.56 **
PHW51	1.71 **	0.69 **	0.14	−0.26	1.32 **	0.10	−0.33 **	−0.70 **	0.48 **
FBLA	−1.75 **	−1.68 **	−2.68 *	−5.34 **	−2.43 **	−0.19 **	−0.41 **	1.49 **	0.42 **
6F629	−1.04 **	−0.27 *	−0.96	−0.37	−1.33 **	1.01 **	−0.11 **	−2.20 **	−0.63 **
6M502A	−1.25 **	2.23 **	7.55 **	6.57 **	1.43 **	1.01 **	−0.41 **	−0.99 **	1.47 **
NL001	−2.79 **	−0.85 **	−2.18	0.09	0.25 *	−0.61 **	0.52 **	−0.25 *	0.51 **
LH181	1.29 **	2.19 **	9.64 **	−0.02	0.48 **	−1.57 **	−0.37 **	2.48 **	0.34 **
LH208	−1.34 **	0.15	3.10 *	−1.36 *	−0.33 *	1.97 **	−0.28 **	0.01	0.89 **
LH212Ht	1.66 **	1.19 **	11.06 **	8.28 **	0.77 **	−0.01	0.01	1.18 **	0.51 **
Lp215D	−1.42 **	−2.02 **	4.17 **	1.17 *	−0.07	−0.42 **	−0.74 **	0.85 **	−0.07
PHJ89	−1.79 **	−1.72 **	−4.54 **	−5.22 **	−0.98 **	0.65 **	−0.07	−0.93 **	−0.11
PHK93	2.62 **	1.53 **	7.38 **	0.57	0.62 **	1.83 **	0.01	1.95 **	−0.19
LSD$_{0.05}$	0.17	0.25	2.58	1.12	0.24	0.13	0.08	0.22	0.21
LSD$_{0.01}$	0.24	0.35	3.66	1.58	0.35	0.18	0.11	0.32	0.30

* and ** in the column represents significance at the 0.05 and 0.01 probability level, respectively.

Number of rows per ear, number of kernel per row, and 100-kernel weight are important factors for grain yield composition. The evaluation of the 100-kernel weight revealed positive and significant GCA values for the USA inbred lines LH191, FBLA, LH181, LH212Ht, Lp215D and PHK93. The GCA effect values of the number of rows per ear were positive and significantly different for LH191, LH192, and NL001, indicating that hybrid combinations obtained by these inbred lines could increase the number of rows per ear; the GCA effect values of kernels per row indicated significant increase due to many inbreds, including LH192, PHN34, PHP76, 6F629, 6M502A, LH208, PHJ89 and PHK93. Hybrid combinations obtained by these inbred lines could increase kernels per row. For total grain yield, the positive and significant GCA coefficient indicated that hybrids developed from the corresponding inbreds may achieved higher than average grain yield (Table 10).

3.5. General Combining Ability (GCA) Effect of China Tester Lines Suitable for Machine-Harvest Indexes and Grain Yield Characters

From Table 11, we can see that the kernel number per row, row number per ear and grain yield GCA effect values of chang7-2 were positively significant. The GCA effects of plant height and ear height of Zheng58 were negatively significant, but the GCA effect of 100-kernel weight was positively significant. The results indicate that Zheng58 would be beneficial in hybrids for mechanized harvest.

The GCA effect values of days to maturity, days to silking, plant height and grain moisture content at harvest were all negative significant, and GCA effect values of kernel number per row were positively significant for the tester four-144. The GCA effect values of days to maturity, days to silking, plant height, ear height and grain moisture content at harvest of four-287 were all significantly negative, and the GCA effect of 100-kernel weight was positively significant. The results showed that the hybrid combination with four-287 was easy to possess the characteristics of earlier maturity, fewer days to silking, low plant height, low ear, low moisture content at harvest, and high 100-kernel weight.

3.6. Specific Combining Ability (SCA) of Hybrid Combination

Among the 64 hybrid combinations, 16 had positive and significant SCA effects for yield. (Table 12). The A × A cis-hybrid combinations with good yield included LH191 × Zheng58, PHN34 × Zheng58, LH208 × Zheng58 and FBLA × four-144. The B × B cis-hybrid combinations producing increased yield included 6M502A × Chang 7-2, LH212Ht × Chang 7-2, Lp215D × Chang 7-2, RS710 × four287, LH181 × four-287 and PHK93 × four-287. Meanwhile, the A × B trans-hybrid combinations with yield increases included PHW51 × Chang 7-2, NL001 × Chang 7-2 and LH208 × four-287. The B × A trans-hybrid combinations with yield increases included 6M502A × Zheng 58, PHP76 × four-144 and 6F629 × four-144.

Conversely, 18 had negative SCA values, which indicated decreased yields (Table 12). The A × A cis-hybrid combination with significantly decreased yield included NL001 × Zheng58 and LH208 × four-144. The B × B cis-hybrid combinations with significantly decreased yield included 6F629 × Chang7-2, LH181 × Chang7-2, PHK93 × Chang7-2, PHP76 × four-287, 6M502A × four-287, LH212Ht × four- 287. The A × B trans-hybrid combinations with significantly decreased yield includes PHN34 × Chang 7-2, FBLA × Chang7-2, PHN34 × four-287, PHW51 × four-287, FBLA × four-287 and NL001 × four-287. Meanwhile, the significantly decreased yield performers in the B × A trans-hybrid combinations included RS710 × Zheng58, LH181 × Zheng58, Lp215D × four-144 and PHJ89 × four-144.

Table 11. The general combining ability, of China tester lines for machine-harvest and yield characteristics.

Line	Days to Maturity	Days to Silking	Plant Height	Ear Height	Grain Moisture at Harvest	Kernels per Row	Kernel Rows	100-Kernel Weight	Grain Yield
Zhen58	1.7**	0.8**	−17.1**	−6.9**	1.0**	−1.0**	−0.3**	0.2**	−0.2**
Chang7-2	2.1**	1.3**	19.8**	19.2**	0.9**	1.8**	0.9**	−0.4**	0.8**
four-144	−1.7**	−0.5**	−0.4	−4.7**	−0.7**	0.5**	−0.03	−2.0**	−0.6**
four-287	−2.1**	−1.6**	−2.3**	−7.6**	−1.2**	−1.2**	−0.6**	2.2**	0.01
$LSD_{0.05}$	0.09	0.09	0.98	0.44	0.07	0.15	0.05	0.15	0.09
$LSD_{0.01}$	0.13	0.12	1.39	0.62	0.10	0.21	0.08	0.21	0.13

** in the column represents significance at the 0.01 probability level, respectively.

Table 12. Analysis of specific combining ability of hybrid combinations for growth and yield parameters.

Hybrid Combinations	Hybrid Pattern	Days to Maturity	Days to Silking	Plant Height	Ear Height	Grain Moisture at Harvest	Grain Yield
RS710 × Zheng58	B × A	-4.25 **	-0.07	-12.84 **	-7.52 **	0.68 *	-0.47 *
LH191 × Zheng58	A × A	-2.91 **	-0.61 **	0.63	-0.09	-0.03	0.57 *
LH192 × Zheng58	A × A	-1.96 **	-1.11 **	-8.85 **	-5.01 **	-0.99 **	-0.14
PHN34 × Zheng58	A × A	-2.54 **	0.06	8.58 **	0.55	-2.68 **	0.87 **
PHP76 × Zheng58	B × A	-1.16 **	0.64 *	2.22	6.94 **	-0.28	0.19
PHW51 × Zheng58	A × A	5.09 **	1.60 **	6.86 *	3.64 **	-1.61 **	0.14
FBLA × Zheng58	A × A	1.71 **	0.47	6.55 *	6.74 **	3.13 **	0.15
6F629 × Zheng58	B × A	2.67 **	1.22 **	3.62	2.43 *	2.59 **	-0.27
6M502A × Zheng58	B × A	1.21 **	0.89 **	7.21 *	1.49	0.73 *	0.50*
NL001 × Zheng58	A × A	0.42 *	-1.19 **	-10.24 **	-7.78 **	-3.18 **	-1.01 **
LH181 × Zheng58	B × A	5.34 **	1.77 **	-1.61	-3.04 *	-0.26	-0.95 **
LH208 × Zheng58	A × A	1.29 **	-0.03	-7.31 **	-3.99 **	0.71 **	0.64 **
LH212Ht × Zheng58	B × A	-1.21 **	-1.57 **	7.94 **	9.37 **	-1.79 **	-0.17
Lp215D × Zheng58	B × A	0.71 **	-0.03	0.53	-0.02	2.22 **	0.12
PHJ89 × Zheng58	B × A	-0.08	-0.32	-8.33 **	-4.97 **	0.99 **	0.10
PHK93 × Zheng58	B × A	-4.33 **	-1.73 **	5.03	1.24	-0.25	-0.26
RS710 × Chang7-2	B × B	-3.47 **	-1.18 **	-11.52 **	-15.97 **	0.59 *	-0.32
LH191 × Chang7-2	A × B	2.03 **	-1.22 **	4.87	7.48 **	-2.25 **	-0.41
LH192 × Chang7-2	A × B	3.16 **	0.94 **	-4.27	9.00 **	0.01	0.26
PHN34 × Chang7-2	A × B	4.08 **	1.11 **	-3.19	3.90 **	-0.08	-0.66 **
PHP76 × Chang7-2	B × B	0.95 **	0.19	-13.36 **	-6.13 **	-0.46	-0.22
PHW51 × Chang7-2	A × B	-3.47 **	-0.18	3.76	-6.91 **	1.51 **	0.60 **
FBLA × Chang7-2	A × B	-1.01 **	0.53 *	0.83	-5.58 **	1.63 **	-1.09 **
6F629 × Chang7-2	B × B	-1.88 **	0.94 **	1.99	-1.89	1.60 **	-0.50 *
6M502A × Chang7-2	B × B	-1.84 **	-0.06	2.65	10.33 **	0.19	0.50 *
NL001 × Chang7-2	A × B	2.03 **	1.03 **	10.24 **	6.15 **	-0.05	1.69 **
LH181 × Chang7-2	B × B	-4.55 **	-1.52 **	-4.85	7.68 **	0.83 **	-0.79 **
LH208 × Chang7-2	A × B	1.08 **	-0.31	3.52	-7.55 **	-0.31	-0.04
LH212Ht × Chang7-2	B × B	2.08 **	-0.02	5.44 *	3.46 **	0.02	1.07 **
Lp215D × Chang7-2	B × B	-0.34 *	0.86 **	1.83	0.99	-2.39 **	0.52*
PHJ89 × Chang7-2	B × B	-0.97 **	-0.93 **	9.02 **	-0.05	-0.91 **	0.27
PHK93 × Chang7-2	B × B	2.12 **	-0.18	-6.98*	-4.92 **	0.21	-0.88 **
RS710 × four-144	B × A	-1.05 **	-2.05 **	6.30 *	-3.32 **	-2.05 **	0.06
LH191 × four-144	A × A	0.78 **	2.74 **	-2.85	-4.01 **	1.24 **	-0.40
LH192 × four-144	A × A	1.24 **	1.08 **	2.20	-3.68 **	-0.20	0.03

Table 12. Cont.

Hybrid Combinations	Hybrid Pattern	Days to Maturity	Days to Silking	Plant Height	Ear Height	Grain Moisture at Harvest	Grain Yield
PHN34 × four-144	A × A	−1.67 **	−0.59 *	8.56 **	5.21 **	3.56 **	0.37
PHP76 × four-144	B × A	4.87 **	1.99 **	8.90 **	3.19 **	0.67 *	0.69 **
PHW51 × four-144	A × A	1.62 **	−0.88 **	−8.83 **	−0.93	1.71 **	0.01
FBLA × four-144	A × A	−0.09	−1.51 **	−4.16	−1.93	−4.25 **	1.48 **
6F629 × four-144	B × A	−2.80 **	−3.26 **	−5.99 *	−0.69	−3.81 **	0.67 **
6M502A × four-144	B × A	0.08	−0.42	−0.58	−4.10 **	−1.10 **	−0.43 *
NL001 × four-144	A × A	−3.72 **	−1.01 **	3.31	−1.55	1.06 **	0.09
LH181 × four-144	B × A	−0.30	0.79 **	−0.33	−1.51	2.41 **	−0.10
LH208 × four-144	A × A	−2.51 **	0.66 *	−0.88	6.08 **	−0.82 **	−1.33 **
LH212Ht × four-144	B × A	0.66 **	0.95 **	−5.34 *	−3.97 **	0.60 *	−0.14
Lp215D × four-144	B × A	−0.42 *	−0.17	−6.22 *	−2.61 *	−0.68 *	−0.47 *
PHJ89 × four-144	B × A	2.78 **	1.04 **	6.76 *	11.94 **	−0.68 *	−0.59 *
PHK93 × four-144	B × A	0.53 *	0.62 *	−0.86	1.90	2.32 **	0.08
RS 710 × four-287	B × B	8.76 **	3.30 **	18.06 **	26.81 **	0.78 **	0.73 **
LH191 × four-287	A × B	0.10	−0.91 **	−2.65	−3.38 **	1.04 **	0.24
LH192 × four-287	A × B	−2.45 **	−0.91 **	10.91 **	−0.31	1.19 **	−0.15
PHN34 × four-287	A × B	0.14	−0.58 *	−13.95 **	−9.66 **	−0.79 **	−0.58 *
PHP76 × four-287	B × B	−4.65 **	−2.83 **	2.24	−4.00 **	0.08	−0.65 **
PHW51 × four-287	A × B	−3.24 **	−0.54 *	−1.79	4.20 **	−1.60 **	−0.74
FBLA × four-287	A × B	−0.61 **	0.51 *	−3.22	0.78	−0.50 *	−0.53 *
6F629 × four-287	B × B	2.01 **	1.09 **	0.38	0.14	−0.36	0.10
6M502A × four-287	B × B	0.55 **	−0.41	−9.29 **	−7.72 **	0.16	−0.57 *
NL001 × four-287	A × B	1.26 **	1.17 **	−3.31	3.18 **	2.17 **	−0.76 **
LH181 × four-287	B × B	−0.49 **	−1.04 **	6.79 *	−3.13 *	−2.98 **	1.84 **
LH208 × four-287	A × B	0.14	−0.33	4.66	5.46 **	0.43	0.74 **
LH212Ht × four-287	B × B	−1.53 **	0.63 *	−8.05 **	−8.86 **	1.18 **	−0.75 **
Lp215D × four-287	B × B	0.05	−0.66 *	3.86	1.65	0.85 **	−0.18
PHJ89 × four2-87	B × B	−1.74 **	0.21	−7.45 **	−6.93 **	0.61 *	0.22
PHK93 × four-287	B × B	1.68 **	1.30 **	2.80	1.78	−2.28 **	1.05 **
LSD p ≤ 0.05		0.34	0.50	5.16	2.23	0.49	0.43
LSD p ≤ 0.01		0.48	0.70	7.31	3.17	0.69	0.60

* and ** in the column represents significance at the 0.05 and 0.01 probability level, respectively.

There were 15 of the hybrid combinations with significantly negative SCA effect values for both days to maturity and days to silking. The A × A cis-hybrid combinations with short growth stage included LH191 × Zheng58, LH192 × Zheng58, PHN34 × four-144 and NL001 × four-144. The B × B cis-hybrid combinations with short growth stage included RS710 × Chang7-2, LH181 × Chang7-2, PHJ89 × Chang7-2, PHP76 × four-287 and LH181 × four-287. The more early maturing A × B trans-hybrid combinations included LH212Ht × Zheng58, PHK93 × Zheng58, RS710 × four-144 and 6F629 × four-144. Meanwhile, the B × A trans-hybrid combinations with rapid development included LH192 × four-287 and PHW51 × four-287.

There were 14 hybrid combinations with significant negative SCA effect values for both plant height and ear height. The A × A cis-hybrid combinations with lower plants and ears included LH192 × Zheng58, NL001 × Zheng58 and LH208 × Zheng58; the B × B cis-hybrid combinations included RS710 × Chang7-2, PHP76 × Chang7-2, PHK93 × Chang7-2, 6M502A × four-287, LH212Ht × four-287 and PHJ89 × four-287; the A × B trans-hybrid combinations included PHN34 × four-287; the B × A trans-hybrid combinations included RS710 × Zheng58, PHJ89 × Zheng58, LH212Ht × four-144 and Lp215D × four-144.

There were 20 of the hybrid combinations with significant negative SCA values for grain moisture content at harvest stage. The A × A cis-hybrid combinations with more rapid grain moisture dry down rate included LH192 × Zheng58, PHN34 × Zheng58, PHW51 × Zheng58, NL001 × Zheng58, FBLA × four-144 and LH208 × four-144. The B × B cis-hybrid combinations with lower grain moisture content included Lp215D × Chang7-2, PHJ89 × Chang7-2, LH181 × four-287 and PHK93 × four-287. The A × B trans-hybrid combinations with decreased grain moisture content at harvest stage included LH191 × Chang7-2, PHN34 × four-287, PHW51 × four-287 and FBLA × four-287. While the B × A trans-hybrid combinations which produced drier grain included LH212Ht × Zheng58, RS710 × four-144, 6F629 × four-144, 6M502A × four-144, Lp215D × four-144 and PHJ89 × four-144.

3.7. Total Combining Effect (TCA) and Control Heterosis (CH) for Yield Trait

As can be seen from Table 13, the TCA value of the yield characters in the worst to best hybrid combinations ranged from −2.62 to 3.03. The TCA effect values of the 30 best-yield and least-yield hybrid combinations were similar to the control heterosis rankings.

Table 13. Ranking of hybrids according to their total combining ability and control heterosis.

Hybrid Combinations	Hybrid Pattern	Female Parent GCA Effects	Male Parent GCA Effects	SCA Effects	TCA Effects	Control Heterosis %	Rank
NL001 × Chang7-2	A × B	0.51 **	0.83 **	1.69 **	3.03	21.48	1
6M502A × Chang7-2	B × B	1.47 **	0.83 **	0.50 *	2.80	19.64	2
LH212Ht × Chang7-2	B × B	0.51 **	0.83 **	1.07 **	2.41	15.93	3
LH181 × four-287	B × B	0.34 **	0.01	1.84 **	2.19	14.05	4
PHW51 × Chang7-2	A × B	0.48 **	0.83 **	0.60 **	1.91	11.60	5
6M502A × Zheng58	B × A	1.47 **	-0.18 **	0.50 *	1.79	11.21	6
LH208 × Chang7-2	A × B	0.89 **	0.83 **	-0.04	1.68	9.79	7
LH208 × four-287	A × B	0.89 **	0.01	0.74 **	1.64	9.50	8
LH208 × Zheng58	A × A	0.89 **	-0.18 **	0.64 **	1.35	6.94	9
Lp215D × Chang7-2	B × B	-0.07	0.83 **	0.52 *	1.28	6.44	10
FBLA × four-144	A × A	0.42 **	-0.64 **	1.48 **	1.26	5.96	11
PHJ89 × Chang7-2	B × B	-0.11	0.83 **	0.27	0.99	3.86	12
6M502A × four-287	B × B	1.47 **	0.01	-0.57 *	0.91	3.36	13
PHK93 × four-287	B × B	-0.19	0.01	1.05 **	0.87	2.96	14
LH192 × Chang7-2	A × B	-0.49 **	0.83 **	0.26	0.60	0.76	15
LH181 × Zheng58	B × A	0.34 **	-0.18	-0.95 **	-0.79	-10.95	50
PHK93 × four-144	B × A	-0.19	-0.64 **	0.08	-0.75	-11.18	51
LH192 × Zheng58	A × A	-0.49 **	-0.18 **	-0.14	-0.81	-11.61	52
PHN34 × four-287	A × B	-0.28 *	0.01	-0.58 *	-0.85	-11.79	53
LH208 × four-144	A × A	0.89 **	-0.64 *	-1.33 **	-1.08	-13.48	54
6F629 × Zheng58	B × A	-0.63 **	-0.18 **	-0.27	-1.08	-13.65	55
LH192 × four-144	A × A	-0.49 **	-0.64 **	0.03	-1.10	-14.19	56
Lp215D × four-144	B × A	-0.07	-0.64 **	-0.47 *	-1.18	-14.32	57
PHP76 × four-287	B × B	-0.56 **	0.01	-0.65 **	-1.2	-14.59	58
RS 710 × four-287	B × B	-1.97 **	0.01	0.73 **	-1.23	-15.12	59
PHJ89 × four-144	B × B	-0.11	-0.64 **	-0.59 *	-1.34	-16.11	60
LH191 × four-144	A × A	-0.31 **	-0.64 **	-0.40	-1.35	-16.17	61
RS710 × Chang7-2	B × B	-1.97 **	0.83 **	-0.32	-1.46	-16.81	62
RS710 × four-144	B × A	-1.97 **	-0.64 **	0.06	-2.55	-26.35	63
RS710 × Zheng58	B × A	-1.97 **	-0.18 **	-0.47 *	-2.62	-26.89	64

* and ** in the column represents significance at the 0.05 and 0.01 probability level, respectively.

Among the top fifteen TCA effect values, there were two A × A cis-hybrid combinations, seven B × B cis-hybrid combinations, five A × B trans-hybrid combinations, and one B × A trans-hybrid combination. The TCA effect values that increased yield can be divided into the following three categories:

(1) Both parental GCA and hybrid SCA effects were large, for hybrid combinations NL001 × Chang7-2, 6M502A × Chang7-2, LH212Ht × Chang7-2, PHW51 × Chang7-2, Lh181 × four-287 and LH208 × four-287.

(2) Complementary parental GCA effects with a positive hybridization combination SCA effect as observed for the hybrids 6M502A × Zheng58, LH208 × Zheng58, Lp215D × Chang7-2, FBLA × four-144, PHJ89 × Chang7-2, PHK93 × four-287 and LH192 × Chang7-2.

(3) Parental GCA effect values were large with small SCA values, as found in the hybrids LH208 × Chang7-2 and 6M502A × four-287.

The 15 hybrids with the lowest TCA effect values can be divided into the following four categories:

(1) The complementary value of parental GCA effect value and the large value for the SCA effect of hybrid combination, such as for RS 710 × four-287.

(2) The complementary value of parental GCA effects and a small value for the SCA effect of hybrid combination, such as LH181 × Zheng58, PHN34 × four-287, LH208 × four-144, PHP76 × four-287, and RS710 × Chang7-2.

(3) Both parents with small GCA effect values and hybrid combinations with positive SCA effect values, such as hybrids PHK93 × four-144, LH192 × four-144, and RS710 × four-144.

(4) Both parents with small GCA and hybrid with low SCA effect values, such as LH192 × Zheng58, 6F629 × Zheng58, Lp215D × four-144, PHJ89 × four-144, LH191 × four-144, and RS710 × Zheng58.

In Table 13, all the control heterosis values were the mean values of two locations, and ranged from −26.89% to 21.48%. There were 15 hybrid combinations with positive heterosis, 2NL001 × chang7-2, 6M502A × chang7-2, LH212Ht × chang7-2, LH181 × four287, PHW51 × chang7-2, 6M502A × Zheng58, LH208 × chang7-2, LH208 × Zheng58, lh215d × chang7-2, FBLA × four-144, PHJ89 × chang7-2, 6M502A × four287, PHK93 × four287, LH192 × chang7-2 (Table 13), indicating that these hybrids had better yield than the standard of Zhengdan 958.

4. Discussion

4.1. Improvement and Utilization of USA Germplasm

The introduction of exotic germplasm was an important way to enrich genetic diversity for China maize crop production. It has been stated that the potential utilization of inbred lines cannot be judged according to the strengths and weaknesses of the inbred plant growth, but needs to be identified based on the analysis of combining ability [11–13]. From Tables 10 and 11, we can see that the North American inbred lines are genetically distinct from the inbred lines of China, and there was a wide regional gap between them. There were significant differences in GCA effect values of the inbred when grown at different locations, which indicates that American inbred lines perform well in comprehensive traits such as yield. In the process of improving, selecting, and matching inbred lines to make improved hybrids, the target traits can be selected according to the GCA, and the grouping of American inbred lines can be determined. On the basis of plant growth, development, and heterotic patterns, according to the principle of complementary characteristics of the same group, successful maize production populations have been constructed with superior inbred lines, and the frequency of superior alleles has been improved by selective repetitive breeding [14–17].

4.2. Classification of USA Germplasm

By analyzing the SCA effect value of 64 hybrid combinations, it was found that there were both cis and trans combinations of heterotic groups in which the yield SCA effect value was positive and significant. Heterosis existed between the USA inbred lines SS group and the China A group, and between the USA inbred lines NSS group and the China B group. This was due to the differences in the Germplasm Foundation of the China A group and the B group, and also the SS group and NSS group in the United States. When using the USA inbred lines, the combining ability of USA inbred lines must be determined on the basis of local indigenous inbred lines. To identify the classification of American inbred lines, and the heterosis group of USA inbreds, in breeding the second-cycle inbred line, the cis hybrid combination is usually used to improve the group.

4.3. Combining Ability and Control Heterosis of USA Germplasm

The TCA effect value of a hybrid was the same trend as its ranking compared to control heterosis. The TCA values of hybrid combinations with yields greater than the control hybrid were all positive, and the TCA values of hybrid combinations yielding less than the control were all negative. The value of TCA can be used as an index in hybrids selection [18,19]. In the top 15 TCA value hybrid combinations, the SCA effect value was mostly positive. There were 8 hybrid combinations with positive GCA values of parental yield, and 8 hybrid combinations with positive and negative GCA values of parental yield. The results showed that hybrid combinations with high heterosis required higher SCA and GCA effect values [20]. The selection of GCA effect value of parental yield should be paid attention to in heterotic crossing combinations, ensuring that at least one parent yield GCA effect value is positive, and SCA effect value should not be too low.

5. Conclusions

The best combiner inbred lines from USA were RS710, PHP76, FBLA, and PHJ89. These materials had great potential for breeding early maturing, had high density tolerance, and were suitable for machine-harvest hybrids. The best USA inbred lines with high GCA in yield traits were 6M502A, LH208, NL001, LH212Ht, PHW51, FBLA, and LH181. These inbred lines had great potential in breeding high-yield hybrids. The use of parental combining ability information will ease the process of making superior hybrids. The inbred line Chang7-2 promoted a high-yield hybrid combination ability. Meanwhile, the inbred line four-287 led to hybrid combinations suitable for machine harvesting. The TCA value could be used as an index to evaluate the heterosis of hybrid combinations without growing the control hybrid for comparison. The best hybrid combination were NL001 × Chang7-2, 6M502A × Chang7-2, LH212Ht × Chang7-2, LH181 × four-287, PHW51 × Chang7-2, 6M502A × Zheng58, LH208 × Chang7-2, LH208 × four-287, LH208 × Zheng58, Lp215D × Chang7-2, FBLA × four-144. Furthermore, these hybrid combinations have potential for further commercial development.

Author Contributions: J.-y.S., J.-l.G. and X.-f.Y. conceived and designed the experiments; J.L., Y.F. and D.W. performed the experiments; Z.-j.S. performed the statistical analysis; J.-y.S. wrote the paper.

Funding: This study was funded by the National Key Research and Development Program of China (2017YFD0300802, 2016YFD0300103), the Maize Industrial Technology System Construction of Modem Agriculture of China (CARS-02-63) and the Fund of Crop Cultivation Scientific Observation Experimental Station in North China Loess Plateau (25204120) of China.

Acknowledgments: We would like to thank the Maize High-Yield and High-Efficiency Cultivation Team for field and data collection, and especially Juliann Seebauer for manuscript revisions.

Conflicts of Interest: The authors declare no conflicts of interest.

References

1. Zhang, S.H.; Tian, Q.Z.; Li, X.H. Advancement of maize germplasm improvement and relevant research. *Maize Sci.* **2006**, *14*, 1–6. [CrossRef]
2. Chen, D.Y.; Jing, X.Q.; Wang, X.J. Discussion on the breeding for mechanical harvesting and density tolerant maize hybrids. *Crops* **2014**, *2*, 13–15. [CrossRef]
3. Li, C.; Qiao, J.F.; Gu, L.M. Analysis of maize biological traits which affected corn kernel mechanically harvesting qualities. *Acta Agric. Boreal. Sin.* **2015**, *30*, 164–169. [CrossRef]

4. Zhang, S.H.; Xu, W.P.; Li, M.S. Challenge and opportunity in maize breeding program. *Maize Sci.* **2008**, *16*, 1–5. [CrossRef]
5. Zhao, W.Y.; Liu, X.; Wang, D.X. Experience and lesson of American maize varieties resource utility. *China Seed Ind.* **2011**, *10*, 50–51. [CrossRef]
6. Li, H.M.; Hh, R.F.; Zhang, S.H. The impacts of US and CGIAR's germplasm on maize production in China. *Sci. Agric. Sin.* **2005**, *38*, 2189–2197.
7. Yong, H.J.; Wang, J.J.; Zhang, D.G. Characterization and potential utilization of maize populations in America region. *Hereditas (Beijing)* **2013**, *35*, 703–713. [CrossRef] [PubMed]
8. Pan, T.Z.; Gao, J.L.; Su, Z.J. Comprehensive evaluation of agronomic traits based on principal component analysis of maize hybrids. *North. Agric.* **2016**, *44*, 1–8. [CrossRef]
9. Federer, W.T.; Wolfinger, R.D. SAS code for recovering inter effect information in experiments with incomplete block and lattice rectangle designs. *Agron. J.* **1998**, *90*, 545–551. [CrossRef]
10. Geraldi, I.O.; Miranda Filho, J.B. An adapted model for the analysis of partial diallel crosses. *Braz. J. Genet.* **1988**, *11*, 419–430.
11. Hallauer, A.R.; Carena, M.J.; Miranda Filho, J.B.; Carena, M.J. *Testers and Combining Ability, Quantitative Genetics in Maize Breeding*, 8th, ed.; Iowa State University Press: Ames-Iowa, IA, USA, 1988; pp. 383–418, ISBN 978-1-4419-0765-3.
12. Duvick, D.N.; Cassman, K.G. Post-green revolution trends in yield potential of temperate maize in the North-Central United States. *Crop Sci.* **1999**, *39*, 1622–1630. [CrossRef]
13. Duvick, D.N. The contribution of breeding to yield advances in maize (*Zea mays* L.). *Adv. Agron.* **2005**, *86*, 91–97. [CrossRef]
14. Mao, J.C.; Zhang, S.H.; Li, X.T. The methodology for maize inbred line with high yield selection and high combining ability. *Sci. Agric. Sin.* **2006**, *39*, 872–878.
15. Gao, X.D.; Zhou, X.M.; Gao, H.M. Combining ability of main agronomic traits and heterosis of European maize germplasm BRC. *Maize Sci.* **2015**, *23*, 28–33. [CrossRef]
16. Sun, F.C.; Feng, Y.; Su, E.H. Analysis of combining ability of yield of fractional inbred line of main corn varieties in inner mongolia and study on their genetic relationship. *Acta Agric. Boreal. Sin.* **2011**, *26*, 119–123.
17. Li, M.S.; Zhang, S.H.; Li, X.H. Study on heterotic groups among maize inbred lines based on SCA. *Sci. Agric. Sin.* **2002**, *35*, 600–605.
18. Chen, C.; Yuan, H.Y.; Lei, Y.T. Analysis of yield combining ability and heterosis of selective lines form the improve maize population MM. *J. Northwest. A F Univ.* **2013**, *41*, 93–98. [CrossRef]
19. Cui, C.; Gao, J.L.; Yu, X.F. Combining ability of traits related to nitrogen use efficiency in eighteen maize inbred lines. *Acta Agon. Sin.* **2014**, *40*, 838–849. [CrossRef]
20. Gou, C.M.; Yu, S.Q.; Huang, N. Analysis on combining ability of 17 inbred lines from maize landraces. *Acta Agric. Sin.* **2015**, *30*, 175–182. [CrossRef]

agronomy

MDPI

Article

Effect of Soybean and Maize Rotation on Soil Microbial Community Structure

Peng Zhang [1,†], Jiying Sun [1,†] , Lijun Li [1,*], Xinxin Wang [2], Xiaoting Li [1] and Jiahui Qu [1]

[1] College of Agronomy, Inner Mongolia Agricultural University, No. 275, Xin Jian East Street,
    Hohhot 010019, China; imauzp@emails.imau.edu.cn (P.Z.); jiying-sun@imau.edu.cn (J.S.)
    lixt1229@163.com (X.L.); nmgqujiahui@163.com (J.Q.)
[2] Chifeng Academy of Agricultural and Animal Husbandry Sciences, Song Shan District,
    Chifeng 024031, China; wxx986@163.com
*   Correspondence: lijun-li@imau.edu.cn; Tel.: +86-158-4815-4170
†   These authors contributed equally to this work.

Received: 29 October 2018; Accepted: 5 January 2019; Published: 22 January 2019

Abstract: Examining the soil microbiome structure has great significance in terms of exploring the mechanism behind plant growth changes due to maize (*Zea mays* L.) and soybean (*Glycine max* Merr.) crop rotation. This study explored the effects of soil microbial community structure after soybean and maize crop rotation by designing nine treatments combining three crop rotations (continuous cropping maize or soybean; and maize after soybean) with three fertility treatments (organic compound fertilizer, chemical fertilizer, or without fertilizer). Soil was sampled to 30 cm depth the second year at approximately the middle of the growing season, and was analyzed for physical, chemical, and phospholipid fatty acid (PLFA) profiles. Bacteria was found to be the predominant component of soil microorganisms, which mainly contained the PLFAs 16:0. Crop rotation with organic compound fertilizer application reduced the percentage of fungi in the soil by 24% compared to continuous maize and soybean with the same fertilizer application. The combination of crop rotation with organic fertilizer can reduce the percentage of fungi/bacteria to the greatest degree. In addition, the content of soil aggregate and organic matter had great influence on Gram-positive bacteria and actinomyces. In conclusion, soybean and maize crop rotation improve the soil nutrient content primarily by influencing the composition of bacterial community, especially the Gram-positive bacteria.

Keywords: crop rotation; maize; soybean; microbial community structure

---

1. Introduction

The quality of soil is one of the most important factors affecting crop growth because it is not only one of the major components of the environment but also is necessary for the survival of field crops. Maize is the world's largest grain crop and thus has high economic value. However, the continuous planting of maize leads to the lack of soil nutrient uniformity and intensifies the occurrence and transmission of soil diseases in the same plot. In order to relieve the land pressure, knowledge of the effects of crop rotation patterns of maize and other crops on soil properties is important. Owing to the long-term economic and ecological benefits of Leguminosae crops and Gramineae crops, crop rotation patterns between the two plant families has long been considered the optimal system for maintaining the soil nutrient cycle. A great deal of research has been done on inserting exotic Leguminosae into various crop rotation patterns. For example, the three-course cropping of ancient Greece and Rome (three plot of field, fallow land, spring sowing land, autumn (winter) sowing land respectively, and rotate in the same order, so in three years, each plot can fallow one time), the British Norfolk 4-year rotation model (red clover-wheat/rye-feeding turnip/sugar-beet-two-rowed barley/red clover) in 1730 and the American 6-year rotation model initiated in 1794 were both examples of legume and non-legume crop rotation [1]. However, there have been few studies on the ecological effects of crop rotation, most of which focus on crop yield and the effect on physical and chemical soil properties. Furthermore, the detection and analysis of the change of soil microbial community structure before and after crop rotation has not been common.

Microorganisms are an important component of the material cycle and energy transformation process in a soil ecosystem. Due to the effect of fertilization, soil microorganisms not only affect the physical and chemical properties of soil but also affect the effectiveness of fertilizers on plants [2]. Fertilization mainly affects soil microorganisms by changing the physical and chemical properties of soil as well as its nutrient contents. Fertilization affects soil microorganisms mainly by changing the physical structure and nutrient content of soil and the amount of root and aboveground litter of crops [3]. The rhizosphere is the microenvironment in which plants come into contact with soil. The soil microbial community composition is an important limiting factor of soil processes, and the composition and activity of a microbial community largely determine biogeochemical cycles, metabolic processes of soil organic matter, and soil fertility and quality [4,5]. In addition, soil microorganisms are closely related to the stability and health of the soil ecosystem. Soil microorganisms are more sensitive to changes in external conditions, such as land use change, management measures, and cultivation than other soil physical and chemical indexes. Therefore, soil microbial biomass, community composition, and diversity are often used as indicators of soil quality changes [6,7].

Phospholipid fatty acid (PLFA) spectrogram technology was used to analyze biological community structures in the 1980s [8,9]. This method of analysis relies on fatty acid spectrograms to quantify the entire microbial community without the need for soil enrichment or cultivation, and therefore is quicker and more reliable than traditional approaches [10]. Although this method cannot identify the specific microbial species at the strain level, PLFA does not depend on the influence of the plant culture system, but can directly provide information and quantitatively describe the whole microbial community. This method also has the advantages of objective and reliable test results, simple operation of test conditions, and multiple test functions and has been used widely in the field of cycle microbiology. In order to clarify the effects of crop rotation and fertilization on soil microbial community structure, PLFA was used to analyze the microbial community composition of soil samples.

2. Materials and Methods

2.1. Study Survey and Design

The experiments were conducted at Chifeng Academy of Agricultural and Animal Husbandry Sciences, Inner Mongolia Autonomous Region, northeastern China ($42°15'$ N, $118°72'$) in 2016 and 2017. The study area has a temperate semi-arid continental climate, with an average annual temperature of 6.5 °C and an average annual precipitation of 380 mm, with the precipitation mainly concentrated in July and August. The sunshine time is generally at least 2800 to 3200 h/year. The main crops grown in the area are maize, buckwheat (*Fagopyrum esculentum* Moench.), and millet (*Panicun italicun* L.). The previous crop in the farmland was maize.

A randomized block design was used in this experiment with three replications. There were nine cropping designs for the study including (1) continuous cropping of maize with organic compound fertilizer (CM + OF), (2) continuous cropping of maize with chemical fertilizer Nitrogen and Phosphorus (CM + NP), (3) continuous cropping maize without fertilizer (CM + 0), (4) maize after soybean with organic compound fertilizer (SM + OF), (5) maize after soybean with chemical fertilizer Nitrogen and Phosphorus (SM + NP), (6) maize after soybean with no fertilizer (SM + 0), (7) soybean continuous cropping with organic compound fertilizer (CS + OF), (8) continuous cropping soybean with chemical fertilizer Nitrogen and Phosphorus (CS + NP), and (9) continuous cropping soybean without fertilizer (CS + 0). The annual planting date was 18 May, and the amount of fertilizer applied per year was consistent. According to the local recommendation, 300 kg/ha of chemical (NP) fertilizer diammonium phosphate (N 13%, P_2O_5 44%) were applied for maize, 150 kg/ha of chemical

(NP) fertilizer diammonium phosphate (N 13%, P_2O_5 44%) were applied for soybean, and the rate of organic compound fertilizer for maize and soybean is 900 kg/ha. The chemical fertilizer and organic compound fertilizer were applied when sowing. The rotation area was planted with soybeans in 2016 and maize in 2017. The field management measures are similar to local management measures. Drip irrigation was carried out after sowing to ensure seedling emergence rate. Beginning on 29 June, every 30 days, field weeds were removed up until the crops were harvested. Sowing rate of soybean and maize rotation in trial plots 2016–2017 is shown in Table 1.

Table 1. Sowing rate of soybean and maize rotation in trial plots 2016–2017.

Plants	Varieties	Sowing Rate	Plant Spacing	Row Spacing
		Plant/ha	cm	cm
maize	Fengdan 189	67,500	33	45
soybean	Red bean 3	210,000	12	40

2.2. Samples Collection

The soil samples of the study area were obtained during the vigorous growth of crops on 8 August 2017. In each plot, three points located 5 cm from the root were randomly selected from the rhizosphere of maize and soybean. Soil samples, which were taken at the soil depth of 0–30 cm, were separated into two samples. One sample was dried to analyze the soil physical structure and chemical properties, the other sample was stored in a freezer at $-20\,°C$ for the determination of soil microbial community structure.

2.3. Soil Physical and Chemical Properties

The content of soil macroaggregates (SA) was measured by mechanical sieving method. Soil pH (pH) in water was measured at a soil/water ratio of 2:5 (*w:v*) after 1 h in suspension for water. Soil available nitrogen (AN) was determined using the alkaline diffusion method; soil available phosphorus (AP) and soil organic matter (SOM) were measured by $NaHCO_3$ leaching molybdenum-antimony anti-absorption spectrophotometry and potassium dichromate volumetric method, respectively [11–13].

2.4. Determination of Soil Microbial Community Structure

Soil microorganisms PLFAs were extracted by Bligh-Dyer modified method and esterified C19:0 was used as the internal standard [14]. Briefly, the processes of extraction, purification, and analysis consisted of measuring 2 g freeze-dried soil, then 20 mL chloroform-methanol-citric acid buffer (1:2:0.8, *v/v/v*) was added to extract total PLFAs of the samples. The extracted PLFAs were subsequently separated by silica gel column (SPE-SI), and consisted of neutral fatty acids, sugar fatty acids, and phosphatidic acid. Phospholipid acid was dissolved in methanol/toluene (1:1, *v/v*) solution, then 0.2 mol/L KOH was added, the solution was esterified at 37 °C for 15 min, then separated by GC-MS (gas chromatograph-mass spectrometry) analyzer, and then separated by bacterial fatty acid standards and commercial MIDI system (Microbial Identification System) to identify and quantify phospholipid fatty acids. Soil microbial phospholipid fatty acid profiles were obtained by analyzing the corresponding microbial communities, and the structural diversity of soil microbial communities were judged by statistical analysis [15]. Phospholipid fatty acids were determined based on Frostegard et al. [16]: [i/a/cy/br/10Me (delspray with methyl in the 10th carbon atom)] X:Y ω Z (OH/cis/t), where X represents the total number of C atoms of fatty acid molecule, Y indicates the number of unsaturated olefin bonds, ω represents the position of the olefinic bond from the carboxyl group, Z represents the position of the olefin bond or cyclopropane chain. The prefix "i" (iso) represents the isomeric methyl branched chain (the third carbon atom from the methyl end), "a" (anteiso) represents the pre-isomeric methyl branched chain (the third carbon atom from the methyl end), "cy" represents the cyclopropyl group, and "br" represents the unknown position of the methyl chain.

The suffixes "cis" and "trans" represent cis and trans isomers, respectively, and the number before "OH" denotes the position of hydroxyl groups (counted from the carboxyl end, the second carbon is alpha, and the third carbon is beta). Characterization of microbial PLFA is shown in Table 2 [17–23].

Table 2. Phospholipid fatty acid (PLFA) characterization of microorganisms.

Microbial Type	Phospholipid Fatty Acid Labelled
Bacteria in general (B)	i14:0, i15:1, i15:0, a15:0, i16:0, i17:0, a17:0, 16:1 ω 7cis, 16:1 ω 9cis, 17:1 ω 7cis, 17:1 ω 8cis, 18:1 ω 7cis, 18:1 ω 5cis, cy17:0, cy19:0, 16:12 OH, 16:0, 18:0
Gram-positive bacteria (G+)	i14:0, i15:1, i15:0, a15:0, i16:0, 16:0, i17:0, a17:0
Gram-negative bacteria (G-)	16:1 ω 7cis, 16:1 ω 9cis, 17:1 ω 7cis, 17:1 ω 8cis, 1:1 ω 7cis, 18:1 ω 7cis, 18:1 ω 5cis, cy17:0, cy19:0, 16:12 OH
Actinomycetes (Act)	10Me16:0, 10Me17:0, 10Me18:0
Fungi (Fug)	16:1 ω 5cis, 18:1 ω 9cis, 18:2 ω 6cis, 18:2 ω 9cis, 18:3 ω 6cis

2.5. Data Analysis

The data in this paper were analyzed by variance analysis, principal component analysis (PCA), and nonlinear dimensionality reduction analysis (RDA) in Excel (Microsoft Office 2016, Microsoft: 2015, Washington, WA, USA), SPSS 25.0 (Statistical Product and Service Solutions. International Business Machines Corporation (IBM), 2017, Chicago, IL, USA), and R (R Foundation for Statistical Computing 3.5.1, R Core Development Team, 2018, Vienna, Austria). Excel was used to calculate the mean and standard deviation of all data. The principal component analysis of phospholipid fatty acid data was carried out using SPSS 25.0. The relationship between soil microbial community structure and characteristics of environmental soil factors was redundantly analyzed by R language vegan package, and the correlation between them was further analyzed by R language corrplot package.

3. Results

3.1. Effects of Each Treatment on the Composition and Content of Phospholipid Fatty Acids

Altogether, 20 kinds of phospholipid fatty acids were checked from the soil samples treated by the different rotation and fertilizer treatments. There were seven kinds of phospholipid fatty acids among the 20 kinds of phospholipid fatty acids with significant effects with regard to the rotation treatment or interaction between rotation treatment and fertilizer treatment. From Table 3, we can see that seven kinds of phospholipid fatty acids were mainly detected in this study, in which 16:0 (Gram-positive bacteria) had the maximum content, accounting for 11.6–12.7% of the total phospholipid fatty acids. These seven kinds of phospholipid fatty acids belonged to Gram-positive bacteria, Gram-negative bacteria, AM Fungi, Actinomycetes, and other bacteria.

The rotation treatment had a significant effect on the content of a15:0, 16:0, 16:1ω5c, and 18:0 at $p \leq 0.01$, had a significant effect on 17:1 ω 7c and 10Me18:0 at $p \leq 0.05$, and did not have a significant effect on i16:0. The fertilizer treatment had no significant effect on the content of the above phospholipid fatty acids, there were significant effect on the content of i16:0 at $p \leq 0.05$ in the interaction between rotation treatment and fertilizer treatment, no significant differences were observed between the others as a result of the interaction between the treatment of rotation and fertilizer.

For a15:0, soybean continuous cropping with chemical fertilizer applied resulted in significantly higher numbers than that maize continuous cropping with organic fertilizer and chemical fertilizer applied. For another Gram-positive bacteria (i16:0), there was significant interaction between fertilizer and cropping system ($p = 0.047$). For 17:1 ω 7c, continuous cropping of maize without fertilizer was remarkably higher than soybean continuous cropping with organic fertilizer and without fertilization.

Cropping system had a significant effect on its content at 0.05 level (p = 0.018). Cropping system had an extremely significant influence on 16:1 ω 5c (AM Fungi (Arbuscular Mycorrhizal Fungi)) at the 0.01 level (p = 0.000); its numbers were significantly lower in soybean continuous cropping than maize continuous cropping and soybean-maize rotation.

The content of 10Me18:0 presented a significant difference under the SM + OF, which displayed significantly higher content than all of the other treatments, and there was no significant difference between the other treatments. The 18:0 displayed a significant difference under the CS + OF, CS + NP, and CS + 0, although there was no significant difference observed between CS + OF, CS + NP, and CS + 0, all of which displayed significantly higher content than all the other treatments.

Table 3. The percentage of phospholipid fatty acids under different treatments in soil (%). Values are the average ± 1 standard error.

Treatment	G+			G-	AMF	Act	OB
	a15:0	16:0	i16:0	17:1 ω 7c	16:1 ω 5c	10Me18:0	18:0
CM + OF	† 4.4 ± 0.2 bc	† 11.7 ± 1.0 bc	† 3.9 ± 0.3 a	† 2.1 ± 0.3 ab	† 3.9 ± 0.1 a	† 1.9 ± 0.2 b	† 2.1 ± 0.2 c
CM + NP	4.4 ± 0.1 c	11.6 ± 0.3 c	3.6 ± 0.1 c	2.0 ± 0.3 ab	3.9 ± 0.1 a	1.8 ± 0.1 b	2.1 ± 0.1 c
CM + 0	4.4 ± 0.0 abc	11.8 ± 0.5 abc	3.7 ± 0.0 abc	2.2 ± 0.4 a	3.9 ± 0.2 a	1.9 ± 0.1 b	2.2 ± 0.2 bc
SM + OF	4.7 ± 0.1 abc	11.6 ± 0.4 c	3.8 ± 0.1 abc	1.9 ± 0.2 ab	3.9 ± 0.1 a	2.2 ± 0.2 a	2.3 ± 0.2 bc
SM + NP	4.6 ± 0.3 abc	11.7 ± 0.2 bc	3.9 ± 0.1 ab	1.9 ± 0.1 ab	3.7 ± 0.1 a	1.9 ± 0.1 b	2.3 ± 0.1 ab
SM + 0	4.7 ± 0.3 ab	12.0 ± 0.4 abc	3.8 ± 0.1 abc	1.9 ± 0.1 ab	3.7 ± 0.2 a	2.1 ± 0.4 ab	2.2 ± 0.2 bc
CS + OF	4.6 ± 0.2 abc	12.7 ± 0.6 a	3.6 ± 0.1 bc	1.8 ± 0.1 b	3.2 ± 0.2 b	1.8 ± 0.1 ab	2.5 ± 0.1 a
CS + NP	4.8 ± 0.1 a	12.7 ± 0.1 ab	3.8 ± 0.1 abc	2.0 ± 0.1 ab	3.1 ± 0.2 b	1.8 ± 0.1 b	2.5 ± 0.1 a
CS + 0	4.6 ± 0.1 abc	12.1 ± 0.4 abc	3.8 ± 0.1 abc	1.8 ± 0.2 b	3.1 ± 0.1 b	1.8 ± 0.1 b	2.5 ± 0.1 a
Cropping system	**	**	—	*	**	*	**
Fertilizer	—	—	—	—	—	—	—
Interaction effect	—	—	*	—	—	—	—

† Means within a column followed by the same letter are not significantly different at $p \leq 0.05$, and the different letters are significantly different at $p \leq 0.05$. '**' Means significant at the 0.01 probability level, '*' means significant at the 0.05 probability level, '—' means not significant. AMF: Arbuscular Mycorrhizal Fungi; G+: Gram-positive bacteria; G-: Gram-negative bacteria; Act: Actinomycetes; OB: other bacteria; CM + OF: continuous cropping of maize with organic compound fertilizer; CM + NP: continuous cropping of maize with chemical fertilizer Nitrogen and Phosphorus; CM + 0: continuous cropping maize without fertilize; SM + OF: maize after soybean with organic compound fertilizer; SM +NP: maize after soybean with inorganic fertilizer Nitrogen and Phosphorus; SM + 0: maize after soybean with no fertilizer; CS + OF: soybean continuous cropping with organic compound fertilizer; CS + NP: continuous cropping soybean with inorganic fertilizer Nitrogen and Phosphorus; CS + 0: continuous cropping soybean without fertilizer.

The rotation treatment had a significant effect on the content of Gram-positive bacteria (G+) and the other bacteria at $p \leq 0.01$; had a significant effect on Gram-negative bacteria (G-), fungi, and fungi/bacteria at $p \leq 0.05$; but did not have a significant effect on Actinomycetes. The fertilizer treatment had no significant effect on the content of the above phospholipid fatty acids, only the content of Gram-positive bacteria (G+) showed significant differences at $p \leq 0.05$ in the interaction between rotation treatment and fertilizer treatment. Table 4 shows that bacteria accounted for 69% to 71% of the total amount of soil microorganisms, and thus represented the main component of the soil microorganisms. While fungi only accounted for about 3% of the total amount of microorganisms, other bacteria accounted for 8.1–9.9% of the total amount. Cropping system had a significant effect on the composition of the soil microbial community and had a significant effect on Gram-positive bacteria (p = 0.008) and other bacteria (p = 0.006). For Gram-positive bacteria, there was an obvious interaction between fertilizer and farming system (p = 0.034).

Analysis of variance for the fertilizer treatment showed that the content of Gram-negative bacteria (G-) was significantly affected by fertilizer application under the condition of continuous maize cropping system at the 0.01 level. The content of Gram-positive bacteria (G+) was significantly affected by fertilizer application under the condition of soybean-maize rotation system at the 0.01 level (Table 5). Under the condition of continuous maize cropping system, the content of Gram-negative bacteria (G-) treated by organic fertilizer and chemical fertilizer application was significantly higher than that of no fertilizer treatment at the level of 0.01, and there was no significant difference between the organic fertilizer and chemical fertilizer application treatments. Under the condition of the soybean-maize rotation system, the content of Gram-positive bacteria (G+) treated by organic fertilizer application and no fertilizer was significantly higher than that of chemical fertilizer application at the 0.01 level, and there was no significant difference between the organic fertilizer application and no fertilizer.

Table 4. Influence of each treatment on soil microbial community structure (%). Values are the average ± 1 standard error.

Treatment	Gram-Positive Bacteria	Gram-Negative Bacteria	Fungi	Actinomycetes	Other Bacteria	Fungi/Bacteria
	G+	G-				
CM + OF	† 37.3 ± 0.5 bc	† 33.2 ± 0.3 ab	† 3.4 ± 0.3 ab	† 17.5 ± 0.8 a	† 8.3 ± 1.3 ab	† 0.049 a
CM + NP	36.2 ± 0.6 c	33.6 ± 0.4 ab	3.4 ± 0.5 a	17.0 ± 1.5 a	9.7 ± 0.4 a	0.050 a
CM + 0	37.0 ± 0.4 bc	32.1 ± 0.6 b	3.0 ± 0.5 abc	18.1 ± 0.3 a	9.9 ± 0.5 a	0.043 ab
SM + OF	38.1 ± 0.8 ab	32.4 ± 0.8 b	2.6 ± 0.1 c	17.7 ± 0.5 a	9.2 ± 0.6 ab	0.036 b
SM + NP	37.5 ± 0.8 ab	32.7 ± 0.5 b	2.9 ± 0.3 abc	17.1 ± 0.6 a	9.7 ± 1.0 a	0.041 ab
SM + 0	38.0 ± 0.9 ab	32.3 ± 0.8 b	2.6 ± 0.3 bc	18.1 ± 0.8 a	9.0 ± 0.6 ab	0.037 b
CS + OF	37.0 ± 0.7 bc	34.4 ± 0.3 a	3.4 ± 0.8 ab	16.9 ± 0.8 a	8.3 ± 0.6 b	0.048 ab
CS + NP	38.5 ± 0.7 a	32.7 ± 1.6 ab	2.6 ± 0.2 bc	18.1 ± 0.5 a	8.1 ± 0.4 b	0.037 b
CS + 0	37.8 ± 0.7 ab	33.6 ± 1.5 ab	2.8 ± 0.5 abc	17.3 ± 1.0 a	8.5 ± 0.4 ab	0.039 ab
Cropping system	**	*	*	—	**	*
Fertilizer	—	—	—	—	—	—
Interaction effect	*	—	—	—	—	—

† Means within a column followed by the same letter are not significantly different at $p \leq 0.05$, and the different letters are significantly different at $p \leq 0.05$. '**' Means significant at the 0.01 probability level, '*' means significant at the 0.05 probability level, '—' means not significant.

Table 5. Influence of fertilizer treatment on Gram-negative bacteria and Gram-positive bacteria (%). Values are the average ± 1 standard error.

Treatment	Gram-Negative Bacteria	Treatment	Gram-Positive Bacteria
	G-		G+
CM + OF	† 33.2 ± 0.3 aA	SM + OF	† 38.1 ± 0.8 aA
CM + NP	33.6 ± 0.4 aA	SM + NP	37.5 ± 0.8 bB
CM + 0	32.1 ± 0.6 bB	SM + 0	38.0 ± 0.9 aA

† Means within a column followed by the same lowercase letter or capital letter are not significantly different at $p \leq 0.05$ and at $p \leq 0.01$, respectively, and the different lowercase and capital letters are significantly different at $p \leq 0.05$ and $p \leq 0.01$, respectively.

3.2. Principal Component Analysis of Microbial Fatty Acids in Soil by Each Treatment

The principal component analysis of the microbial fatty acids in soil is presented in Figure 1, where explanation variances of the first principal component (PC1) and the second principal component are 34.76% and 28.98%, respectively (Figure 1). Indeed, PC1 is highly related with i17:0, a17:0, and i15:0, which all belong to Gram-positive bacteria, and their component matrix coefficients are 0.932, 0.911, and 0.876, respectively (Figure 1, Table 6). Meanwhile, PC2 displays the more obvious correlations with 18:1 ω 9c, cy17:0 ω 7c, and 16:1 ω 5c, with corresponding component matrix coefficients of 0.866, 0.846, and 0.802, respectively (Figure 1, Table 6). However, there are some lower correlations between PC1 and 18:0, 16:0, and 17:1 ω 7c, which have corresponding component matrix coefficients of −0.053, −0.030, and −0.080, respectively. Moreover, the component matrix coefficients of i17:0, a17:0, and 10Me16:0 to PC2 are also lower, with corresponding values of 0.064, 0.054, and 0.062, respectively (Figure 1, Table 6).

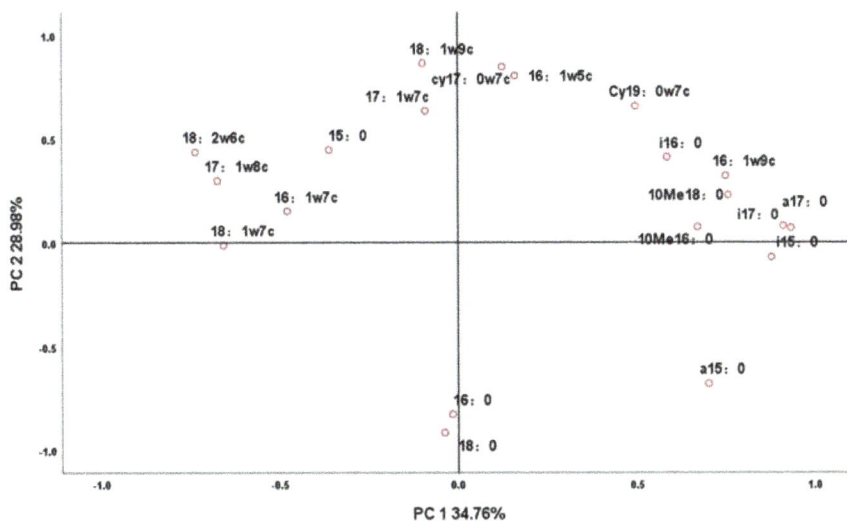

Figure 1. Principal component analysis (PCA) of microbial fatty acids in soil. This figure is a principal component analysis based on the phospholipid fatty acid structure nomenclature. PC: Principal Component.

Table 6. Principal component analysis component matrix.

Phospholipid Fatty Acid	Component	
	1	2
a17:0	0.932	0.054
i17:0	0.911	0.064
i15:0	0.876	−0.086
10Me18:0	0.757	0.216
16:1 ω9c	0.752	0.309
18:2 ω6c	−0.728	0.448
10Me16:0	0.670	0.062
17:1 ω8c	−0.667	0.310
18:1 ω7c	−0.655	−0.003
i16:0	0.592	0.401
16:1 ω7c	−0.475	0.161
18:0	−0.053	−0.917
18:1 ω9c	−0.083	0.866
cy17:0 ω7c	0.138	0.846
16:0	−0.030	−0.829
16:1 ω5c	0.173	0.802
a15:0	0.687	−0.695
cy19:0 ω7c	0.507	0.648
17:1 ω7c	−0.080	0.636
15:0	−0.352	0.453

3.3. Relationship between Soil Microbial Community Structure and Soil Properties

Relationships among soil microbial community structure and soil properties were investigated by redundancy analysis (RDA) and correlation analysis in Figure 2, where AP showed a significant negative effect on OB, and inversely positively affected GP with R^2 = 0.80 and 0.51 ($p < 0.05$), respectively. There were significant correlativity between SA with Fug and GN, as well as GP ($p < 0.05$), with R^2 values of −0.92, −0.65, and 0.95, respectively (Figure 2). GP was also dramatically positively influenced by SOM ($R^2 = 0.78$, $p < 0.05$) and AN ($R^2 = 0.68$, $p < 0.05$), besides SA and AP. There were no soil properties significantly affecting Act ($p < 0.05$).

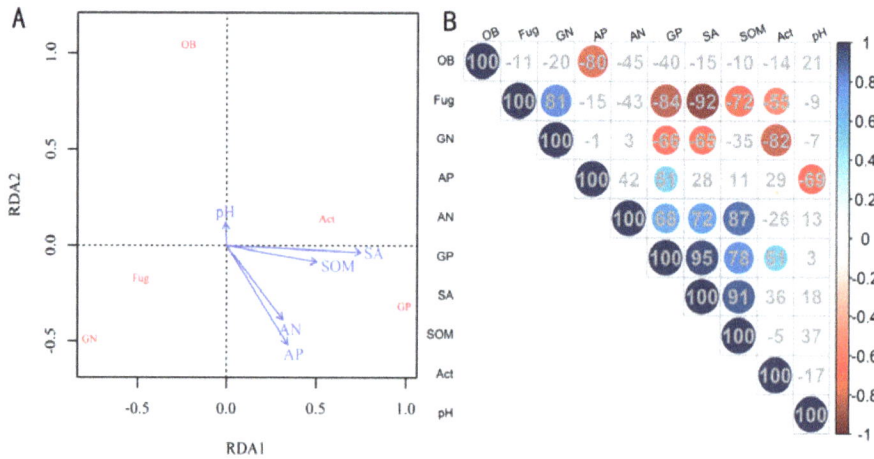

Figure 2. Effects of environmental factors on soil microbial community structure. (**A**) Redundancy analysis (RDA) of environmental factors on soil microbial community structure. (**B**) Correlation analysis of soil environmental factors and soil microbial community. GP: Gram-positive bacteria; GN: Gram-negative bacteria; Fug: Fungi; Act: Actinomycetes; OB: other bacteria. SA: soil macroaggregate content; AP: available phosphorus; AN: soil available nitrogen; SOM: soil organic matter.

4. Discussion

Microorganisms play a vital role in the material cycle and energy transformation of the soil ecosystem, and the soil microbial community structure represents an important aspect to consider when investigating the effect of different crop planting patterns on soil properties. In this study, the results showed that bacteria were the main components of the soil microbial community, which mainly included Gram-positive bacteria and Gram-negative bacteria under the three crop planting patterns (Tables 3 and 4). Some previous research that suggested that bacteria are the main content of soil microorganisms is in agreement with our results [24]. Many types of bacteria have high environmental resistance, such as Gram-positive bacteria and Gram-negative bacteria, which have spores and thicker cell walls and adjust their cell wall structure to adapt to environmental changes [24–26]. Hence variation of bacteria community structure has a significant effect on soil properties [25]. However, some differences in the bacteria community structure were observed under the three crop planting patterns (see Table 4).

Crop rotation has a significant impact on the composition of soil microbial community. Previous studies have shown that complex crop rotation can improve soil quality and crop productivity, including studies on perennial plants [27], and cover crops such as oat, radish, and vetch could increase the bacterial content in PLFA, especially the content of Gram-positive bacteria [28]. In this study, there were some obvious variations observed with Gram-positive bacteria. Gram-positive bacteria presented a significant negative correlation with Gram-negative bacteria (Figure 2). The percentage of Fungi content was reduced in other treatments in comparison to maize continuous cropping (Table 4). Bin Zhang's study had a similar conclusion as the results of this study: short-time rotation has a significant impact on soil fungi community structure, especially arbuscular mycorrhizal fungi, short-time rotation can significantly reduce the content of soil fungi, which resulted in higher biomass of fungi in continuous cropping of maize than rotation. Bacteria account for a large proportion of the soil microbial community, and continuous cropping can increase the content of fungi in soil, which leads to the intensification of soil-borne diseases, but fungi play an important role in the decomposition of recalcitrant compounds in corn residues at later stages [29]. Chavarría et al. suggested that microbial community structure would not change during short-time rotations, which is not supported from the results of this research [28].

Fertilizer application had a significant impact on the content of Gram-negative bacteria (G-) under the cropping system of continuous maize, and showed no significant difference under the cropping system of soybean-maize rotation and continuous soybean. It is possible that the function of nitrogen fixation by soybean fertilized the soil, which covered the response of the Gram-negative bacteria (G-) to the fertilizer application. Fertilizer application had a significant impact on Gram-positive bacteria (G+) under the soybean-maize rotation system, but no significant differences were observed under either the continuous maize or continuous soybean cropping systems. Under the soybean-maize rotation system, the organic fertilizer application treatment and no fertilizer treatment showed a higher content of Gram-positive bacteria (G+) than that of the chemical fertilizer application treatment. The reason for this may be that the Gram-positive bacteria (G+) can be promoted by the organic fertilizer, and be restrained by the chemical fertilizer, due to the supplement of organic fertilizer by nitrogen fixation of no fertilizer application treatment, the effect of promotion and restraining was affected by the soil condition of rotation system, continuous maize system and continuous soybean system. The mechanism(s) behind the response of Gram-negative bacteria (G-) and Gram-positive bacteria (G+) to soil characteristics and fertilizer application should be explored in future research.

The relationship between environmental factors and soil microbial community structure is rather complicated. In our exploration, bacteria (Gram-positive bacteria) primarily improve soil nutrient content, which includes soil available phosphorus, soil available nitrogen, and soil organic matter (Figure 2), Our results also indicated that the Gram-positive bacteria is primarily composed of i17:0, a17:0, and i15:0 (Figure 1). The principal reason for this phenomenon is microbial decomposition. The role of specific species of bacteria needs further study.

5. Conclusions

This study clearly demonstrates the effects of soil microbial community structure under three crop planting patterns on soil properties. The specific conclusions are as follows: (1) bacteria were the main components of the soil microbial community, which mainly included 16:0 (Gram-positive bacteria) under the three crop planting patterns; (2) crop rotation changed the microbial community structure, especially Gram-positive bacteria, which was significantly impacted by the interaction between crop rotation and fertilizer application; (3) Gram-positive bacteria have significant beneficial effects on the soil nutrient content, including the soil available phosphorus, soil available nitrogen, and soil organic matter after soybean and maize crop rotation. In conclusion, soybean and maize crop rotation can improve the soil nutrient content primarily by influencing the composition of bacterial community.

Author Contributions: P.Z., J.S., and L.L. conceived and designed the experiments; X.W., X.L., and J.Q. performed the experiments; P.Z. and L.L. wrote the paper.

Funding: This study was funded by the National Natural Science Foundation of China (31660374), National Commonwealth Industry (Agriculture) Program of China [201503120], National Key Research and Development Program of China [2017YFD0300802].

Acknowledgments: We would like to thank the Oat Research Team for field and data collection, and especially Juliann Seebauer for manuscript revisions.

Conflicts of Interest: The authors declare no conflicts of interest.

References

1. Olsen, R.J.; Hensler, R.F.; Attoe, O.J.; Witzel, S.A.; Peterso, L.A. Fertilizer Nitrogen and Crop Rotation in relation to movement of Nitrate Nitrogen through Soil Profiles. *Soil Sci. Soc. Am. J.* **1970**, *34*, 448–452. [CrossRef]
2. Jenkinson, D.S.; Ladd, J.N. Microbial Biomass in Soil: Measurement and Turnover. *Soil Biochem.* **1981**, *5*, 415–471.

3.   Wu, Q.; Lu, K.; Mao, X.; Qin, H.; Wang, H. Responses of Soil Nutrients and Microbial Biomass and Community Composition to Long-term Fertilization in Cultivated Land. *Chin. Agric. Sci. Bull.* **2015**, *31*, 150–156.

4.   Cavigelli, M.A.; Robertson, G.P. The functional significance of denitrifire community composition in a terrestrial ecosystem. *Ecology* **2000**, *81*, 1402–1414. [CrossRef]

5.   Zelles, L. Fatty acid patterns of phospholipids and lipopolysaccharides in the characterisation of microbial communities in soil: A review. *Biol. Fertil. Soils* **1999**, *29*, 111–129. [CrossRef]

6.   Miller, M.; Dick, R.P. Dynamics of soil C and microbial biomass in whole soil and aggregates in two cropping systems. *Appl. Soil Ecol.* **1995**, *2*, 253–261. [CrossRef]

7.   Bucher, A.E.; Lanyon, L.E. Evaluating soil management with microbial community-level physiological profiles. *Appl. Soil Ecol.* **2005**, *29*, 59–71. [CrossRef]

8.   White, D.C.; Davis, W.M.; Nickels, J.S.; King, J.D.; Bobbie, R.J. Determination of the sedimentary microbial biomass by extractible lipid phosphate. *Oecologia* **1979**, *40*, 51–62. [CrossRef]

9.   Tunlid, A.H.; Baird, B.B.; Trexler, M.; Olsson, S.; Findlay, R.; Odham, G.C.; White, D. Determination of phospholipid ester-linked fatty acids and poly β-hydroxybutyrate for the estimation of bacterial biomass and activity in the rhizosphere of the rape plant *Brassica napus* (L.). *Can. J. Microbiol.* **1985**, *31*, 1113–1119. [CrossRef]

10.   Roslev, P.; Iversen, N.; Henriksen, K. Direct fingerprinting of metabolically active bacteria in environmental samples by substrate specific radiolabelling and lipid analysis. *J. Microbiol. Methods* **1998**, *31*, 99–111. [CrossRef]

11.   ISSCAS. *Physical and Chemical Analysis Methods of Soils*; Institute of Soil Sciences, Chinese Academy of Sciences; Shanghai Science and Technology Press: Shanghai, China, 1978; pp. 7–59. (In Chinese)

12.   Hong, S.B.; Piao, S.L.; Chen, A.P.; Liu, Y.W.; Liu, L.L.; Peng, S.S.; Sardans, J.; Sun, Y.; Penuelas, J.; Zeng, H. Afforestation neutralizes soil pH. *Nat. Commun.* **2018**, *9*, 7. [CrossRef] [PubMed]

13.   Chen, X.H.; Duan, Z.H. Changes in soil physical and chemical properties during reversal of desertification in Yanchi County of Ningxia Hui autonomous region, China. *Enviorn. Geol.* **2009**, *57*, 975–985. [CrossRef]

14.   Bossio, D.A.; Scow, K.M. Impacts of Carbon and Flooding on Soil Microbial Communities: Phospholipid Fatty Acid Profiles and Substrate Utilization Patterns. *Microb. Ecol.* **1998**, *35*, 265–278. [CrossRef] [PubMed]

15.   Wu, J.; Jiang, Y.; Wu, Y.; Xu, J. Effect of complex heavy metal pollution on biomass and community structure of soil microbes in paddy soil. *Acta Pedol. Sin.* **2008**, *45*, 1102–1109. [CrossRef]

16.   Frostegård, Å.; Bååth, E.; Tunlio, A. Shifts in the structure of soil microbial communities in limed forests as revealed by phospholipid fatty acid analysis. *Soil Biol. Biochem.* **1993**, *25*, 723–730. [CrossRef]

17.   Federle, T.W. Microbial distribution in the soil-new techniques. In *Perspectives in Microbial Ecology, Proceedings of the International Symposium of Microbial Ecology IV, Ljubljana, Slovenia, January 1986*; Megusar, F., Gantar, M., Eds.; Slovene Society for Microbiology: Ljubljana, Slovenia, 1986; pp. 493–498.

18.   Hamel, C.; Vujanovic, V.; Jeannotte, R.; Nakano-Hylander, A.; St-Arnaud, M. Negative feedback on a perennial crop: Fusarium crown and root rot of asparagus is related to changes in soil microbial community structure. *Plant Soil* **2005**, *268*, 75–87. [CrossRef]

19.   Tunlid, A.; Hoitink, J.H.A.; Low, C.; White, C.D. Characterization of bacteria that suppress rhizoctonia damping-off in bark compost media by analysis of Fatty Acid biomarkers. *Appl. Environ. Microbiol.* **1989**, *55*, 1368–1374.

20.   Zogg, G.P.; Zak, D.R.; Ringelberg, D.B.; White, D.C.; MacDonald, N.W.; Pregitzer, K.S. Compositional and Functional Shifts in Microbial Communities Due to Soil Warming. *Soil Sci. Soc. Am. J.* **1997**, *61*, 475. [CrossRef]

21.   Blume, E.; Bischoff, M.; Reichert, J.M.; Moorman, T.; Konopka, A.; Turco, R.F. Surface and subsurface microbial biomass, community structure and metabolic activity as a function of soil depth and season. *Appl. Soil Ecol.* **2002**, *20*, 171–181. [CrossRef]

22.   Brant, J.B.; Myrold, D.D.; Sulzman, E.W. Root controls on soil microbial community structure in forest soils. *Oecologia* **2006**, *148*, 650–659. [CrossRef]

23.   Xu, Y.; Cai, X.; Zhu, X. Comparative analysis of microbial community structures in soils from rice-upland crop rotation fields by PLFA profile technique. *Acta Agric. Zhejiangensis* **2013**, *25*, 1056–1061. [CrossRef]

24.   Song, Y.S.P.; Deng, V. Acosta-Martínez, E. Katsalirou. Characterization of redox-related soil microbial communities along a river floodplain continuum by fatty acid methyl ester (fame) and 16s rrna genes. *Appl. Soil Ecol.* **2008**, *40*, 499–509. [CrossRef]

25. Zhang, X.; Xin, X.; Zhu, A.; Yang, W.; Zhang, J.; Ding, S.; Mu, L.; Shao, L. Linking macroaggregation to soil microbial community and organic carbon accumulation under different tillage and residue managements. *Soil Tillage Res.* **2018**, *178*, 99–107. [CrossRef]

26. Mccann, M.C.; Shi, J.; Roberts, K.N.; Carpita, C. Changes in pectin structure and localization during the growth of unadapted and nacl-adapted tobacco cells. *Plant J.* **2010**, *5*, 773–785. [CrossRef]

27. Kiani, M.; Hernandez-Ramirez, G.; Quideau, S.; Smith, E.; Janzen, H.; Larney, F.J.; Puurveen, D. Quantifying sensitive soil quality indicators across contrasting long-term land management systems: Crop rotations and nutrient regimes. *Agric. Ecosyst. Environ.* **2017**, *248*, 123–135. [CrossRef]

28. Chavarría, D.N.; Verdenelli, R.A.; Serri, D.L.; Restovich, S.B.; Andriulo, A.E.; Meriles, J.M.; Vargas-Gil, S. Effect of cover crops on microbial community structure and related enzyme activities and macronutrient availability. *Eur. J. Soil Biol.* **2016**, *76*, 74–82. [CrossRef]

29. Zhang, B.; Li, Y.; Ren, T.; Tian, Z.; Wang, G.; He, X. Short-term effect of tillage and crop rotation on microbial community structure and enzyme activities of a clay loam soil. *Biol. Fertil. Soils* **2014**, *50*, 1077–1085. [CrossRef]

agronomy

MDPI

Article

Hybrid Selection and Agronomic Management to Lessen the Continuous Corn Yield Penalty

Alison M. Vogel and Frederick E. Below *

Department of Crop Sciences, University of Illinois, Urbana, IL 61801, USA; amvogel2@illinois.edu
* Correspondence: fbelow@illinois.edu; Tel.: +1-217-333-9745

Received: 11 September 2018; Accepted: 12 October 2018; Published: 16 October 2018

Abstract: Yield reductions occur when corn (*Zea mays* L.) is continuously grown compared to when it is rotated with soybean [*Glycine max* (L.) Merr.]; primarily due to soil nitrogen availability, corn residue accumulation, and the weather. This study was conducted to determine if a combination of agronomic practices could help overcome these causative factors of the continuous corn yield penalty (CCYP) to obtain increased corn yields. Field experiments conducted during 2014 and 2015 at Champaign, IL, U.S.A. assessed the yield penalty associated with continuous corn verses long-term corn following soybean. Agronomic management was assessed at a standard level receiving only a base rate of nitrogen fertilizer, and compared to an intensive level, which consisted of additional N, P, K, S, Zn, and B fertility at planting, sidedressed nitrogen fertilizer, and a foliar fungicide application. Two levels of plant population (79,000 verses 111,000 plants ha^{-1}) and eight different commercially-available hybrids were evaluated each year. Across all treatments, the CCYP was 1.53 and 2.72 Mg ha^{-1} in 2014 and 2015, respectively. Intensive agronomic management improved grain yield across rotations (2.17 Mg ha^{-1} in 2014 and 2.28 Mg ha^{-1} in 2015), and there was a 40 to 60% greater yield response to intensive management in continuous corn verses the corn-soybean rotation, suggesting intensified management as a method to mitigate the CCYP. With select hybrids, intensive management reduced the CCYP by 30 to 80%. Agronomic management and hybrid selection helped alleviate the CCYP demonstrating continuous corn can be managed for better productivity.

Keywords: continuous corn yield penalty (CCYP); corn-soybean rotation; hybrid; intensive management; maize; population

1. Introduction

Crop rotation is a decision that can affect the productivity and profitability of agriculture production systems. Global trade tentions and crop demand can alter commodity prices that can allow grain price to offset typical lost productivity of corn monocropping. The grain yield reduction when corn is grown continuously (corn grown after previous-crop corn, i.e., continuous corn) compared to when it is rotated with soybean has been widely reported [1–10]. Factors primarily contributing to the continuous corn yield penalty (CCYP) are soil nitrogen availability or immobilization, residue accumulation, and the weather [9]. The consequence of adverse environmental effects are more determental on continuous corn grain yield than corn grown in rotation with soybean [9,11,12]. Environments with minimal rainfall have been documented to increase the magnitude of the CCYP [1,3,13,14], along with cooler than average spring temperatures [12], and excessive warmth during the summer [9,12]. Although weather cannot be controlled, there are many crop inputs that increase yields, and may mitigate the CCYP, including hybrid selection, plant population, fertilizer, and foliar fungicides.

Yield potential is greater with modern corn hybrids as a result of improved tolerance to the stresses, such as those associated with increased plant population, reduced soil nitrogen, and low soil moisture [15–20]. Hybrids vary in their growth and yield response to different management factors, including crop rotation [2,21]. Yet, the greatest yield potential cannot be achieved with newer corn genetics unless grown at higher plant populations than older corn genetics [18,22]. Nitrogen (N) and phosphorus (P) use efficiency [23,24], water-use efficiency [25], and the value of fungicide and insecticide applications, have been shown to improve with increased planting population. Future improvement of corn yield will focus on increased tolerance to even higher plant populations, due to corn's inadequate input use at lower plant populations [17]. However, increased plant population results in a more stressful environment, which could exacerbate the yield-reducing effects of continuously-grown corn.

Nitrogen is the nutrient required in the greatest quantities for corn [26] and is the most frequently limited nutrient for corn production [27]. After N, the second highest quantity of mineral nutrient acquired by corn during the growing season is potassium (K) [26]. Additionally, phosphorus (P) is the least mobile macronutrient and least available in the soil [28]; however, P has the highest nutrient removal rate from the field at harvest with corn grain [26]. Other nutrients found to limit U.S. Corn Belt yields are sulfur (S), zinc (Zn), and boron (B) [29–34].

A more recent tool for increasing grain yields is through foliar fungicide applications [35,36]. Strobilurin fungicides are effective against fungal pathogens that induce foliar fungal diseases in susceptible corn germplasm [37]. Corn residue on the soil surface from previous crops can serve as an overwintering inoculum for several important foliar diseases, such as grey leaf spot (*Cercospora zeae-maydis*) and northern leaf blight (*Exserohilum turcicum*) [38]. Residue accumulation can increase through continuous corn rotations [2,9], no- or reduced-tillage [39], higher plant density [40], and greater grain yields [41]. Furthermore, foliar protection by a strobilurin fungicide has also been documented to increase grain yield even when fungal disease is not present [35].

Intelligent intensification of agronomic management, including hybrid selection, and additional plant population, fertilizer, and fungicide application, may offset the negative causative effects of continuously grown corn [5,42] and promote greater yields [43]. The objectives of this research were to (i) demonstrate the CCYP and quantify the impact of different crop management practices on the reduction of the CCYP, (ii) determine the effect of these management factors on in-season biomass accumulation and plant health, and (iii) assess the effect of these practices on yield components to ascertain when these yield responses are occurring. To achieve these objectives, multiple corn hybrids were grown under two crop rotations (previous crop corn verses soybean), at two population densities and crop management levels (standard verses intensive). In this trial, intensive management (i.e., high input) encompassed additional nitrogen fertilizer, broadcast (i.e., K and B source) and banded (i.e., P, S, Zn and N source) fertility, and a foliar fungicide.

2. Materials and Methods

2.1. Agronomic Practices

Field experiments were conducted in 2014 and 2015 at Champaign, IL, U.S.A. using a long-term site dedicated to crop rotation. Due to the rotation treatment in this study, two comparable field sites of approximately 2 ha each were established within 4.5 km of each other and predominantly (>75%) consisted of a Flanagan silt loam (a fine, smectitic, mesic Aquic Argiudoll) with 0 to 2% slope. The sites were tile drained and unirrigated. The preplanting soil properties at the 0- to 15-cm depth for 2014 and 2015 included, respectively, 39 and 41 g kg^{-1} organic matter, pH 6.1 and 5.5, 19 and 37.1 mg kg^{-1} P, and 101 and 126 mg kg^{-1} K. The minerals P and K were extracted using Mehlich III solution. The study alternated between the two field sites each year, generating for this study 11th (2014) or 13th (2015) year continuous corn vs. long-term corn following soybean rotation. The 11th and 13th year continuous corn were considered as similar treatments in line with other rotational experiments [3,14]. The setup site (the site not used for the current year) established the replicated blocks of corn and soybean that served as the previous crop for the following year's experiment. The corn and soybean blocks in the setup site were maintained with minimal crop management inputs through maturity, harvested, and tilled in preparation for the upcoming year's study. Individual experimental plots consisted of four rows, 5.3-m in length with 76-cm spacing, and planted with a precision ALMACO SeedPro 360 research plot planter (Nevada, IA, USA). Treatments were arranged in a split-split plot in a randomized complete block design with four replications; crop rotation was the main plot and the subplot was hybrid with a factorial arrangement input level and population at the sub-sub plot level.

The hybrids evaluated represented a range of maturities (106- to 113-day relative maturity; RM), as well as two seed brands, that represented varying genetic backgrounds and potential tolerance to continuous corn. In 2014, the hybrids grown included DKC58-87SSRIB (108 RM), DKC60-67RIB (110 RM), DKC62-08 RIB (112 RM), DKC64-87RIB (114 RM), DKC63-33RIB (113 RM), 209-53STXRIB (109 RM), 212-86STXRIB (112 RM), and DKC63-55RIB (113 RM) [Bayer, Leverkusen, Germany]; hybrids grown in 2015 included 5415SS (106 RM), 5887VT3P (108 RM), 5975VT3P (109 RM), 6110SS (110 RM), 6065SS (111 RM), 6265SS (112 RM), 6594SS (113 RM), and 6640VT3P (113 RM) [WinField United, LLC., Arden Hills, MN, USA].

Tillage included a chisel plow in fall with field cultivations in spring for entire seedbed preparation. Plots were planted on 27 April 2014 and 24 April 2015 to achieve an approximate final stand of 79,000 or 111,000 plants ha^{-1}, denoted as standard and high density, respectively. All plots received an in-furrow application of tefluthrin ((1S,3S)-2,3,5,6-tetrafluoro-4-methylbenzyl 3-((Z)-2-chloro-3,3,3-trifluoroprop-1-en-1-yl)-2,2-dimethylcyclopropanecarboxylate) at a rate of 0.11 kg a.i. ha^{-1} for additional control of seedling insect pests. Weed control consisted of a pre-emergence application of S-metolachlor (2-chloro-N-(2-ethyl-6-methylphenyl)-N-(2-methoxy-1-methylethyl)acetamide), atrazine (6-chloro-N-ethyl-N'-(1-methylethyl)-1,3,5-triazine-2,4-diamine), and mesotrione ([2-[4-(methylsulfonyl)-2-nitrobenzoyl]- 1,3-cyclohexanedione), and a post-emergence application of glyphosate [N-(phosphonomethyl)glycine].

One week before planting, 202 kg N ha^{-1} as urea ammonium nitrate was applied to all plots and incorporated by shallow cultivation. The standard management treatment only received this N fertilizer. The additional products utilized in the intensive management system included additional fertilizer (containing N, P, K, S, Zn, and B) and a foliar fungicide application. Immediately prior to planting the intensive management plots, 112 kg P_2O_5 ha^{-1} was banded (10–15 cm beneath the row) by a toolbar fitted with Dawn Equipment 6000 Series Universal Fertilizer Applicators (Dawn Equipment, Sycamore, IL, USA) as MESZ [MicroEssentials SZ; 12-40-0-10S-1Zn] (The Mosaic Company, Plymouth, MN, USA) supplying an additional 34 kg N ha^{-1}, 28 kg S ha^{-1}, and 2.8 kg Zn ha^{-1}. Additionally at planting, 84 kg K_2O ha^{-1} was broadcast applied (Aspire, 0-0-58-0.5B, The Mosaic Company, Plymouth, MN, USA) supplying an additional 0.4 kg B ha^{-1}. At the V6 growth stage (six fully formed leaves), a side-dress application of 67 kg N ha^{-1} was applied to these plots as urea with urease inhibitors [$CO(NH_2)_2$ + n-(n-butyl) thiophosphoric triamide; Agrotain urea; 46-0-0] (Koch Agronomic Services, LLC, Wichita, KS, USA) on 6 June 2014 and 4 June 2015. When plants were approximately between the VT to R1 growth stages (tasseling to silk emergence) (6 June 2014 and 13 July 2015), intensive management plots received an application of Headline AMP (BASF, Florham Park, NJ, USA), a product containing pyraclostrobin (carbamic acid, [2-[[[1-(4-chlorophenyl)-1H-pyrazol-3-yl]oxy]methyl]phenyl]methoxy-, methyl ester) and metconazole (5-[(4-chlorophenyl)methyl]-2,2-dimethyl-1-(1H-1,2,4-triazol-1-ylmethyl)cyclopentanol), at the labeled rates of 0.15 and 0.06 kg a.i. ha^{-1}, respectively. The fungicide was applied using a CO_2-pressurized backpack sprayer via an aqueous suspension at 140 L H_2O ha^{-1} and mixed with the surfactant MasterLock (WinField Solutions, LLC, St. Paul, MN, USA) at 0.45 kg ha^{-1}.

2.2. Plant Biomass Samplings, Health Assessment, and Harvest

To evaluate seasonal aboveground biomass, plants were sampled at two growth stages: V6 (six leaves with collars visible) and R6 (physiological maturity) [44]. Corn tissue sampling was conducted on 2 June 2014 (V6), 3 June 2015 (V6), 9 September 2014 (R6), and 31 August 2015 (R6). Sampling consisted of manually excising plants from the outer two rows at V6 (10 plants per plot), and from the center two rows of each plot at R6 (6 plants per plot) to determine biomass. Plants at the V6 growth stage were dried to 0% moisture and weighed. The plants at R6 were partitioned into grain and stover (including husk) components, and biomass was determined by weighing the total fresh stover then processing it through a wood chipper (BC600XL, Vermeer Corporation, Pella, IA, USA) to obtain representative stover subsamples. The stover subsamples were immediately weighed to determine aliquot fresh weight, and then weighed again after drying to 0% moisture in a forced air oven at 75 °C, to determine subsample aliquot dry weight and calculate dry biomass. Corn ears were dried and then weighed to obtain grain and cob weight. The grain was removed using a corn sheller (AEC Group, St. Charles, IA, USA) and analyzed for moisture content using a moisture reader (Dickey John, GSF, Ankeny, IA, USA). Cob weight was obtained by difference, and dry stover and cob weights were summed to calculate the overall R6 stover biomass. All biomass and grain weight measurements are presented on a 0 g kg^{-1} moisture concentration basis.

To assess treatment effects on plant health, leaf greenness was measured at the R2 growth stage (kernel blister) (29 July 2014 and 20 July 2015) using a Minolta SPAD-502 chlorophyll meter (Spectrum Technologies, East-Plainfield, IL, USA), on the lamina at the midleaf region of five ear leaves (with no lesions) per plot. SPAD values were used to estimate differences in leaf N concentration among treatments, given that N is a main element in chlorophyll molecules, and therefore related to leaf greenness [45].

Plant stand counts were tallied to confirm plant populations at the R6 plant growth stage. The center two rows of each plot were mechanically harvested for determination of grain yield at physiological maturity, and yield values are presented at 0% moisture. Grain subsamples from each plot were collected from the plot combine at harvest and 300 randomly selected kernels were weighed to estimate average individual kernel weight, also expressed at 0% moisture. Kernel number was estimated by dividing grain yield by the average individual kernel weight of each plot.

2.3. Statistical Analysis

Biomass accumulation, leaf greenness, grain yield, and yield Fcomponents (kernel number and kernel weight) were analyzed using the PROC MIXED procedure [46]. All units are expressed on a 0 g kg^{-1} moisture concentration basis. Rotation, hybrid, agronomic management, and population were included as fixed effects and replication as a random effect. Due to differences in years of continuous corn and hybrid, years were analyzed separately. Least square means were separated using the PDIFF option of LSMEANS in SAS PROC MIXED. Unless indicated, fixed effects were considered significant in all statistical calculations if $p \leq 0.05$. Pearson's correlation coefficient was used to evaluate the linear association between grain yield and measured parameters across all treatments and within each rotation, using the CORR procedure of SAS.

3. Results and Discussion

3.1. Temperature and Precipitation

The weather conditions of 2014 and 2015 in Champaign, IL resulted in varied temperatures and levels of precipitation (Table 1). In 2014, temperatures were below-average with above-average precipitation, particularly in June and July. Temperatures were at or below normal in July, August, and September of 2014. Illinois experienced a warm April and May in 2015, with a cooler than average June, July, and August. Rainfall in May 2015 was slightly above average in Champaign, but in June, the whole state of Illinois experienced rainfall amounts breaking records that date back to 1886. Champaign received 122.7 mm of rainfall above the 30-year average. Pollination and grain-filling conditions were good with a drier July and August. Overall, the 2014 and 2015 production years experienced very little weather-induced heat or moisture stress. As a result, conditions were generally conducive to favorable grain yields.

Table 1. Monthly weather data for Champaign, IL, USA during the production seasons of 2014 and 2015. Temperature is the average daily air temperature and precipitation is the average monthly accumulated rainfall. Values were obtained from the U.S. National Oceanic and Atmospheric Administration and values in parentheses are the deviations from the 30-year average (1981–2010).

Year	Month					
	April	May	June	July	Aug.	Sept.
2014						
Temperature, °C	11.5 (0.4)	17.7 (0.8)	22.8 (0.4)	21.0 (−2.8)	23.0 (0.0)	18.1 (−0.9)
Precipitation, mm	100.1 (6.6)	111.3 (−13)	208.5 (98.3)	221.2 (101.9)	38.6 (−61.2)	87.4 (7.9)
2015						
Temperature, °C	12.0 (1.1)	18.6 (1.7)	22.2 (−0.1)	23.0 (−0.8)	22.1 (−0.9)	21.0 (2.0)
Precipitation, mm	91.9 (−1.5)	154.2 (30.0)	232.9 (122.7)	107.2 (−12.2)	80.3 (−19.6)	163.6 (84.1)

3.2. Plant Biomass Accumulation and Plant Health Assessment

Agronomic input level significantly impacted early season (V6 growth stage) biomass accumulation (Table 2). Compared to standard input, intensive management led to 42 to 56% greater aboveground biomass accumulation when averaged across both planting densities and rotations (Table 3). A significant increase in corn early season biomass accumulation with increased fertilizer inputs is well known [27,47]. Since this sampling was completed immediately prior to the additional sidedressed N, management responses were primarily from the broadcasted K and B and banded P, N, S, and Zn supplied at planting.

Table 2. Tests of fixed sources of variation on early and late season biomass accumulation, in-season leaf greenness, final grain yield, and yield components for the continuous corn trial conducted at Champaign, IL during 2014 and 2015. Rotation (R), hybrid (H), management (M), and population (P) served as fixed effects.

Year/Fixed Effect	V6 Biomass	R2 SPAD	R6 Stover	Grain Yield	Kernel Number	Kernel Weight
			p > F			
2014						
Rotation (R)	0.3407	0.0004	0.1487	<0.0001	0.0670	0.0121
Hybrid (H)	0.0015	0.0186	0.0827	<0.0001	<0.0001	<0.0001
R × H	0.0270	0.4260	0.1861	0.0057	0.0205	0.1536
Management (M)	<0.0001	<0.0001	<0.0001	<0.0001	<0.0001	<0.0001
R × M	0.4083	<0.0001	0.6366	<0.0001	0.0003	0.1309
H × M	0.1088	0.5741	0.4072	0.0971	0.4142	0.3174
R × H × M	0.0302	0.1897	0.7955	0.0665	0.1201	0.4127
Population (P)	<0.0001	<0.0001	0.0053	0.0093	0.0104	<0.0001
R × P	0.4215	0.0133	0.8181	0.2644	<0.0001	<0.0001
H × P	0.2080	0.7575	0.6459	0.0998	0.0635	0.5584
R × H × P	0.8140	0.7686	0.6102	0.8172	0.0474	0.0786
M × P	0.1084	0.2811	0.0919	0.6485	0.2868	0.4939
R × M × P	0.2410	0.0795	0.1966	0.6369	0.9993	0.7101
H × M × P	0.5074	0.5361	0.9915	0.6744	0.8588	0.9917
R × H × M × P	0.0442	0.6956	0.9135	0.1519	0.3386	0.6266
2015						
Rotation (R)	0.0469	0.0099	0.1655	0.0018	0.0016	0.0022
Hybrid (H)	0.0150	0.1512	0.0278	0.0453	<0.0001	<0.0001
R × H	0.5081	0.6912	0.7316	0.1713	0.0587	0.1215
Management (M)	<0.0001	<0.0001	<0.0001	<0.0001	<0.0001	<0.0001
R × M	0.1449	<0.0001	0.2669	0.0015	0.0013	0.0228
H × M	0.0223	0.0154	0.6895	0.4599	0.1188	0.0017
R v H × M	0.1975	0.3037	0.1246	0.2247	0.0265	0.6136
Population (P)	<0.0001	<0.0001	<0.0001	0.0026	<0.0001	<0.0001
R × P	0.4649	0.3326	0.5698	0.7829	0.9614	0.0960
H × P	0.5189	0.3828	0.5145	0.0997	0.1563	<0.0001
R × H × P	0.8911	0.2649	0.3833	0.0915	0.3801	0.2294
M × P	0.7745	0.5340	0.2152	0.0140	0.0025	0.6258
R × M × P	0.9055	0.6813	0.9914	0.2584	0.5078	0.3064
H × M × P	0.5882	0.3390	0.5763	0.9153	0.7468	0.7456
R × H × M × P	0.7129	0.1925	0.9442	0.1882	0.0234	0.3173

Table 3. Aboveground biomass accumulation as influenced by crop rotation, hybrid, agronomic input level, and population at Champaign, IL in 2014 and 2015. All values are reported at 0 g kg^{-1} moisture concentration.

		V6 Shoot Biomass				R6 Stover Biomass			
		Input Level §							
		Standard		Intensive		Standard		Intensive	
	Year/	Planting Density (× 1000 plants ha^{-1}) ¶							
Rotation †	Hybrid ‡	79	111	79	111	79	111	79	111
		kg ha^{-1}							
Cont. Corn	**2014**								
	209-53STX	382	513	557	789	9561	11,180	12,710	12,118
	212-86STX	332	516	651	855	11,224	12,942	12,750	12,650
	DKC58-87	366	452	565	754	10,552	12,009	12,176	11,483
	DKC60-67	316	548	653	683	9276	10,774	11,711	12,037
	DKC62-08	391	570	852	864	12,038	11,061	12,357	10,543
	DKC63-33	407	417	741	874	9144	10,334	11,638	11,321
	DKC63-55	362	480	657	763	10,337	12,096	12,150	12,923
	DKC64-87	423	525	630	836	9170	11,226	11,331	13,034
	2014 Means	372	503	663	802	10,163	11,453	12,103	12,014
	2015								
	5415SS	279	332	661	931	8073	9337	9973	11,581
	5887VT3P	315	372	614	890	7728	8655	8743	9034
	5975VT3P	290	372	793	827	8500	8523	9303	10,623
	6065SS	310	379	719	986	9400	9705	11,717	11,205
	6110SS	333	400	843	1025	8628	7933	9640	10,652
	6265SS	421	587	963	1017	8799	8813	10,327	11,404
	6594SS	485	565	680	1036	7648	9130	9570	10,853
	6640VT3P	332	474	789	940	6849	7684	10,070	11,100
	2015 Means	346	435	758	957	8203	8722	9918	10,806
Corn-Soybean	**2014**								
	209-53STX	357	473	630	817	9082	9416	11,814	12,731
	212-86STX	321	402	671	809	10,769	11,096	11,438	12,962
	DKC58-87	322	407	559	684	10,771	11,601	11,637	12,728
	DKC60-67	315	370	598	751	9745	11,940	10,921	12,088
	DKC62-08	361	413	569	712	10,943	12,799	13,470	12,799
	DKC63-33	395	486	712	751	10,206	10,363	12,890	12,599
	DKC63-55	295	431	618	819	10,761	11,618	11,901	13,019
	DKC64-87	423	541	706	887	11,523	11,356	12,211	12,268
	2014 Means	349	440	633	779	10,475	11,274	12,035	12,649
	2015								
	5415SS	327	376	812	785	9210	9814	10,489	10,709
	5887VT3P	217	316	658	958	8525	8262	10,581	10,324
	5975VT3P	287	317	747	880	8264	8435	9160	11,466
	6065SS	269	376	651	829	9512	9942	11,232	11,547
	6110SS	336	361	747	921	8323	8907	8787	10,009
	6265SS	314	361	785	913	9018	9484	10,535	10,544
	6594SS	404	432	673	952	9054	8815	11,254	11,270
	6640VT3P	296	412	693	760	8896	9977	9882	11,789
	2015 Means	306	369	721	875	8850	9205	10,240	10,957

† Rotation V6 LSD ($p \leq 0.05$) = nonsignificant (NS) in 2014 and 54 kg ha^{-1} in 2015; Rotation R6 LSD ($p \leq 0.05$) = NS.
‡ Hybrid V6 LSD ($p \leq 0.05$) = 48 kg ha^{-1} in 2014 and 68 kg ha^{-1} in 2015; Hybrid R6 LSD ($p \leq 0.05$) = NS in 2014 and 900 kg ha^{-1} in 2015. § Input level V6 LSD ($p \leq 0.05$) = 19 kg ha^{-1} in 2014 and 34 kg ha^{-1} in 2015; Input level R6 LSD ($p \leq 0.05$) = 455 kg ha^{-1} in 2014 and 291 kg ha^{-1} in 2015. ¶ Plant density V6 LSD ($p \leq 0.05$) = 19 kg ha^{-1} in 2014 and 34 kg ha^{-1} in 2015; Plant density R6 LSD ($p \leq 0.05$) = 455 kg ha^{-1} in 2014 and 292 kg ha^{-1} in 2015.

When assessed at V6, accumulation of biomass was comparable, but tended to be greater in the continuous corn rotation relative to the corn-soybean rotation, similar to previous reports [48]. Regardless of rotation, the 40% greater planting population of the high population treatment increased the early season biomass accumulation per area by 19 to 20% in both years. Hybrid selection influenced early season biomass accumulation; there was a difference of 108 kg ha^{-1} and 127 kg ha^{-1} in 2014 and 2015, respectively, between the smallest and largest hybrids at V6.

At the R2 growth stage, the corn-soybean rotation led to enhanced ear leaf greenness compared to continuous corn (Table 4). In 2014 and 2015, leaf greenness was 59.1 vs. 52.9 and 62.5 vs. 57.9 SPAD relative units from corn-soybean rotation vs. continuous corn, respectively (Table 4). Intensive management increased leaf greenness in continuous corn, but not of those plants grown in the corn-soybean rotation. Increased population reduced the leaf greenness levels. When averaged across the hybrids, the least leaf greenness was measured in continuous corn cultivated with standard agronomic management and the higher planting density.

Table 4. Leaf greenness for hybrids as influenced by crop rotation, agronomic input level, and population at the R2 growth stage of the ear leaf. Hybrids were grown in continuous corn and following soybean rotations at Champaign, IL in 2014 and 2015.

	Crop Rotation ‡							
	Continuous Corn				Corn-Soybean			
	Input Level §							
	Standard		Intensive		Standard		Intensive	
Year/	Plant Density (plant ha^{-1}) ¶							
Hybrid †	79,000	111,000	79,000	111,000	79,000	111,000	79,000	111,000
	——————————————— SPAD relative unit ———————————							
2014								
209-53STX	58.1	50.1	60.6	56.2	61.9	59.4	63.0	59.4
212-86STX	56.1	47.1	60.0	57.1	62.4	57.9	63.0	59.4
DKC58-87	50.5	46.4	59.1	54.1	59.5	57.8	58.6	57.2
DKC60-67	54.1	50.5	57.6	55.1	59.6	55.9	60.0	57.5
DKC62-08	51.4	44.2	51.3	50.3	60.6	58.2	60.8	57.4
DKC63-33	51.5	43.8	57.3	55.0	59.7	59.0	59.5	57.7
DKC63-55	48.7	44.5	57.1	51.1	58.1	57.5	59.3	56.2
DKC64-87	52.9	49.6	58.0	52.4	60.8	55.9	61.6	56.0
2014 Means	**52.9**	**47.0**	**57.6**	**53.9**	**60.3**	**57.7**	**60.7**	**57.6**
2015								
5415SS	59.3	55.6	65.3	60.4	64.7	62.8	65.6	64.1
5887VT3P	57.3	53.5	62.4	55.3	64.7	62.5	64.3	62.4
5975VT3P	54.6	50.1	62.0	57.7	63.0	58.1	62.2	60.4
6065SS	61.4	55.8	62.1	60.7	63.6	63.2	64.3	60.6
6110SS	60.7	57.9	60.0	58.9	65.0	60.7	62.4	61.1
6265SS	54.0	53.8	60.7	58.3	64.0	60.5	62.8	60.1
6594SS	55.7	57.2	61.1	58.5	62.5	59.8	63.5	60.8
6640VT3P	56.9	49.2	59.2	57.2	65.4	61.1	64.2	61.0
2015 Means	**57.5**	**54.1**	**61.6**	**58.4**	**64.1**	**61.1**	**63.7**	**61.3**

† Hybrid LSD ($p \leq 0.05$) = 2.7 SPAD unit in 2014 and nonsignificant (NS) in 2015. ‡ Rotation LSD ($p \leq 0.05$) = 0.7 SPAD unit in 2014 and 2.5 SPAD unit in 2015. § Input level LSD ($p \leq 0.05$) = 0.8 SPAD unit in 2014 and 0.6 SPAD unit in 2015. ¶ Plant density LSD ($p \leq 0.05$) = 0.8 SPAD unit in 2014 and 0.6 SPAD unit in 2015.

Hybrid selection also impacted these R2 measurements, thirteen of the sixteen hybrids had significantly reduced ear leaf greenness when grown following corn rather than after soybean, while the other hybrids exhibited that tendency. Greater leaf chlorophyll concentrations and boosted levels of plant N have also been found when corn was rotated with soybean compared to grown continuously, which have been attributed to the greater N availability observed in non-continuous corn systems [49]. Leaf chlorophyll concentration, photosynthetic potential of the plant, and leaf N nutrient status are closely related [50–52]. These treatment-induced differences in leaf chlorophyll resulting from cropping system and management level changes suggest that N uptake and N availability play a key role in the continuous corn yield penalty and indicate potential ways to mitigate it.

Stover biomass accumulation at the R6 growth stage was 11% and 17% greater from the intensive management when compared to the standard input level, in 2014 and 2015 respectively (Table 4). On an individual plant basis, the intensive input level led to an additional 17 g plant^{-1} of dry weight in 2014 and 18 g plant^{-1} of dry weight in 2015 (data not shown). The increased population treatment provided an additional 32,000 plants ha^{-1} and resulted in increased overall biomass production per area (Table 4). Conversely, individual plants' R6 stover accumulation at the two populations were 139 and 125 g plant^{-1} when grown at 79,000 plants ha^{-1} compared to 110 and 100 g plant^{-1} when grown at 111,000 plants ha^{-1} in 2014 and 2015, respectively (data not shown). Previous crop minimally ($p = 0.15$ and $p = 0.17$), or slightly increased final stover biomass with alternating rotation. It has been previously reported that 75% of the time, corn grown after soybean produced greater dry matter than when grown following corn [53]. Combined with the data presented here, these results indicate that corn stover grown with crop rotation will often produce at least similar, if not greater, stover biomass, than when grown continuously.

3.3. Grain Yield and Yield Components

Rotation, hybrid, management, and population treatments significantly influenced grain yield (Table 2). When averaged across all treatment combinations, the CCYP associated with continuous corn compared to first year corn following soybean was 1.53 Mg ha^{-1} (-13%; $p < 0.0001$) in 2014 and 2.72 Mg ha^{-1} (-22%; $p = 0.0018$) in 2015 (Table 5). Although increased planting densities decreased yield by an average of 0.19 and 0.36 Mg ha^{-1} in 2014 and 2015, respectively, the continuous corn rotation did not magnify this response as originally predicted. Unexpectedly, the yield reduction associated with increased planting densities tended to be greater in the corn-soybean rotation vs. continuous corn. The increased inter-plant competition of higher planting densities tended to reduce corn yield more when grown under standard management (-0.22 Mg ha^{-1} in 2014 and -0.65 Mg ha^{-1} in 2015) compared to when grown under the high input management (-0.16 Mg ha^{-1} in 2014 and -0.06 Mg ha^{-1} in 2015, non-significant) when averaged across rotations. The lowest yield was observed when corn was grown after corn with standard agronomic management and the higher plant density.

Intensive agronomic management significantly improved grain yield when averaged across crop rotations (2.17 and 2.28 Mg ha^{-1} in 2014 and 2015, respectively), but the effect was 40–60% greater in continuous corn vs. the corn-soybean rotation (2.65 vs. 1.69 Mg ha^{-1} in 2014 and 2.67 vs. 1.90 Mg ha^{-1} in 2015) (Table 5). These findings are consistent with other studies that found additional fertilizer inputs are needed to achieve continuous corn yields that approach or are similar to rotated corn yields [1,27,54]. These data indicate that the continuous corn yield penalty can be ameliorated with agronomic management. Although the highest yields were consistently achieved in the corn-soybean rotation using intensive management and low planting densities, individual hybrids were found to respond differently to management. Select hybrids, for example, were able to nearly overcome the CCYP when grown with intensive management (Figure 1). The CCYP was reduced by 0.89 to 1.93 Mg ha^{-1} with intensive management for seven hybrids: 6265SS (34%), DKC58-87 (37%), 6640VT3P (38%), DKC64-87 (54%), 212-86STX (72%), 209-53STX (75%), and DKC63-55 (77%).

Table 5. Corn grain yield for hybrids as influenced by crop rotation, agronomic input level, and population. Hybrids were grown in continuous corn and following soybean rotations at Champaign, IL in 2014 and 2015. All values are reported at 0 g kg^{-1} moisture concentration.

	Crop Rotation [‡]							
	Continuous Corn				Corn-Soybean			
	Input Level [§]							
	Standard		Intensive		Standard		Intensive	
Year/	Plant Density (plant ha^{-1}) [¶]							
Hybrid [†]	79,000	111,000	79,000	111,000	79,000	111,000	79,000	111,000
	Mg ha^{-1}							
2014								
209-53STX	9.65	9.81	12.60	12.62	11.59	10.82	12.93	13.02
212-86STX	8.71	8.52	11.67	12.13	10.69	11.27	12.99	12.11
DKC58-87	8.44	8.60	11.20	11.23	10.75	11.16	12.38	13.13
DKC60-67	9.46	9.00	11.76	12.11	11.10	10.61	12.58	12.63
DKC62-08	8.00	8.10	9.84	9.27	10.32	9.59	11.86	11.50
DKC63-33	9.04	8.80	11.59	11.23	11.56	11.10	13.23	12.89
DKC63-55	8.75	8.55	11.53	11.04	10.48	10.05	11.90	11.39
DKC64-87	9.25	9.07	12.37	12.04	11.88	11.05	13.48	13.05
2014 Means	**8.92**	**8.81**	**11.57**	**11.46**	**11.04**	**10.71**	**12.67**	**12.47**
2015								
5415SS	9.46	8.05	11.42	11.40	10.74	10.88	13.03	12.97
5887VT3P	9.15	7.08	10.57	9.66	11.63	9.83	13.07	12.80
5975VT3P	8.24	7.97	11.23	10.44	11.44	10.16	13.49	12.94
6065SS	9.82	9.62	11.24	12.27	11.68	11.06	13.37	13.30
6110SS	10.01	9.09	11.68	12.19	12.05	12.58	13.47	13.53
6265SS	7.73	6.75	10.71	10.01	12.02	12.31	13.46	13.72
6594SS	7.85	8.58	11.10	11.69	12.08	10.48	13.84	13.74
6640VT3P	6.91	5.99	9.14	10.21	11.51	11.43	13.33	12.20
2015 Means	**8.65**	**7.89**	**10.89**	**10.98**	**11.64**	**11.09**	**13.38**	**13.15**

[†] Hybrid LSD ($p \leq 0.05$) = 0.35 Mg ha^{-1} in 2014 and 1.05 Mg ha^{-1} in 2015. [‡] Rotation LSD ($p \leq 0.05$) = 0.06 Mg ha^{-1} in 2014 and 0.81 Mg ha^{-1} in 2015. [§] Input level LSD ($p \leq 0.05$) = 0.14 Mg ha^{-1} in 2014 and 0.23 Mg ha^{-1} in 2015. [¶] Plant density LSD ($p \leq 0.05$) = 0.14 Mg ha^{-1} in 2014 and 0.23 Mg ha^{-1} in 2015.

Grain yield is derived from yield components (i.e., kernel number and individual kernel weight) that may be altered by changes in fertility, planting population, and germplasm [23,55,56]. The improved grain yields as a result of intensified agronomic management increased both kernel number and kernel weight (Table 6). Similarly, the consistently greater yields resulting from the corn-soybean rotation compared to the continuously grown corn yields were derived from a combination of increased kernel number and kernel weight.

Table 6. Grain yield components as influenced by crop rotation, hybrid, agronomic input level, and population at Champaign, IL in 2014 and 2015. All values are reported at 0 g kg^{-1} moisture concentration.

		Kernel Number				Kernel Weight			
					Input Level [§]				
		Standard		Intensive		Standard		Intensive	
Rotation [†]	Year/ Hybrid [‡]	Planting Density ($\times$ 1000 plants ha^{-1}) [¶]							
		79	111	79	111	79	111	79	111
		——— m^{-2} ———				——— mg kernel^{-1} ———			
Cont. Corn	**2014**								
	209-53STX	4394	4736	4779	5346	219	207	264	236
	212-86STX	3984	4169	4618	5200	218	204	253	233
	DKC58-87	4047	4307	4632	5032	208	199	241	223
	DKC60-67	4146	4455	4533	5031	228	202	259	241
	DKC62-08	3398	3852	3629	3866	235	210	271	240
	DKC63-33	4155	4800	4897	5512	219	183	237	204
	DKC63-55	4341	4433	4986	5331	201	193	231	207
	DKC64-87	4597	4704	5374	5388	201	192	230	226
	2014 Means	**4133**	**4432**	**4681**	**5088**	**216**	**199**	**248**	**226**
	2015								
	5415SS	4749	5102	4962	5485	199	171	230	207
	5887VT3P	4441	4522	4584	5015	206	177	230	209
	5975VT3P	4129	4670	4978	5252	199	169	226	198
	6065SS	4307	4741	4555	5378	228	202	247	228
	6110SS	4320	4649	4782	5603	230	195	244	217
	6265SS	3352	3197	4129	4120	230	209	258	241
	6594SS	4335	4750	5117	5923	184	180	217	196
	6640VT3P	3424	3226	4095	5203	199	183	220	195
	2015 Means	**4132**	**4357**	**4650**	**5247**	**209**	**186**	**234**	**212**
Corn-Soybean	**2014**								
	209-53STX	5282	4844	5101	5027	220	227	253	263
	212-86STX	4390	4612	4923	4460	243	246	264	273
	DKC58-87	4280	5022	4873	5549	251	222	254	237
	DKC60-67	4442	4375	4355	4583	250	244	289	277
	DKC62-08	4025	3752	4526	4203	257	259	263	274
	DKC63-33	5375	4784	5084	4936	215	234	260	261
	DKC63-55	4512	4514	4958	4731	232	222	240	241
	DKC64-87	5690	4912	5585	5598	209	227	244	234
	2014 Means	**4750**	**4602**	**4926**	**4886**	**235**	**235**	**258**	**257**
	2015								
	5415SS	5009	5410	5189	5738	214	201	251	227
	5887VT3P	4779	4567	4865	5566	243	215	269	230
	5975VT3P	4861	5078	4936	5851	236	199	273	221
	6065SS	4636	4721	4988	5499	252	234	268	242
	6110SS	4797	5858	5084	5729	251	213	265	236
	6265SS	4224	4558	4507	4799	284	270	298	286
	6594SS	5488	5316	5786	6251	220	197	239	220
	6640VT3P	4646	5222	5143	5275	247	219	258	230
	2015 Means	**4805**	**5091**	**5062**	**5589**	**243**	**218**	**265**	**236**

[†] Rotation kernel number LSD ($p \leq 0.05$) = nonsignificant (NS) in 2014 and 161 m^{-2} in 2015; Rotation kernel weight LSD ($p \leq 0.05$) = 9.0 mg kernel^{-1} in 2014 and 10.2 mg kernel^{-1} in 2015. [‡] Hybrid kernel number LSD ($p \leq 0.05$) = 198 m^{-2} in 2014 and 297 m^{-2} in 2015; Hybrid kernel weight LSD ($p \leq 0.05$) = 8.7 mg kernel^{-1} in 2014 and 11.2 mg kernel^{-1} in 2015. [§] Input level kernel number LSD ($p \leq 0.05$) = 98 m^{-2} in both 2014 and 2015; Input level kernel weight LSD ($p \leq 0.05$) = 4.4 mg kernel^{-1} in 2014 and 2.3 mg kernel^{-1} in 2015. [¶] Plant density kernel number LSD ($p \leq 0.05$) = 99 m^{-2} in both 2014 and 2015; Plant density kernel weight LSD ($p \leq 0.05$) = 4.4 mg kernel^{-1} in 2014 and 2.3 mg kernel^{-1} in 2015.

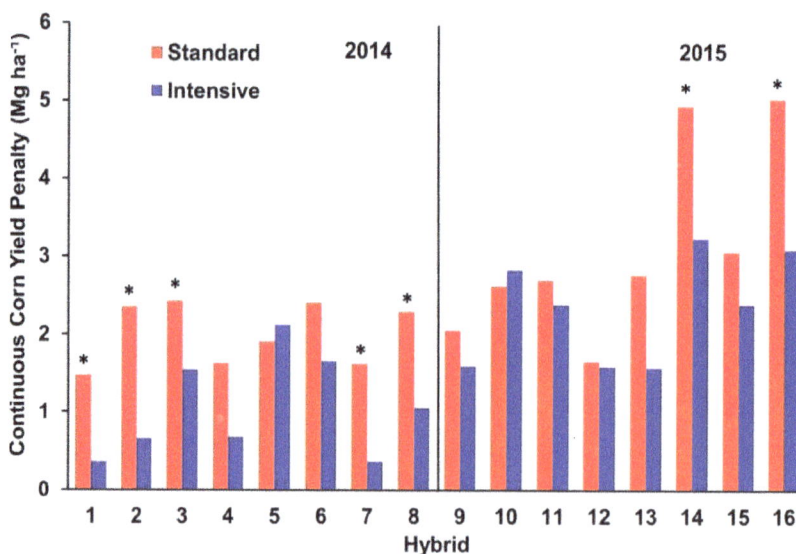

Figure 1. The yield penalty (yield difference between corn-soybean and continuous corn rotation) as influenced by two levels of management (standard vs. intensive) at Champaign, IL during 2014 (hybrids 1–8) and 2015 (hybrids 9–16). Hybrids 1–16 follow the order hybrids were presented in Tables 3–6. Values represent the average of two planting populations. * CCYP (continuous corn yield penalty) significantly different at $p \leq 0.05$, due to crop management for each hybrid.

When combined, the 40–60% greater yield response in continuous corn vs. corn-soybean rotation when grown with high input management was linked to a greater production in the amount and weight of those kernels. When plants in continuous corn were cultivated with intensive management, kernel weight was equivalent to that of the corn rotated with soybean managed with standard input levels. It has been previously documented that corn in rotation with soybean, regardless of if they were nodulated or non-nondulated, resulted in both larger and more numerous kernels compared to when grown continuously [57]. These results indicate that throughout much of the growing season corn in rotation was more successful at setting and maintaining yield potential than corn following corn. As early as the V5 (five leaf) growth stage, the number of kernel rows is determined, followed by spikelet pairs that produce kernels at V6, with the number of ovules (potential kernels) and the size of the ear set at V12 (12 leaves) [44]. Kernel number can be altered by the quality of pollination or through kernel abortion in response to any stress from environmental conditions or plant competition [58]. Later in the season, the size of the individual kernels is set (R2) followed by the expansion and filling of those kernels with starch [44]. Rotation of corn with soybean increased the grain-filling period or rate of grain-filling that resulted in heavier kernels. Part of this response can be attributed to the additional N availability in rotated corn compared to corn on corn [9,49], which influences both the production and size of kernels [59].

Increased planting populations resulted in minimal yield reductions regardless of the previous crop (Table 5). Under high input management, the yield penalty from the continuous corn rotation was not magnified with the higher planting density. Regardless of rotation, the increased kernel numbers produced per area from higher planting densities was offset by lesser kernel weights (Table 6). These compensatory patterns resulted in no overall yield advantage from the increased planting population. Kernel number produced per area was greater at the plant population that resulted in more grain yield. While individual kernel number per plant has been found to be reduced as plant population was increased, there was, however, a greater kernel number per unit area produced as a result of more harvestable ears at the higher plant population [60].

3.4. Correlations between Crop Growth and Final Grain Yield

Early season plant growth assessments at the V6 growth stage, had a stronger positive correlation to final grain yield in the corn-soybean rotation than in continuous corn (Table 7). Leaf greenness at the R2 growth stage was strongly positively correlated to final grain yield in continuous corn. Similar to previous findings, leaf greenness had this stronger correlation to grain yield when assessed in continuous corn compared to corn in rotation with soybean [61]. Kernel number had a strong to very strong positive correlation to grain yield in the continuous corn plots. Setting the highest potential kernels and decreasing kernel abortion is essential in maintaining and improving grain yield [43]. When corn was rotated with soybean, kernel weight was moderately correlated to grain yield and the correlation was strong when grown continuously. Harvest index, the ratio of grain to total aboveground biomass, was strongly correlated to grain yield in continuous corn. Overall, these correlations show the importance of interactions within the crop throughout the growing season to maintain grain yield potential; with kernel number being determined earlier in the growing season and kernel weight later in crop development.

Table 7. Pearson correlation coefficients and associated significance level for final grain yield between selected corn growth parameters as influenced by crop rotation and averaged across all other treatments.

Corn Parameter	2014		2015	
	CC [†]	CS	CC	CS
V6 Biomass	0.69 ***	0.76 ***	0.42 ***	0.57 ***
R2 SPAD	0.70 ***	0.12	0.72 ***	0.09
Harvest Index	0.64 ***	0.46 ***	0.65 ***	0.36 ***
Kernel Number	0.76 ***	0.59 ***	0.84 ***	0.49 ***
Kernel Weight	0.55 ***	0.24 *	0.62 ***	0.56 ***

*** Significant at the 0.001 probability level. * Significant at the 0.05 probability level. [†] CC, Continuous Corn; CS, Corn-Soybean Rotation.

4. Conclusions

In central Illinois, cropping rotation, hybrid selection, agronomic management, and plant population all significantly influenced the measured parameters in corn, with numerous interactions. The highest yields of this study were achieved in the corn-soybean rotation grown with intensive management and at the standard planting density. The data presented here suggest that the CCYP can be mitigated with intensified management. Without enhanced fertility (i.e., standard management) continuous corn production yielded significantly less grain than corn grown following soybean. Intensive agronomic management increased grain yield by enhancing both kernel number and kernel weight. Through growth responses both pre-and post-pollination, there was a 40–60% greater yield response to intensive management in continuous corn compared to the corn-soybean rotation. As a result of certain genetic predispositions, corn germplasm varied in growth and yield response and magnitude of responses to rotation, input level, and population, emphasizing the importance of hybrid selection in continuous corn acres. When population was increased, continuous corn grain yields were maintained when treated with the high input level. Improvement in crop health (i.e., leaf greenness and biomass accumulation) and productivity was made using both crop rotation and intensive management. Enhanced fertility and leaf protection (i.e., intensive management level) in combination with select hybrids resulted in a multifaceted approach to reduce the CCYP and increase yields.

Author Contributions: Performed experiment, A.M.V.; formal analysis, A.M.V.; writing—original draft, A.M.V.; writing—review and editing, F.E.B.; project administration, F.E.B.

Funding: This research was made possible with partial funding from the National Institute of Food and Agriculture project NC1200 "Regulation of Photosynthetic Processes" and the Illinois AES project 802-908. We also greatly appreciate the support from Crop Sciences, a division of Bayer, Winfield United, and the Mosaic Company.

Acknowledgments: We would like to thank the Crop Physiology Lab personnel for field and data collection, and especially Juliann Seebauer for manuscript preparation.

Conflicts of Interest: The authors declare no conflict of interest.

References

1. Peterson, T.A.; Varvel, G.E. Crop yield as affected by rotation and nitrogen rate. III. Corn. *Agron. J.* **1989**, *81*, 735–738. [CrossRef]
2. Meese, B.G.; Carter, P.R.; Oplinger, E.S.; Pendleton, J.W. Corn/soybean rotation effect as influenced by tillage, nitrogen, and hybrid/cultivar. *J. Prod. Agric.* **1991**, *4*, 74–80. [CrossRef]
3. Porter, P.M.; Lauer, J.G.; Lueschen, W.E.; Ford, J.H.; Hoverstad, T.R.; Oplinger, E.S.; Crookston, R.K. Environment affects the corn and soybean rotation effects. *Agron. J.* **1997**, *89*, 442–448. [CrossRef]
4. Howard, D.D.; Chambers, A.Y.; Lessman, G.M. Rotation and fertilization effects on corn and soybean yields and soybean cyst nematode populations in a no-tillage system. *Agron. J.* **1998**, *90*, 518–522. [CrossRef]
5. Katsvairo, T.W.; Cox, W.J. Tillage x rotation x management interactions in corn. *Agron. J.* **2000**, *92*, 493–500. [CrossRef]
6. Pedersen, P.; Lauer, J.G. Corn and soybean response to rotation sequence, row spacing, and tillage system. *Agron. J.* **2003**, *95*, 965–971. [CrossRef]
7. Pikul, J.L., Jr.; Hammack, L.; Riedell, W.E. Corn yield, N use and corn rootworm infestation of rotations in the northern Corn Belt. *Agron. J.* **2005**, *97*, 854–863. [CrossRef]
8. Stanger, T.F.; Lauer, J.; Chavas, J. Long-term cropping systems: The profitability and risk of cropping systems featuring different rotations and nitrogen rates. *Agron. J.* **2008**, *100*, 105–113. [CrossRef]
9. Gentry, L.F.; Ruffo, M.L.; Below, F.E. Identifying factors controlling the continuous corn yield penalty. *Agron. J.* **2013**, *105*, 295–303. [CrossRef]
10. Al-Kaisi, M.M.; Archontoulis, S.V.; Kwaw-Mensah, D.; Miguez, F. Tillage and crop rotation effects on corn agronomic response and economic return at seven Iowa locations. *Agron. J.* **2015**, *107*, 1411–1424. [CrossRef]
11. Varvel, G.E. Monoculture and rotation system effects on precipitation use efficiency in corn. *Agron. J.* **1994**, *86*, 204–208. [CrossRef]
12. Wilhelm, W.W.; Wortmann, C.S. Tillage and rotation interactions for corn and soybean grain yield as aff ected by precipitation and air temperature. *Agron. J.* **2004**, *96*, 425–432. [CrossRef]
13. Raimbault, B.A.; Vyn, T.J. Crop rotation and tillage effects on corn growth and soil structural stability. *Agron. J.* **1991**, *83*, 979–985. [CrossRef]
14. Seifert, C.A.; Roberts, M.J.; Lobell, D.B. Continuous corn and soybean yield penalties across hundreds of thousands of fields. *Agron. J.* **2017**, *109*, 541–548. [CrossRef]
15. Carlone, M.R.; Russell, W.A. Response to plant densities and nitrogen levels for four maize cultivars from different eras of breeding. *Crop Sci.* **1987**, *27*, 465–470. [CrossRef]
16. Sangoi, L.; Gracietti, M.A.; Rampazzo, C.; Bianchetti, P. Response of Brazilian maize hybrids from different eras to changes in plant population. *Field Crops Res.* **2002**, *79*, 39–51. [CrossRef]
17. Tollenaar, M.; Lee, E.A. Yield potential, yield stability and stress tolerance in maize. *Field Crops Res.* **2002**, *75*, 161–169. [CrossRef]
18. Tokatlidis, I.S.; Koutroubas, S.D. A review of maize hybrids' dependence on high plant populations and its implications for crop yield stability. *Field Crops Res.* **2004**, *88*, 103–114. [CrossRef]
19. Duvick, D.N. The contribution of breeding to yield advances in maize (*Zea mays* L.). *Adv. Agron.* **2005**, *86*, 83–145.
20. Hammer, G.L.; Dong, Z.; McLean, G.; Doherty, A.; Messina, C.; Schussler, J.; Zinselmeier, C.; Paszkiewicz, S.; Cooper, M. Can changes in canopy and/or root system architecture explain historical maize yield trends in the U.S. Corn Belt? *Crop Sci.* **2009**, *49*, 299–312. [CrossRef]
21. Mastrodomenico, A.T.; Haegele, J.W.; Seebauer, J.R.; Below, F.E. Yield stability differs in commercial maize hybrids in response to changes in plant density, nitrogen fertility, and environment. *Crop Sci.* **2018**, *58*, 230–241. [CrossRef]

22. Tollenaar, M. Physiological basis of genetic improvement of maize hybrids in Ontario from 1959 to 1988. *Crop Sci.* **1991**, *31*, 119–124. [CrossRef]

23. Boomsma, C.R.; Santini, J.B.; Tollenaar, M.; Vyn, T.J. Maize morphological responses to intense crowding and low nitrogen availability: An analysis and review. *Agron. J.* **2009**, *101*, 1426–1452. [CrossRef]

24. Clay, S.A.; Clay, D.E.; Horvath, D.P.; Pullis, J.; Carlson, C.G.; Hansen, S.; Reicks, G. Corn response to competition: Growth alteration vs. yield limiting factors. *Agron. J.* **2009**, *101*, 1522–1529. [CrossRef]

25. Kuchenbuch, R.O.; Gerke, H.H.; Buczko, U. Spatial distribution of maize roots by complete 3D soil monolith sampling. *Plant Soil* **2009**, *315*, 297–314. [CrossRef]

26. Bender, R.R.; Haegele, J.W.; Ruffo, M.L.; Below, F.E. Nutrient uptake, partitioning, and remobilization in modern, transgenic insect-protected maize hybrids. *Agron. J.* **2013**, *105*, 161–170. [CrossRef]

27. Ciampitti, I.A.; Vyn, T.J. Physiological perspectives of changes over time in maize yield dependency on nitrogen uptake and associated nitrogen efficiencies: A review. *Field Crops Res.* **2012**, *133*, 48–67. [CrossRef]

28. Kovar, L.K.; Claasen, N. Soil–root interactions and phosphorus nutrition of plants. In *Phosphorus: Agriculture and the Environment*; Sims, J.T., Sharpley, A.N., Eds.; Agronomy Monograph Series No. 46; ASA, CSSA, SSSA: Madison, WI, USA, 2005; pp. 379–414.

29. David, M.B.; Gentry, L.E.; Mitchell, C.A. Riverine response of sulfate to declining atmospheric sulfur deposition in agricultural watersheds. *J. Environ. Qual.* **2016**, *45*, 1313–1319. [CrossRef] [PubMed]

30. Camberato, J.; Casteel, S. Sulfur Deficiency. Purdue Univ. Dep. of Agronomy, Soil Fertility Update. 2017. Available online: http://www.agry.purdue.edu/ext/corn/news/timeless/sulfurdeficiency.pdf (accessed on 16 July 2018).

31. Bell, R.W.; Dell, B. *Micronutrients in Sustainable Food, Feed, Fibre and Bioenergy Production*; International Fertilizer Industry Association: Paris, France, 2008; ISBN 978-2-9523139-3-3.

32. Alloway, B.J. Soil factors associated with zinc deficiency in crops and humans. *Environ. Geochem. Health* **2009**, *31*, 537–548. [CrossRef] [PubMed]

33. Karlen, D.L.; Flannery, R.L.; Sadler, E.J. Aerial accumulation and partitioning of nutrients by corn. *Agron. J.* **1988**, *80*, 232–242. [CrossRef]

34. Berger, K.C.; Heikkinen, T.; Zube, E. Boron deficiency, a cause of blank stalks and barren ears in corn. *Soil Sci. Soc. Am. J.* **1957**, *21*, 629–632. [CrossRef]

35. Jeschke, M.; Doerge, T. Management of foliar diseases in corn with fungicides. *Crop Insights* **2010**, *18*, 1–4.

36. Bartlett, D.W.; Clough, J.M.; Godwin, J.R.; Hall, A.A.; Hamer, M.; Parr-Dobrzanski, B. The strobilurin fungicides. *Pest Manag. Sci.* **2002**, *58*, 649–662. [CrossRef] [PubMed]

37. Grossmann, K.; Retzlaff, G. Bioregulatory effects of the fungicidal strobilurin kresoxim-methyl in wheat (*Triticum aestivum*). *Pestic. Sci.* **1997**, *50*, 11–20. [CrossRef]

38. Wise, K.; Mueller, D. Are fungicides no longer just for fungi? An analysis of foliar fungicide use in corn. *Am. Phytopathol. Soc.* **2011**. [CrossRef]

39. Bockus, W.W.; Shroyer, J.P. The impact of reduced tillage on soilborne plant pathogens. *Annu. Rev. Phytopathol.* **1998**, *36*, 485–500. [CrossRef] [PubMed]

40. Shapiro, C.A.; Wortmann, C.S. Corn response to nitrogen rate, row spacing, and plant density in eastern Nebraska. *Agron. J.* **2006**, *98*, 529–535. [CrossRef]

41. Graham, R.L.; Nelson, R.; Sheehan, J.; Perlack, R.D.; Wright, L.L. Current and potential U.S. corn stover supplies. *Agron. J.* **2007**, *99*, 1–11. [CrossRef]

42. Riedell, W.E.; Schumacher, T.E.; Clay, S.A.; Ellsbury, M.M.; Pravecek, M.; Evenson, P.D. Corn and soil fertility responses to crop rotation with low, medium, or high inputs. *Crop Sci.* **1998**, *38*, 427–433. [CrossRef]

43. Ruffo, M.L.; Gentry, L.F.; Henninger, A.S.; Seebauer, J.R.; Below, F.E. Evaluating management factor contributions to reduce corn yield gaps. *Agron. J.* **2015**, *107*, 495–505. [CrossRef]

44. Abendroth, L.J.; Elmore, R.W.; Boyer, M.J.; Marlay, S.K. *Corn Growth and Development*; PMR 1009; Iowa State University: Ames, IA, USA, 2011.

45. Daughtry, C.S.; Walthall, C.L.; Kim, M.S.; De Colstoun, E.B.; McMurtrey, J.E., III. Estimating corn leaf chlorophyll concentration from leaf and canopy reflectance. *Remote Sens. Environ.* **2000**, *74*, 229–239. [CrossRef]

46. *Statistical Analysis System (SAS) Version SAS/STAT 9.4*; SAS Institute Inc.: Cary, NC, USA, 2012.

47. Riedell, W.E.; Pikul, J.L.; Jaradat, A.A.; Schumacher, T.E. Crop rotation and nitrogen input effects on soil fertility, maize mineral nutrition, yield, and seed composition. *Agron. J.* **2009**, *101*, 870–879. [CrossRef]

48. Crookston, R.K.; Kurle, J.E.; Copeland, P.J.; Ford, J.H.; Jueschen, W.E. Rotational cropping sequence affects yield of corn and soybean. *Agron. J.* **1991**, *83*, 108–113. [CrossRef]

49. Ennin, S.A.; Clegg, M.D. Effect of soybean plant population in a soybean and maize rotation. *Agron. J.* **2001**, *93*, 396–403. [CrossRef]

50. Filella, I.; Serrano, L.; Serra, J.; Penuelas, J. Evaluating wheat nitrogen status with canopy reflectance indices and discriminant analysis. *Crop Sci.* **1995**, *35*, 1400–1405. [CrossRef]

51. Moran, J.A.; Mitchell, A.K.; Goodmanson, G.; Stockburger, K.A. Differentiation among effects of nitrogen fertilization treatments on conifer seedlings by foliar reflectance: A comparison of methods. *Tree Physiol.* **2000**, *20*, 1113–1120. [CrossRef] [PubMed]

52. Hatfield, J.L.; Gitelson, A.A.; Schepers, J.S.; Walthall, C.L. Application of spectral remote sensing for agronomic decisions. *Agron. J.* **2008**, *100*, S117–S131. [CrossRef]

53. Sindelar, A.J.; Schmer, M.R.; Jin, V.L.; Wienhold, B.J.; Varvel, G.E. Long-term corn and soybean response to crop rotation and tillage. *Agron. J.* **2015**, *107*, 2241–2252. [CrossRef]

54. Varvel, G.E.; Wilhelm, W.W. Soybean nitrogen contribution to corn and sorghum in western Corn Belt rotations. *Agron. J.* **2003**, *95*, 1220–1225. [CrossRef]

55. D'Andrea, K.E.; Otegui, M.E.; Cirilo, A.G. Kernel number determination differs among maize hybrids in response to nitrogen. *Field Crops Res.* **2008**, *105*, 228–239. [CrossRef]

56. Cox, W.J. Whole-plant physiological and yield responses of maize to plant density. *Agron. J.* **1996**, *88*, 489–496. [CrossRef]

57. Bergerou, J.A.; Gentry, L.E.; David, M.B.; Below, F.E. Role of N 2 fixation in the soybean N credit in maize production. *Plant Soil* **2004**, *262*, 383–394. [CrossRef]

58. Ritchie, S.W.; Hanway, J.J.; Benson, G.O. *How a Corn Plant Develops*; Special Report No. 48; Iowa State Univ.: Ames, IA, USA, 1986.

59. Pearson, C.J.; Jacobs, B.C. Yield components and nitrogen partitioning of maize in response to nitrogen before and after anthesis. *Aust. J. Agric. Res.* **1987**, *38*, 1001–1009. [CrossRef]

60. Hashemi-Dezfouli, A.; Herbert, S.J. Intensifying plant density response of corn with artificial shade. *Agron. J.* **1992**, *84*, 547–551. [CrossRef]

61. Attia, A.; Shapiro, C.; Kranz, W.; Mamo, M.; Mainz, M. Improved yield and nitrogen use efficiency of corn following soybean in irrigated sandy loams. *Soil Sci. Soc. Am. J.* **2015**, *79*, 1693–1703. [CrossRef]

![agronomy logo] *agronomy*

MDPI

Article

Agronomic Comparisons of Conventional and Organic Maize during the Transition to an Organic Cropping System

William J. Cox * [ORCID] and **Jerome H. Cherney**

School of Integrated Plant Sciences, Unit of Soil and Crop Sciences, Cornell University, Ithaca, NY 14853, USA; jhc5@cornell.edu
* Correspondence: wjc3@cornell.edu; Tel.: +1-607-255-1758

Received: 22 May 2018; Accepted: 2 July 2018; Published: 5 July 2018

Abstract: Maize producers transitioning to an organic cropping system must grow crops organically without price premiums for 36 months before certification. We evaluated conventional and organic maize with recommended and high seeding and N rates in New York to identify the best organic management practices during the transition. Conventional versus organic maize management differences included a treated (fungicide/insecticide) Genetically Modified (GM) hybrid versus a non-treated non-GM isoline; side-dressed synthetic N versus pre-plow composted manure; and Glyphosate versus mechanical weed control, respectively. Organic versus conventional maize yielded 32% lower as the entry crop (no previous green manure crop). Grain N% and weed densities explained 72% of yield variability. Organic and conventional maize, following wheat/red clover in the second year, yielded similarly. Organic maize with high inputs following wheat/red clover and conventional maize with high inputs following soybean in the third year yielded the highest. Grain N% and maize densities explained 54% of yield variability. Grain crop producers in the Northeast USA who do not have on-farm manure and forage equipment should plant maize after wheat/red clover with additional N (~56 kg N/ha) at higher seeding rates (~7%) during the transition to insure adequate N status and to offset maize density reductions from mechanical weed control.

Keywords: organic cropping system; maize; maize densities; weed densities; grain N%; yield components

1. Introduction

Recent downward trends in crop prices have prompted some cash crop producers, who practice maize (*Zea mays* L.)-soybean {*Glycine max* (L.) Merr.} or maize-soybean-wheat/red clover (*Triticum aestivum* L./*Trifolium pratense* L.) rotations, to contemplate transitioning from a conventional to an organic cropping system. The United States Department of Agriculture (USDA), however, requires a 36-month transition period that prohibits the use of GM crops, synthetic fertilizer, pesticides, and so on before a field can be certified as organic and eligible for the organic price premium [1]. Furthermore, comprehensive survey data indicate that organic maize, despite higher profits because of the price premium, had lower yields and higher per-hectare production costs when compared with conventional maize [2]. Consequently, a major deterrent for potential organic crop producers is a loss in profit during the transition because of higher production costs, lower yields, and the absence of a price premium. Organic maize has proved particularly challenging during the transition because of its high N requirement and marginal competitiveness with weeds [3,4]. The identification of best management practices for organic maize could help grain crop producers minimize yield and profit losses during the transition period.

Organic compared with conventional maize yielded 34% lower during the transition years in a maize-soybean rotation in a Minnesota study established in 2002 [3]. Organic maize yielded lower mostly because of the lack of available soil N, associated with low N content of the solid dairy manure applied to organic maize. In another Minnesota study, organic compared with conventional maize in a maize-soybean rotation yielded 24% lower from 1993–2007 [5]. In the same study, however, organic maize in a four-year oat/alfalfa-alfalfa-maize-soybean rotation compared with conventional maize in a maize-soybean rotation yielded ~8% lower during the transition years [6], but similarly from 1993–2007 [5]. Both authors concluded that with a diversified rotation, organic compared with conventional maize can have comparable yields. A study in Iowa confirmed this conclusion as organic maize in a more diversified maize-soybean-oat/alfalfa-alfalfa rotation yielded similarly compared with conventional maize in a maize-soybean rotation during the transition period [7] and in the second phase of the study [8].

A meta-analysis study indicated that organic crop yields are low in the first years after conversion and gradually increase over time, owing to improvements in soil fertility and management skills [9]. In a cropping system study established in Maryland, however, organic maize in a maize-soybean-wheat/hairy vetch (*Vicia velossa*) rotation yielded 28% lower compared with conventional no-till (NT) maize in a maize-soybean-wheat/soybean rotation during the transition years, and 40% lower after the transition, mostly because of low soil N availability [10]. Also, in a long-term Wisconsin study, conventional maize in a NT maize-soybean rotation had a ~150 kg/ha/year yield trend compared with only a ~100 kg/ha/year yield trend for organic maize in the organic maize-soybean-wheat rotation [11]. The difference in yield trends was attributed to either technology advances in the conventional cropping system and/or increased weed competition in the organic cropping system [12]. Another meta-analysis study indicated that organic compared with conventional maize yields were typically ~25% lower [13]. Furthermore, the Agricultural Resource Management Survey (ARMS) data for maize (794 conventional and 451 organic farms) in 2010 reported that organic maize in diversified rotations compared with conventional maize yielded 27% lower [2]. The use of diversified rotations thus may not eliminate the yield gap between organic and conventional maize.

A major deterrent to adoption of organic crop production is the uncertainty associated with selection of the best entry crop and subsequent rotation during the 36-month transition period during which organic premiums do not exist [3]. Another deterrent is that novice organic crop producers are uncertain of the best organic management practices to use during the transition and beyond [4]. Two objectives of this study are as follows: (1) to compare organic and conventional maize in different sequences of the maize-soybean-wheat/red clover rotation to identify the best year to plant maize during the transition period and (2) to evaluate recommended and high input management practices (high seeding and high N rates) to determine if high input management increases weed competitiveness and improves soil N availability for organic maize.

2. Materials and Methods

We initiated a cropping system study at a Cornell University research farm near Aurora, New York, (42°44′ N, 76°40′ W) in 2015 to evaluate three sequences of the maize-soybean-wheat/red clover rotation. Three contiguous experimental fields (220 m × 40 m) with similar tile-drained silt loam soil (fine-loamy, mixed, mesic, Glossoboric Hapludalfs) but different previous crops in 2014 (spring barley, maize, and soybean) were used in the study. The experimental design is a split-split plot (four replications) with previous crops as whole plots, cropping systems (conventional and organic) as sub-plots, and management inputs (recommended and high inputs) as sub-sub plots. The entire 40 m lengths were planted to maize, soybean or winter wheat in each field, but plot length was shortened to 30 m to allow for 5 m borders on the north and south sides of the plots. Also, 3 m borders were inserted between sub-plots (cropping systems) to minimize spray drift or fertilizer movement from conventional into organic plots. Likewise, 3 m border plots were inserted between each sub-subplot to minimize border effects from each crop, which differed in height. Whole plot dimensions were 216 m wide and 30 m long, sub-plot dimensions were 27 m wide and 30 m long, and sub-subplot dimensions were 3 m wide and 30 m long.

Winter wheat was not planted in the fall of 2013 before the onset of the study so red clover was not inter-seeded in the spring of 2014. Instead, red clover was seeded in mid-July of 2015 into bare soil to ensure a green manure crop for the 2016 maize crop. In addition, soybean developed green stem and did not shed all its leaves in the fall of 2016 delaying harvest until 9 November, which is too late to plant winter wheat in this environment. Consequently, maize in 2017 followed the intended wheat crop (planted after soybean harvest in the fall of 2015, inter-seeded with red clover in March of 2016, and harvested in July of 2016) as well as an unintended soybean crop. Our three sequences from 2015 to 2017 thus included red clover-maize-soybean, soybean-wheat/red clover-maize, and maize-soybean-maize. This paper will focus exclusively on maize in each year.

The fields were moldboard plowed from 16–19 May in all three years, followed by secondary tillage the following day. Maize was planted in 0.76 m row spacing immediately after secondary tillage in all three years. The maize planting date, which was delayed so some early-season weeds could emerge before plowing in the organic cropping system, remained within the optimum planting date range (25 April–20 May) at this site [14]. We used different rates of composted poultry manure (5-4-3 N, P, K analysis, respectively), depending upon the year and previous crops, as an N source for organic maize. The composted manure was applied one day before plowing. We estimated that 50% of the N from the composted poultry manure would be mineralized and available to organic maize.

Table 1 lists the management inputs for maize for the 3 years. Major differences between conventional and organic maize include (a) a treated (insecticide/fungicide seed treatment) GM hybrid versus the non-treated, non-GM isoline, (b) starter fertilizer of 10-20-20 (N, P, K analysis) versus composted manure (5-4-3), (c) injected-side-dressed liquid N (32-0-0 N, P, K analysis) versus composted poultry manure applied pre-plow as the N source, and (d) Glyphosate application versus mechanical weed control, respectively. Seeding rates of ~73,110 kernels/ha were used in recommended input and ~87,810/ha in high input management of both cropping systems. Nitrogen rates in the recommended and high input management varied according to previous crops and years (Table 1). We selected a non-GM isoline for organic maize instead of an organically developed and produced hybrid so we could determine how management practices (and not hybrid selection) affected yield and yield components.

Table 1. Planting rate, seed treatment, hybrid, starter fertilizer, N fertilizer, and weed control practices for conventional and organic maize with recommended (Rec.) and high input management at Aurora, New York in 2015, 2016, and 2017.

Practices	2015		2016		2017	
	Rec.	High	Rec.	High	Rec.	High
			Conventional			
Planting rate (seeds/ha)	73,110	87,810	73,110	87,810	73,110	87,810
Seed Treatment			Fungicide/insecticide			
Hybrid			P9675AMXT			
Starter Fert. (kg/ha)			305 kg/ha (10-20-20, N, P, K analysis)			
N fertilizer-side-dress (kg N/ha)	135 kg N/ha (liquid)	180 kg N/ha (liquid)	None	56 kg N/ha (liquid)	56 kg N/ha (following wheat/RC) and 111 kg N/ha (following soybean)	111 kg N/ha (following wheat/RC) and 155 kg N/ha (following soybean)
Weed Control			Glyphosate (Single Post-application)			
			Organic			
Planting rate (kernels/acre)	73,110	87,810	73,110	87,810	73,110	87,810
Seed Treatment			None			
Hybrid			P9675			
Starter Fertilizer			365 kg/ha composted poultry manure (5-4-3)			
Pre-plant N fertilizer (kg N/ha)	135 kg N/ha composted manure	180 kg N/ha composted manure	None	56 kg N/ha composted manure	56 kg N/ha (following wheat/red clover and 111 kg N/ha (following soybean) composted manure	111 kg N/ha (following wheat/red clover and 155 kg N/ha (following soybean) composted manure
Weed Control			Rotary hoe + close cultivation + in-row cultivations (3×)			

Red clover biomass was estimated a few days before plowing in 2016 and 2017 by sampling three regions of each sub-subplot with a quadrat (0.8 m^2). The samples were oven-dried for three days at 60 °C, ground, and then analyzed for total N by combustion (LECO CN628 Nitrogen Analyzer, LECO Corporation, St. Joseph, MI, USA). Maize densities were taken immediately before rotary hoeing (~1–2 days after 90% emergence) and again at the ninth leaf stage (V9 stage, [15]), after the completion of mechanical weed control practices, by counting all the plants along the 30 m plot length of the two harvest rows. The first maize density measurement was taken to determine if the treated GM maize hybrid and non-treated non-GM maize isoline differed in emergence rates and plant establishment. The second measurement was taken to determine the extent of maize damage by mechanical weed control practices (rotary hoeing, a close cultivation, and three in-row cultivations) in organic maize. Weed densities were also determined by counting all the weeds taller than 5 cm in height along the 30 m length of the two harvest rows at the V14 stage, the end of the critical weed-free period in maize in this environment [16]. Predominant weed species, which did not differ among previous crops or between cropping systems, included *Polygonum convovulus* L., *Chenopodium album* L., *Echinochloa crus-galli* (L.) Beauv., *Polygonum pensylvanicum* L., *Setaria vidis* L., *Ambrosia artemisiifolia* L., and *Amaranthus retroflexus* L.

Yield components were determined a few days before harvest by hand-harvesting all the plants in a 1 m length of the two harvest rows every 10 m along the 30 m-length of the sub-subplot for a total of three sampling regions or 25–35 plants. Whole plants were air-dried in a greenhouse for a few weeks; counted (reported as plants/m^2) and weighed; ears were removed and counted; kernels were hand-threshed and counted with a seed counter (Old Mill Co., Savage, MD, USA); kernels were weighed; kernels were then ground and brought to the lab to determine grain N concentrations by combustion (LECO CN628 Nitrogen Analyzer, LECO Corporation, St. Joseph, MI, USA). Total kernel weight was divided by total kernel number (3000–20,000) to determine individual kernel weight; and divided by total plant weight to determine harvest index (HI) values.

The three 10 m lengths in each sub-subplot were harvested with a small plot Almaco combine (Nevada, IA, USA) in late October or early November in each year when grain moistures were ~18%. The three yields in each sub-subplot were then pooled and averaged. An approximate 1000 g sample was collected from each sub-subplot to determine grain moisture. Yields were adjusted to 15.5% moisture. Grain moisture differences were less than 1% between cropping systems, and thus will not be reported.

Maize had different previous crops in 2015 (small grain, maize, and soybean) compared with 2016 (red clover) and 2017 (wheat/red clover and soybean), which resulted in different N application rates across years and within a year (2017). Consequently, we analyzed each year separately. Previous crop (2014 crops), cropping systems (conventional and organic), and management inputs (recommended and high) were considered fixed and replications random for statistical analyses for individual years using the REML function in the MIXED procedure of SAS (version 9.4; SAS Institute Inc., Cary, NC, USA). Previous crops showed significance for yield, grain N%, and kernel weight in 2015 (higher in the field following soybean compared with maize), but did not have significant two-way or three-way interactions in any of the years (Table 2). Consequently, the data will be pooled across previous crops (the three contiguous fields) for each year. Least square means of the main effects (cropping system and management inputs) were computed and means separations were performed on significant effects using Tukey's studentized range test (HSD) test, with statistical significance set at $p < 0.05$. Differences among least square means for cropping system interactions were calculated also using Tukey's HSD test. Two-way interactions (cropping system by management inputs) were detected for some variables so the interaction comparisons will be presented. Simple correlations (Pearson) among all measurements within each year were calculated using CORR in SAS. Also, the PROC STEPWISE REG SAS procedure was used to build statistical models to explain yield variability using data from the entire plot (maize densities, weed densities, and grain N% concentrations) or from the sampling area (plants/m^2, ears/plant, kernels/ear, kernel weight, and HI in each year and across years).

Table 2. Significance for grain yield, maize densities before (DEN1) and after (DEN2) rotary hoeing and cultivating operations, weed density, grain N% concentration, plants/m², ears/plant, kernels/ear (Kern./ear), kernel weight (Kwt.), and harvest index (HI) in 2015, 2016, and 2017 at Aurora, New York.

Variable	Yield	DEN1	DEN2	Weeds	Grain N	Plants/m²	Ears/Plant	Kern./Ear	Kwt.	HI
					2015					
Previous Crop	* +	NS	NS	NS	***	NS	NS	NS	*	NS
Cropping System	***	NS	***	***	***	**	*	**	*	*
PC × CS	NS ++	NS	NS	NS	NS	NS	NS	NS	NS	NS
Inputs	*	NS	***	NS	NS	***	NS	**	***	NS
PC × I	NS	NS	NS	NS	NS	NS	NS	NS	NS	NS
CS × I	NS	NS	NS	NS	NS	NS	NS	NS	NS	**
PC × CS × I	NS	NS	NS	NS	NS	NS	NS	NS	NS	NS
					2016					
Previous Crop	NS	*	*	NS	NS	NS	NS	NS	NS	NS
Cropping System	NS	NS	***	**	*	*	*	NS	NS	NS
PC × CS	NS	***	NS	NS	NS	NS	NS	NS	NS	NS
Inputs	NS	***	***	*	***	*	NS	**	NS	NS
PC × I	NS	NS	NS	NS	NS	NS	NS	NS	NS	NS
CS × I	NS	NS	NS	NS	NS	NS	NS	*	*	NS
PC × CS × I	NS	NS	NS	NS	NS	NS	NS	NS	NS	NS
					2017					
Previous Crop	NS	*	*	NS	NS	NS	NS	NS	NS	*
Cropping System	***	NS	***	**	*	**	NS	NS	***	**
PC × CS	NS	*	NS	NS	NS	NS	NS	NS	NS	NS
Inputs	***	***	***	**	***	***	NS	NS	NS	NS
PC × I	NS	NS	NS	NS	NS	NS	NS	NS	NS	NS
CS × I	*	NS	NS	*	***	*	NS	NS	**	NS
PC × CS × I	NS	NS	NS	NS	NS	NS	NS	NS	NS	NS

+ * = significant at 0.05, ** at 0.01, *** at 0.001, ++ NS = not significant at 0.05.

3. Results

3.1. 2015

The 2015 growing season had the second wettest 1 May through 30 June period (Table 3) on record at the experimental site (61 years of records, http://climod.nrcc.cornell.edu/climod/rank/). Conditions became exceedingly dry for the remainder of the growing season as the 2015 growing season had the fourth driest 1 July through 9 September period at the site (http://climod.nrcc.cornell.edu/climod/rank/). Late spring and early summer conditions were cool, especially during June and July, so maize did not attain the silking stage until ~25 July. Maize experienced some drought stress, as indicated by premature leaf senescence, from the early grain-filling stage (~15 August) until physiological maturity (~10 September).

Table 3. Monthly and total precipitation and growing degree days (30–10 °C system) at Aurora, New York from 1 May through August during the 2015, 2016, and 2017 growing seasons.

Month	Precipitation			Growing Degree Days		
	2015	2016	2017	2015	2016	2017
	mm			°C		
May	141	63	133	255	262	261
June	201	28	97	244	268	267
July	80	48	186	328	374	341
August	35	116	38	307	396	305
Total	457	255	454	1134	1300	1174

Cropping system and input management significantly affected yield, and there was no cropping system by management input interaction (Table 2). Organic compared with conventional maize yielded 32% lower, when averaged across management inputs (Table 4). The yield data agree with a previous study that had 34% lower organic maize yields during the first transition year when no green manure crop was in place and solid manure was the primary N source [3]. When averaged across cropping systems, high input compared with recommended input management yielded 3.5% higher, which probably was not an economical response to higher seeding and N rates.

Table 4. Grain yield, maize densities before (Density1) and after (Density2) rotary hoeing and cultivating operations, weed densities at the 14th leaf stage (V14), grain N%, plants/m^2, ears/plant, kernels/ear, kernel weight (kwt.), and harvest index (HI) under conventional and organic management at recommended and high inputs in 2015 at Aurora, New York. Averages are provided to compare main effects of cropping systems when there are no cropping systems × input management interactions.

Treatments	Yield	Density1	Density2	Weeds	Grain N
	kg/ha	Plants/ha	Plants/ha	No./m^2	%
Conventional					
Recommended	10,321	72,608	72,158	0.47	1.33
High input	10,545	86,635	86,391	0.39	1.32
Ave.	10,357	79,621	79,275	0.43	1.32
Organic					
Recommended	6905	69,875	64,750	2.41	1.05
High input	7281	83,882	80,819	2.13	1.06
Ave.	7093	76,879	72,875	2.27	1.05
HSD 0.05	829 [+]	NS	1898 [+]	0.55 [+]	0.05 [+]
	Plant/m^2	Ears/Plant	Kernels/Ear	Kwt.	HI
Conventional	No./m^2	No./plant	No./ear	mg	no.
Recommended	7.28	1.0	572	262	0.59
High input	8.62	1.0	542	247	0.60
Ave.	7.95	1.0	557	254	
Organic					
Recommended	6.63	1.03	506	247	0.59
High input	7.40	1.03	472	236	0.58
Ave.	7.02	1.03	489	242	
LSD 0.05 [+]	0.80 [+]	0.03 [+]	51 [+]	9 [+]	0.01 [++]

[+] Compares means of cropping systems. [++] Compares means of cropping system × input management interactions.

Organic compared with conventional maize had similar plant densities shortly after emergence but ~8% lower plant densities at the V9 stage, probably due to mechanical weed control practices (Density 2, Table 4). A previous study also reported lower organic maize compared with conventional maize densities because of rotary hoe damage [7]. Despite the close and repeated cultivations, organic compared with conventional maize had more than five times higher weed densities (Table 4). Nevertheless, weed densities in organic maize averaged a relatively low 2.27 weeds/m^2. Weed densities had negative correlations with maize densities at the V9 stage ($r = -0.41$, $n = 48$, Table 5) and grain N% ($r = -0.81$), but high seeding and N rates did not significantly reduce weed densities in organic maize.

Organic maize had very low grain N% concentrations (1.05%) compared with conventional maize (1.32%, Table 4). Excessive precipitation (276 mm) from planting to the silking stage may have leached or denitrified a considerable amount of the N in the pre-plow application of composted poultry manure. In contrast, the experimental site received 98 mm of precipitation from the side-dressed N application (26 June) to the silking stage, which was probably not sufficient to leach or denitrify much of the side-dressed N. Lower organic maize yields were observed in a study using poultry compost litter as the N source because of low N status associated with increased immobilization of N [17]. Organic maize also had low grain N% concentrations (1.07%) during the first transition year in another study, but without a yield reduction [7].

Grain yield had a strong positive correlation with grain N% concentrations ($r = 0.80$, $n = 48$, Table 5) and a strong negative correlation with weed densities ($r = -0.78$). Stepwise regression analyses indicated that linear and quadratic weed density coefficients and a quadratic grain N% coefficient explained 72% of the yield variability ($n = 48$, Table 6). This agrees with results from a previous study that reported lower organic maize yields mostly because of low soil N availability (73%) and weed competition (23%) with only 4% associated with lower maize densities [10].

Table 5. Correlations (*r*-values, *n* = 48) among grain yield, maize densities before (DEN1) and after (DEN2) rotary hoeing and cultivating operations, weed density, grain N% concentration, plants/m^2, ears/plant, kernels/ear, kernel weight (Kwt.), and harvest index (HI) in 2015 at Aurora, New York.

Variable	Yield	DEN1	DEN2	Weeds	Grain N	Plants/m^2	Ears/Plant	Kernels/Ear	Kwt.	HI
Yield	-	NS	0.42	−0.78	0.8	0.49	−0.30	0.49	0.52	0.44
DEN1	NS [++]	-	0.88	NS	NS	0.48	NS	NS	NS	NS
DEN2	0.42	0.7	-	−0.41	NS	0.65	NS	NS	NS	NS
Weeds	−0.78	NS	−0.41	-	−0.81	−0.34	0.41	−0.48	−0.34	−0.29
Grain N%	0.81	NS	NS	−0.81	-	0.29	−0.31	0.68	0.52	0.6
Plants/m^2	0.49	0.48	0.65	−0.34	0.29	-	NS	NS	NS	NS
Ears/Plant	−0.30	NS	NS	0.41	−0.31	NS	-	NS	NS	NS
Kernels/ear	0.49	NS	NS	−0.48	0.68	NS	NS	-	0.55	0.73
Kwt.	0.52	NS	NS	−0.34	0.52	NS	NS	0.55	-	0.57
HI	0.44	NS	NS	−0.29	0.6	NS	NS	0.73	0.57	-

[++] Not Significant at 0.05.

Table 6. Model (*n* = 48) significance (*p*-value), adjusted R^2 and C(*p*) values, and parameter estimates, of maize density (after mechanical weed control operations), weed density, and grain N% from stepwise regression equations predicting maize yields at Aurora, New York in 2015, 2016, and 2017, and averaged over 2015–2017.

Variables	*p*	Adj. R^2	$\hat{\beta_0}$	$\hat{\beta 1}$	$\hat{\beta 2}$	C(*p*)
			kg/ha			
			2015			
Model	<0.0001	0.72				5.2
Intercept	0.001 [+]		8402			
Maize density	NS [++]					
Maize density^2	NS					
Weed density	0.002			−2350		
Weed density^2	0.02				406	
Grain N%	NS					
Grain N%^{2}	0.04				1735	
			2016			
Model	0.001	0.21				2.32
Intercept	0.001		3683			
Maize density	0.001			0.06		
Maize density^2	NS					
Weed density	NS					
Weed density^2	NS					
Grain N%	NS					
Grain N%^{2}	NS					
			2017			
Model	<0.0001	0.53				2.27
Intercept	<0.0001		−2797			
Maize density	NS			0.74		
Maize density^2	0.0004				−0.0000036	
Weed density	NS					
Weed density^2	NS					
Grain N%	<0.0001		9157			
Grain N%^{2}	NS					
			2015–2017			
Model	<0.0001	0.56				5.26
Intercept	<0.0001		−6593			
Maize density	<0.0001			0.74		
Maize density^2	<0.0001				−0.000005	
Weed density	NS					
Weed density^2	NS					
Grain N%	<0.0001			67,615		
Grain N%^{2}	<0.0001				−23,797	

[+] *p*-values. [++] Not significant at 0.05.

Yield component analyses from the sampling area indicated that organic compared with conventional maize had 11.7% lower plants/m^2, 12% lower kernel number, and 4.7% lower kernel weight (Table 4). Kernel number and kernel weight typically increase as maize densities decrease [18,19] so the lower kernel number and kernel weight in organic maize was somewhat surprising. The low N status in maize, however, can also lower kernel number and kernel weight [20,21]. Grain N% did have positive correlations with kernels/ear (r = 0.68, n = 48, Table 5) and kernel weight (r = 0.52). The three yield components also had significant positive correlations (~0.50) with yield. Stepwise regression analyses indicated that a linear plant density coefficient, linear and quadratic kernels/ear coefficients, and a quadratic kernel weight coefficient explained 73% of the yield variability (n = 48, Table 7).

Table 7. Model (n = 48) significance (p-value), adjusted R^2 and C(p-values), and parameter estimates of plants/m^2, ears/plant, kernels/ear, and kernel weight from stepwise regression equations predicting maize yields at Aurora, New York in 2015, 2016, and 2017, and averaged over 2015–2017.

Variables	p	Adj. R^2	$\hat{\beta}_0$	$\hat{\beta}1$	$\hat{\beta}2$	C(p)
			kg/ha			
			2015			
Model	<0.0001	0.73				4.61
Intercept	0.56		2377			
Plants/m^2	<0.0001			1089		
(Plants/m^2)^2	NS[+]					
Ears/plant	NS[++]			−2350		
(Ears/plant)^2	NS				406	
Kernels/ear	0.02			−41.7		
(Kernels/ear)^2	0.005				0.06	
Kernel weight	NS					
(Kernel weight)^2	0.005				0.05	
			2016			
Model	0.03	0.10				2.32
Intercept	<0.0001		3683			
Plants/m^2	0.03			247		
(Plants/m^2)^2	NS					
Ears/plant	NS					
(Ears/plant)^2	NS					
Kernels/ear	NS					
(Kernels/ear)^2	NS					
Kernel weight	NS					
(Kernel weight)^2	NS					
			2017			
Model	<0.0001	0.35				7.30
Intercept	0.24		8042			
Plants/m^2	0.006			2892		
(Plants/m^2)^2	0.03				−157	
Ears/plant	0.06			2227		
(Ears/plant)^2	NS					
Kernels/ear	<0.01			−76.3		
(Kernels/ear)^2	0.01				0.07	
Kernel weight	<0.0001			28.3		
(Kernel weight)^2	NS					
			2015–2017			
Model	<0.0001	0.63				5.82
Intercept	<0.0001		−8277			2
Plants/m^2	<0.0001		827			
(Plants/m^2)^2	<0.0004				0.0000036	
Ears/plant	NS					
(Ears/plant)^2	NS					

[+] p-values. [++] Not significant at 0.05.

3.2. 2016

The 2016 growing season had the second driest 1 May through 18 July period on record at the experimental site with only 53 mm of precipitation recorded from planting until the silking stage (http://climod.nrcc.cornell.edu/climod/rank/). Soils are typically shallow in the Northeast USA, resulting in an effective rooting depth of only 0.75 m [22]. Consequently, dry climatic conditions result in significant crop stress in this environment [22]. Conditions improved during the remainder of the growing season with 160 mm of precipitation recorded from the silking stage (~18 July) until physiological maturity (~3 September).

Cropping system and management inputs did not affect yield and there was no cropping system × management input interaction (Table 2). The exceedingly dry conditions from planting until silking contributed to low maize yields, which probably negated yield responses to cropping systems and management inputs. Maize densities in both cropping systems were very low before rotary hoeing because dry soil conditions reduced emergence (Table 8). Organic compared with conventional maize had similar plant densities before rotary hoeing but 8% lower plant densities at the V9 stage probably because of crop damage from mechanical weed control practices. Maize densities in both cropping systems were much lower than the threshold final plant density (~67,000 plants/ha) for maximum yield in this environment, even in dry years [23]. Consequently, yield had a positive correlation with maize densities ($r = 0.45$, $n = 48$, Table 9).

Table 8. Grain yield, maize densities before (Density1) and after (Density2) rotary hoeing and cultivating operations, weed densities at the 14th leaf stage (V14), grain N%, plants/m^2, ears/plant, kernels/ear, kernel weight (kwt.), and harvest index (HI) under conventional and organic management at recommended and high inputs in 2016 at Aurora, New York. Averages are provided to compare main effects for cropping systems when there are no cropping systems x input management interactions.

Treatments	Yield	Density1	Density2	Weeds	Grain N
	Kg/ha	Plants/ha	Plants/ha	No./m^2	%
Conventional					
Recommended	7783	58,784	56,566	0.27	1.68
High input	7156	69,663	65,606	0.18	1.56
Ave.	7469	64,225	61,086	0.22	1.62
Organic					
Recommended	7093	58,080	51,472	0.99	1.61
High input	7156	69,602	60,648	0.64	1.51
Ave.	7124	63,842	56,059	0.82	1.56
LSD 0.05 [+]	NS	NS	2034	0.27	0.04
	Plants/m^2	Ears/Plant	Kernels/Ear	Kwt.	HI
Conventional	No./m^2	No./plant	No./ear	mg	No.
Recommended	6.08	1.06	394	309	0.64
High input	7.00	1.06	359	305	0.63
Ave.	6.54	1.06	377	307	0.64
Organic					
Recommended	5.55	1.12	381	312	0.65
High input	5.83	1.19	346	309	0.64
Ave.	5.69	1.15	363	310	0.65
HSD 0.05 [+]	0.56	0.07	NS	NS	NS

[+] Compares means of cropping systems.

Cropping system and management inputs affected weed densities and there was no cropping system by input treatment interaction (Table 2). Weed densities were higher in organic compared with conventional maize, but densities were less than 1.0 weed/m^2 (Table 8). Dry soil conditions probably reduced weed emergence. Input management also influenced weed densities (Table 2), which had a weak negative correlation with maize densities at the V9 stage ($r = -0.38$) but no correlation with grain N%. Grain N% concentrations were greater in conventional compared with organic management, but values in both cropping systems exceeded 1.50%, which indicates sufficient N. Consequently, grain yield did not correlate with weed densities nor grain N% concentrations (Table 9).

Table 9. Correlations (*r*-values, *n* = 48) among grain yield, maize densities before (DEN1) and after (DEN2) rotary hoeing and cultivating operations, weed density, grain N concentration, plants/m^2, ears/plant, kernels/ear, kernel weight (Kwt.), and harvest index (HI) in 2016 at Aurora, New York.

Variable	Yield	DEN1	DEN2	Weeds	Grain N%	Plants/m^2	Ears/Plant	Kernels/Ear	Kwt	HI
Yield	-	0.27	0.45	NS	NS	0.32	NS	NS	NS	NS
DEN1	0.27	-	0.82	NS	NS	0.35	NS	−0.33	NS	NS
DEN2	0.45	0.82	-	−0.38	0.3	0.53	NS	NS	NS	NS
Weeds	NS [+]	NS	−0.38	-	NS	NS	NS	NS	NS	NS
Grain N%	NS	NS	0.3	NS	-	NS	NS	NS	NS	NS
Plants/m^2	0.32	0.35	0.53	NS	NS	-	−0.35	NS	NS	NS
Ears/Plant	NS	NS	NS	NS	NS	−0.35	-	−0.46	NS	NS
Kernels/ear	NS	−0.33	NS	NS	NS	NS	−0.46	-	NS	NS
Kwt.	NS	NS	NS	NS	NS	NS	NS	NS	-	NS
HI	NS	NS	NS	NS	NS	NS	NS	NS	NS	-

[+] Not Significant at 0.05.

Organic compared with conventional maize had ~13% fewer plants/m^2 in the sampling area, a few days before harvest (Table 8). Organic and conventional maize had similar kernels/ear and kernel weight. Organic compared with conventional maize, however, did have greater ears/plant. Apparently, the greater number of ears/plant compensated for the lower plant densities, resulting in similar yields between organic and conventional maize in the exceedingly dry growing season.

3.3. 2017

The 2017 growing season had the second wettest (tied with 2015) 1 May through 31 July period on record at the experimental site (http://climod.nrcc.cornell.edu/climod/rank/, Table 3). As in 2015, conditions became dry in August with the 2017 growing season having the fourth driest August on record (http://climod.nrcc.cornell.edu/climod/rank/). Despite excessively wet antecedent moisture conditions, premature leaf senescence was observed in maize in late August and early September. Silking was observed on ~22 July and physiological maturity on ~8 September so some drought stress occurred during the late kernel filling stage.

Yield had significant cropping system and management input effects but there was a cropping system × management input interaction (Table 2). Organic maize following wheat/red clover or soybean and conventional maize following soybean showed ~15% to 19% yield responses to high input management (Table 10). Conventional maize following wheat/red clover, however, showed only an 8.6% response. Organic maize following wheat/red clover with high inputs and conventional maize following soybean with high inputs yielded the highest. Conventional compared with organic maize following soybean with high inputs yielded ~4% higher. In contrast, organic compared with conventional maize following wheat/red clover with high inputs yielded ~15% higher. Overall, organic maize in a soybean-wheat/red clover-maize rotation compared with a maize-soybean-maize rotation yielded ~9% higher, which supports previous findings that organic maize performs best in a more complex rotation [5–8].

Organic compared with conventional maize had similar maize densities before rotary hoeing for the third consecutive year (Table 2), which indicates that the lack of an insecticide/fungicide treatment and the GM genes in organic maize did not hinder plant establishment in this study. Organic compared with conventional maize, however, had 9% fewer plants at the V9 stage probably because of crop damage with mechanical weed control practices (Table 10). Plant densities in organic maize with recommended inputs averaged only ~60,000 plants/ha, much lower than the threshold plant density for maximum yield in this environment. Maize densities once again had a positive correlation (*r* = 0.46, *n* = 96, Table 11) with yield.

Table 10. Grain yield, maize densities before (Density1) and after (Density2) rotary hoeing and cultivating operations, weed densities at the 14th leaf stage (V14), grain N%, plants/m^2, ears/plant, kernels/ear, kernel weight (kwt.), and harvest index (HI) under conventional and organic management at recommended and high inputs in 2017 at Aurora, New York. Averages are provided to compare main effects for cropping systems when there are no cropping systems × input management interactions.

Treatment/Previous Crop	Yield	Density1	Density2	Weeds	Grain N
Conventional	Kg/ha	Plants/ha	Plants/ha	Weeds/m^2	%
Recommended-wheat/RC	10,145	63,693	65,964	1.26	1.33
Recommended-soybean	10,556	65,131	66,448	1.15	1.33
High Input-wheat/RC	11,014	73,502	75,851	0.90	1.34
High Input-soybean	12,547	75,905	76,807	0.96	1.43
Ave.		69,558	71,200		
Organic					
Recommended-wheat/RC	11,301	63,790	59,364	0.67	1.37
Recommended-soybean	10,294	64,595	60,379	2.48	1.26
High Input-wheat/RC	12,952	75,374	70,896	0.55	1.43
High Input-soybean	12,001	75,992	68,757	2.28	1.38
Ave.		69,937	64,849		
HSD 0.05	451 [++]	NS	1607 [+]	0.52 [++]	0.05 [++]
	Plants/m^2	Ears/plant	Kernels/ear	Kwt.	HI
Conventional	no./m^2	no./plant	no./ear	mg	no.
Recommended-wheat/RC	7.43	1.02	545	271	0.46
Recommended-soybean	7.41	1.02	550	275	0.48
High Input-wheat/RC	8.11	1.03	517	270	0.47
High Input-soybean	7.87	1.04	556	282	0.46
Ave.		1.03			0.47
Organic					
Recommended-wheat/RC	6.41	1.03	561	325	0.48
Recommended-soybean	6.12	1.02	528	291	0.51
High Input-wheat/RC	6.92	1.08	556	316	0.51
High Input-soybean	7.80	1.02	531	294	0.52
Ave.		1.04			0.51
HSD 0.05	0.41 [++]	NS	29 [++]	12 [++]	0.02 [+]

[+] Compares means of cropping systems. [++] Compares means of cropping system × input management interactions.

Table 11. Correlations (*r*-values, *n* = 96) among grain yield, maize densities before (DEN1) and after (DEN2) rotary hoeing and cultivating operations, weed density, grain N concentration, plants/m^2, ears/plant, kernels/ear, kernel weight (Kwt.), and harvest index (HI) in 2017 at Aurora, New York.

Variable	Yield	DEN1	DEN2	Weeds	Grain N%	Plants/m^2	Ears/Plant	Kernels/Ear	Kwt	HI
Yield	-	0.66	0.46	−0.2	0.68	NS	NS	NS	0.39	NS
DEN1	0.66	-	0.8	NS	0.34	0.41	NS	−0.33	NS	0.39
DEN2	0.46	0.88	-	NS	0.32	0.47	NS	NS	−0.33	NS
Weeds	−0.2	NS	NS	-	−0.41	NS	NS	−0.29	−0.24	0.24
Grain N%	0.68	0.34	0.32	−0.41	-	0.22	NS	0.22	0.32	NS
Plants/m^2	NS [+]	0.41	0.54	NS	0.22	-	−0.26	−0.22	−0.45	0.42
Ears/Plant	NS	NS	NS	NS	NS	−0.26	-	NS	NS	0.42
Kernels/ear	NS	NS	NS	−0.29	0.22	−0.22	NS	-	0.46	0.21
Kwt.	0.39	NS	−0.33	−0.24	0.32	−0.45	NS	0.46	-	0.22
HI	NS	0.34	NS	−0.24	NS	NS	0.42	0.21	0.22	-

[+] Not Significant at 0.05.

Weed densities had a significant cropping system × management input interaction (Table 2). Conventional maize showed a ~23% reduction in weed densities with high input management compared with only a~10% reduction in organic maize. Interestingly, organic maize following wheat/red clover, regardless of input management1, had lower weed densities compared with conventional maize following wheat/red clover or soybean with recommended inputs (Table 10). Likewise, organic maize following wheat/red clover compared with following soybean had approximately three times lower weed densities. A previous study also reported fewer weeds in an organic soybean-wheat/red clover-maize rotation compared with a maize-soybean rotation [24]. Weed densities, however, had a weak negative correlation (*r* = −0.20, *n* = 96) with yield. Weed densities of fewer than 2.5 weeds/m^2 in organic maize following soybean may not have affected yield greatly because of the exceedingly wet conditions through the early grain-filling period. Other studies have also reported higher weed densities in organic compared with conventional maize with limited impacts on yield [7,25]. Weed densities did not correlate with maize densities at the V9 stage but did have a negative correlation with grain N% ($r = -0.41$). Weed densities in organic maize trended lower in high input management in all three years but weed densities were generally low in this study so yield effects were probably limited.

Grain N% showed a cropping system × management input interaction (Table 2). Grain N% showed a 0.06 to 0.12% grain N% increase in the high versus recommended input treatments (even with significant yield increases), except in conventional maize when following red clover (1.33 and 1.34% N, respectively, Table 10). Red clover, which was frost-seeded into conventional winter wheat in early March of 2016, averaged only ~3400 kg/ha of biomass (about 25% grasses were in the sample) with a 3.0% N concentration compared with ~5600 kg/ha of biomass with a 3.85% N concentration in the organic cropping system. For some unknown reason, the ammonium nitrate applied to conventional wheat in April of 2016 resulted in less red clover emergence and/or growth compared to red clover in organic wheat. Conventional maize following wheat/red clover in the recommended treatment (56 kg N/ha side-dressed because of low biomass and N concentration of red clover) yielded 10% lower than the recommended organic maize treatment, which received no additional N and relied totally on plowed in red clover for its N supply. In a previous study [26], the green manure crop, hairy vetch, did not provide adequate N to organic maize when biomass was below a critical value (4630 kg/ha) so the low red clover biomass before planting conventional maize most likely did not provide adequate N.

Red clover, however, decomposes rapidly with estimates of 35% release four weeks after incorporation and complete release about 10 weeks after incorporation [27]. A considerable amount of N was thus released by late June and early July (six to seven weeks after incorporation) when maize was not taking up large amounts of N (V5 to V8 stage of growth from 25 June–5 July). This probably resulted in some leaching of the released N from red clover incorporation (17 May) until 5 July (310 mm). Consequently, red clover +56 kg N/ha side-dressed did not provide adequate N to conventional maize as indicated by the 9% yield increase in the high input treatment (red clover +100 kg N/ha, side-dressed). Likewise, red clover alone probably did not provide adequate N to organic maize as indicated by the 15% yield increase in the high input treatment (an additional 56 kg N/ha of pre-plow composted manure), although higher maize densities undoubtedly also contributed to the yield increase.

Grain N% concentrations increased by ~0.1% in conventional and organic maize following soybean with high compared with recommended inputs, despite the 16 to 19% yield increases (Table 10). Again, the positive yield and grain N% responses to high inputs following soybean are probably associated with leaching of some of the pre-plow composted manure or side-dressed N. Grain N% concentrations had a positive correlation (0.68, $n = 96$) with grain yield. Stepwise regression analyses indicated that a linear grain N% coefficient and a quadratic maize density coefficient explained 53% of the yield variability ($n = 96$, Table 6). Another study also found that low soil N availability and low plant densities contributed to lower organic maize yields when compared with conventional maize [17].

Organic compared with conventional maize had ~11.5% fewer plants/m^2 at harvest, but there was a cropping system × input interaction (Table 2). Conventional maize with high inputs showed only an ~8% increase compared with a ~17% increase in plants/m^2 in organic maize with high input management (Table 10). Ears/plant as well as kernels/ear, was similar between cropping systems. Kernel weight also had a cropping system × input interaction as indicated by ~1% increase of conventional maize to high inputs and ~1% decrease in organic maize to high inputs. Overall, organic compared with conventional maize had ~11.5% higher kernel weight, which apparently compensated for the lower maize densities as indicated by higher yields when following wheat/red clover. Kernel weight had negative correlations with plant density ($r = -0.45$, $n = 96$) and positive correlations with grain N% concentration ($r = 0.32$, Table 11), which agrees with previous studies [18,19,21]. Kernel weight also had a positive correlation with yield ($r = 0.39$). Stepwise regression analyses indicated that a linear plants/m^2 coefficient and a linear kernel weight coefficient explained 31% of the yield variability (Table 7).

4. Conclusions

Maize as an entry crop in the transition period to an organic cropping system proved problematic when a green manure crop was not in place as indicated by grain N% concentrations of only 1.05% and 32% lower yield in organic compared with conventional maize. Organic maize, which yielded similarly in the dry second year and 15% greater in the wet third year compared with conventional maize when following wheat/red clover, appears viable as second or third transition year crops when following wheat/red clover in this environment. Interestingly, red clover, which is typically inter-seeded into wheat to provide N to the subsequent maize crop, also appeared to reduce weed densities in the third year, which bodes well for the sustainability of an organic soybean-wheat/red clover-maize rotation. Our fields, however, did not have problematic weeds at the initiation of the study so weed interference was not a major factor in this study. In fields with high densities of problematic weeds, organic maize may not have yielded as well or 15% higher in the second and third years, respectively.

Maize N status and maize densities appeared to be the major factors explaining yield variability in this study (the linear and quadratic maize density and grain N% coefficients explained 56% of the yield variability when averaged across the three years, $n = 192$, Table 5). Transitioning cash crop producers in the Northeast USA who do not have an available supply of manure nor equipment for perennial forage production should either plant wheat/red clover the year before transitioning, plant wheat/red clover as the entry crop followed by maize as the second-year transition crop, or plant soybean as the entry crop, wheat/red clover in the second year, followed by organic maize in the third year. Transitioning cash crop producers should also apply additional N (~56 kg N/ha) and increase maize seeding rates, not to improve weed competitiveness, but rather to offset the potential loss of N associated with rapid red clover decomposition in wet springs and the ~10% maize density reduction with mechanical weed control. Maize seeding rates may only have to increase by ~7% because yield component compensation via increased ears/plant or increased kernel weight can mitigate some of the yield reduction associated with low final plant densities.

Author Contributions: W.J.C. designed and conducted the experiment, and wrote the initial draft of the manuscript. J.H.C. conducted the statistical analyses and reviewed the manuscript.

Funding: This project was partially supported by the U.S. Department of Agriculture Cooperative State Research, Education, and Extension Service through New York Hatch Project 1257322.

Conflicts of Interest: The authors have no conflict of interests.

Abbreviations:

MN	Minnesota, USA
IA	Iowa, USA
MD	Maryland, USA
WI	Wisconsin, USA
NY	New York, USA
MI	Michigan, USA
NC	North Carolina
GM	genetically modified

References

1. USDA. Organic. 2012. What Is Organic Certification? Available online: http://www.ams.usda.gov/sites/default/files/media/What%20is%20Organic%20Certification.pdf (accessed on 8 May 2018).

2. USDA. ERS. 2015. The Profit Potential of Certified Organic Field Crop Production. Available online: http://www.ers.usda.gov/media/1875176/err188_summary.pdf (accessed on 8 May 2018).
3. Archer, D.W.; Jaradat, A.A.; Johnson, J.M.F.; Lachnicht-Weyers, S.; Gesch, R.W.; Forcella, F.; Kludze, H.K. Crop Productivity and Economics during the Transition to Alternative Cropping Systems. *Agron. J.* **2007**, *99*, 1538–1547. [CrossRef]
4. Archer, D.W.; Kludze, H. Transition to organic cropping systems under risk. In Proceedings of the American Agricultural Economics Association Annual Meeting, Long Beach, CA, USA, 23–26 July 2006.
5. Coulter, J.A.; Delbridge, T.A.; King, R.P.; Sheaffer, C.C. Productivity, economic, and soil quality in the Minnesota Variable-Input Cropping Systems Trial. *Crop Manag.* **2013**. [CrossRef]
6. Porter, P.M.; Huggins, D.R.; Perillo, C.A.; Quiring, S.R.; Crookston, R.K. Organic and other management strategies with two and four year crop rotations in Minnesota. *Agron. J.* **2003**, *95*, 233–244. [CrossRef]
7. Delate, K.; Cambardella, C.A. Agroecosystem Performance during Transition to Certified Organic Grain Production. *Agron. J.* **2004**, *96*, 1288–1298. [CrossRef]
8. Delate, K.; Cambardella, C.; Chase, C.; Johanns, A.; Turnbull, R. The Long-Term Agroecological Research (LTAR) experiment supports organic yields, soil quality, and economic performance in Iowa. *Crop Manag.* **2013**. [CrossRef]
9. Seufert, V.; Ramankutty, N.; Floey, J.A. Comparing the yields of organic and conventional agriculture. *Nature* **2012**, *485*, 229–232. [CrossRef] [PubMed]
10. Cavigelli, M.A.; Teasdale, J.R.; Conklin, A.E. Long-term agronomic performance of organic and conventional field crops in the mid-Atlantic region. *Agron. J.* **2008**, *100*, 785–794. [CrossRef]
11. Chavas, J.P.; Posner, J.L.; Hedtcke, J.L. Organic and Conventional Production Systems in the Wisconsin Integrated Cropping Systems Trial: II. Economic and Risk Analysis 1993–2006. *Agron. J.* **2009**, *101*, 288–295. [CrossRef]
12. Baldock, J.O.; Hedtcke, J.L.; Posner, J.L.; Hall, J.A. Organic and Conventional Production Systems in the Wisconsin Integrated Cropping Systems Trial: III. Yield Trends. *Agron. J.* **2014**, *106*, 1509–1522. [CrossRef]
13. Hossar, L.; Archer, D.W.; Bertrand, M.; Colnenne-David, C.; Debaeke, P.; Ernfors, M.; Helene Jeuffroy, M.; Munier-Jolain, N.; Nilsson, C.; Sanford, G.R.; et al. Meta-Analysis of Maize and Wheat Yields in Low-Input vs. Conventional and Organic Systems. *Agron. J.* **2016**, *108*, 1155–1167. [CrossRef]
14. Cox, W.J.; Atkins, P. Planting a Full-Season Hybrid at the ~1.5 to 2.0 Inch Depth from ~May 15 to ~May 20 Resulted in Maximum Yield (but Higher Grain Moisture) on a Silt Loam Soil in the Finger Lakes Region. What's Cropping Up? Newsletter. Cornell Univ.: Ithaca, NY, 2015. Available online: http://blogs.cornell.edu/whatscroppingup/2015/01/08/planting-a-full-season-hybrid-at-the-1-5-to-2-0-inch-depth-from-may-15-to-may-20-resulted-in-maximum-yield-but-higher-grain-moisture-on-a-silt-loam-soil-in-the-finger-lakes-region/ (accessed on 8 May 2018).
15. Ritchie, S.W.; Hanway, J.J.; Benson, G.O. *How a Corn Plant Develops*; Special Report No. 48; Iowa State University Cooperative Extension Service: Ames, IA, USA, 1993.
16. Hall, M.R.; Swanton, C.L.; Anderson, G.W. The critical period of weed control in grain corn (*Zea mays*). *Weed Sci.* **1992**, *40*, 441–447.
17. Clark, K.M.; Boardman, D.L.; Staples, J.S.; Easterby, S.; Reinbott, T.M.; Kremer, R.J.; Kitchen, N.R.; Veum, K.S. Crop Yield and Soil Organic Carbon in Conventional and No-Till Organic Systems on a Claypan Soil. *Agron. J.* **2017**, *109*, 588–599. [CrossRef]
18. Borras, L.; Slafer, G.; Otegui, M.E. Seed dry weight response to source-sink manipulations in wheat, maize, and soybean. A quantitative reappraisal. *Field Crops Res.* **2004**, *86*, 131–146. [CrossRef]
19. Van Roekel, R.J.; Coulter, J. Agronomic responses of corn to planting date and plant density. *Agron. J.* **2011**, *103*, 1412–1422. [CrossRef]
20. Boosma, C.R.; Santini, J.B.; Tollenaar, M.; Vyn, T.J. Maize morphophysiological responses to intense crowding and low nitrogen availability: An analysis and review. *Agron. J.* **2009**, *101*, 1426–1452. [CrossRef]
21. Ciampitti, I.A.; Vyn, T.J. A compressive study of plant density consequences on nitrogen uptake dynamics of maize plants from vegetative to reproductive stages. *Field Crops Res.* **2011**, *121*, 2–18. [CrossRef]
22. Singer, J.S.; Cox, W.J. Corn Growth and Yield under Different Crop Rotation, Tillage and Management Systems. *Crop Sci.* **1998**, *38*, 996–1003. [CrossRef]
23. Cox, W.J.; Cherney, J.H. Lack of Hybrid, Nitrogen and Seeding Rate Interactions for Corn Growth and Yield. *Agron. J.* **2012**, *104*, 945–952. [CrossRef]

Agronomy **2018**, *8*, 113

24. Teasdale, J.R.; Mangum, R.W.; Radhakrishnan, J.; Cavigelli, M.A. Weed Seedbank Dynamics in Three Organic Farming Crop Rotations. *Agron. J.* **2004**, *96*, 1429–1435. [CrossRef]
25. Smith, R.G.; Barbercheck, M.E.; Mortensen, D.A.; Hyde, J.; Hulting, A.J. Yield and Net Returns during the Transition to Organic Feed Production. *Agron. J.* **2011**, *103*, 51–59. [CrossRef]
26. Spargo, J.T.; Cavigelli, M.A.; Mirsky, S.B.; Meisinger, J.J.; Ackroyd, V.J. Organic Supplemental Nitrogen Sources for Field Corn Production after a Hairy Vetch Cover Crop. *Agron. J.* **2016**, *108*, 1992–2002. [CrossRef]
27. Gaudin, A.C.M.; Westra, S.; Louks, C.E.S.; Janovicek, K.; Martin, R.C.; Deen, W. Improving Resilience of Northern Field Crop Systems Using Inter-Seeded Red Clover: A Review. *Agronomy* **2013**, *3*, 148–180. [CrossRef]

agronomy

MDPI

Article

Non-Structural Carbohydrate Metabolism, Growth, and Productivity of Maize by Increasing Plant Density

Jairo O. Cazetta [1,*] and Marcos D. Revoredo [2]

[1]   Department of Technology, Unesp—São Paulo State University, Jaboticabal SP 14884-900, Brazil
[2]   Department of Technology, Soil Science Graduation Program, Unesp—São Paulo State University,
    Jaboticabal SP 14884-900, Brazil; mrevoredo@alltech.com
*   Correspondence: cazetta@fcav.unesp.br or jairo.cazetta@unesp.br; Tel.: +55-16-997041137

Received: 8 September 2018; Accepted: 30 October 2018; Published: 2 November 2018

Abstract: Increasing plant density seems to improve the productivity of maize crops, and the understanding of how the metabolism of non-structural carbohydrates is affected in plants under high crop density is critical. Thus, with the objective of further clarifying this issue, maize plants were subjected to densities from 30,000 to 90,000 plants ha^{-1}, and the plant growth, soluble sugars and starch contents, invertase and sucrose synthase activities, and plant production were evaluated. We found that the stalk is more sensitive to the increasing plant density than leaves and kernels. The dry weight of the stalk and leaves per single plant decreased more drastically from low to intermediate plant densities, while grain production was reduced linearly in all plant density ranges, leading to higher values of harvest index in intermediate plant densities. The sucrose concentration did not change in leaves, stalk, or kernels of plants subjected to increasing plant densities at the R4 stage. Also, the specific activity of soluble invertase, bound invertase, and sucrose synthase did not change in leaf, stalk, or kernels of plants subjected to increased plant density. The productivity was increased with the increase in plant density, using narrow row (0.45 m) spacing.

Keywords: *Zea mayz*, L.; plant density; invertase; sucrose synthase; carbohydrates; production components

1. Introduction

Humanity has the great challenge of achieving continued improvements in agricultural production to face the growing demand for food, feed, and renewable fuels, likely against unfavorable global climatic changes [1]. Grasses still constitute the most productive and widely grown crop family which provides the bases for human life across the world [2]. In this context, maize plays important role, with growing relevance. Since 2001, maize surpassed other cereals to be the second most important commodity produced in the world, just after sugarcane [3]. Considering that sugarcane is used to produce mainly water-soluble carbohydrates and biomass for energy, it is reasonable to say that maize is the most important crop grown to obtain non-soluble carbohydrates. Nevertheless, the kernels are used mainly as food and feed. There are also expectations of using maize for biomass and sugar production [4].

To maintain paired demand and production, multiple approaches will be required, with an immediate effort to understand the gaps between potential plant production and actually obtained yield [1,5]. Among other approaches, increasing plant density is an important way to improve productivity [6–8]. While it is clear that intra-specific competition for water, nutrients, and light is increased by increasing plant density, it is not completely clear how maize crop yield tends to increase even above relatively high plant densities [9–11]. Probably the small increases in productivity obtained by an additional plant population, above relatively high plant densities, is no longer compensated by additional cultural costs needed for plant protection, by increased risks of plant lodging and drought stress, among other problems that may occur under those conditions [6]. However, the understanding of how plant physiology works to perform those small increases is very important to drive new technologies for productivity enhancement. As a result, this issue has recently attracted the attention of many researchers in this area [11–17]. There are suggestions that even though carbohydrate reserves do not directly increase yield potential in maize plants under normal conditions for growth, they contribute to yield stability when plants are challenged by environmental or biotic stresses [2,6]. Also, it is considered that controlling plant carbohydrate reserves could be a way to avoid yield loss in maize grown under environmentally unpredictable regions, or grown as a rain-fed crop in which plants are commonly subjected to some detrimental conditions such as low nutrient availability or water stress [18]. In this context, it is well known that sucrose is the main photosynthetic compound in leaves, which is partitioned from source leaves to the sink organs of maize plants [19–23]. Also, it is already known that invertases are enzymes that irreversibly catalyze the cleavage of sucrose into reducing sugars (glucose + fructose), while sucrose synthase is a cytosolic enzyme that reversibly catalyzes sucrose cleavage, but whose reaction relies on the production of ADP- and UDP-glucose for starch and cell wall polysaccharide synthesis [23]. Therefore, it is expected that these compounds, and the activity of the related enzymes, would change in plants challenged by increasing plant density. So, it is supposed that, if plant non-structural carbohydrate reserves are important to stabilize yield under the mentioned situations [2], these reserves also could be responsible for sustaining maize productivity even when plants are grown under relatively high densities. However, the involvement of non-structural carbohydrate reserves, and related enzyme activity, on the process of increasing kernel productivity of maize subjected to increasing plant density is not completely clear. Taking into account the aforementioned information, there is no doubt that maize is one of the most important crops in the world, and that yield improvement can be reached by increasing plant density. Also, there is no doubt that non-structural carbohydrates play important roles in this situation. Therefore, improving our knowledge on non-structural carbohydrate metabolism in whole plants subjected to increasing density is essential to understand as well as circumvent related problems, with the aim of maximizing maize productivity. Thus, this study was designed with the objective of gaining a better understanding of how non-structural carbohydrates levels, and the activities of some related enzymes, change in plants subjected to increasing density. Also, this study aims to understand the relationship of these variables with plant production under real rain-fed field conditions.

2. Materials and Methods

2.1. Characterization of Experimental Conditions, Treatments, and Assay Design

This research was carried out under field conditions at Jaboticabal, SP, Brazil, located at 48°18′58″ W and 21°15′22″ S, at an altitude of 575 m. The climate of this area is classified as Cwa (humid subtropical climate, with dry winters) according to the classification of Köppen. The climate characteristics of the experimental period are presented in Figure 1.

The soil of the experimental area is a typical eutrophic red oxisol with a clay texture. In this area, soybean crops were grown for two years before the development of this assay. The chemical analysis of the soil collected from the upper layer of 0–20 cm was accomplished according to method of Raij et al. [24], and revealed the following values: pH (in $CaCl_2$) = 6.1; organic matter = 25 g dm^{-3}; P (extracted by resin) = 84 mg dm^{-3}; K = 3.7 $mmol_c$ dm^{-3} (milimol of charge per cubic decimeter); Ca = 63 $mmol_c$ dm^{-3}; Mg = 50 $mmol_c$ dm^{-3}; H + Al = 20 $mmol_c$ dm^{-3}; sum of bases = 116.7 $mmol_c$ dm^{-3}; total cationic exchange capacity = 136.7 $mmol_c$ dm^{-3}; and base saturation = 85%.

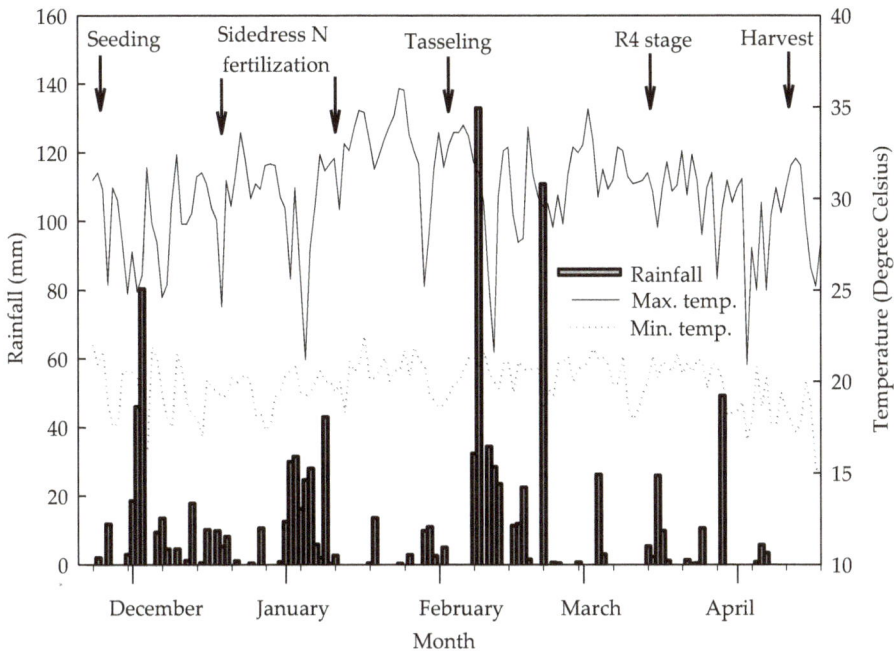

Figure 1. Rainfall, maximum temperature (Max. temp.), and minimum temperature (Min. temp.) during the assay development, as well as indications date of seeding, N fertilization, tasseling, R4 stage, and final harvest.

Treatments consisted of five densities of plants (30,000, 45,000, 60,000, 75,000, and 90,000 plants per hectare) and four replications, in a complete randomized block design, resulting in 20 experimental units (plots). Each experimental unit consisted of 10 rows of plants, spaced 0.45 m apart, and with a length of 6 m. A border of 1 m was maintained along all sides of the plot, so the central part was considered for samplings. To obtain the five treatments (plant densities) mentioned above, plants were allocated at 0.74 m, 0.49 m, 0.37 m, 0.30 m, and 0.25 m apart. The sowing was conducted manually and two seeds were sown in each sowing point. At the V3 stage, thinning was accomplished in order to leave a single plant per sowing point.

The simple hybrid AG 9010 (early cycle, relatively small plant height, and erect leaves) was used, grown under rain-fed conditions (Figure 1).

Before the experiment assembly, the soil was prepared by using the conventional system (plowing once and harrowing twice). Based on the soil chemical analysis and recommendations of Raij and Cantarella [25], planting fertilization of all experimental unities was performed in the sowing furrow by applying 16 kg ha^{-1} of N, 48 kg ha^{-1} P_2O_5, 32 kg ha^{-1} K_2O, 22 kg ha^{-1} S, and 9.6 kg ha^{-1} Zn.

Nitrogen side-dress fertilization was performed with 134 kg ha^{-1} N (ammonium sulfate), with 60% of this total being applied in the V4 stage and the remaining 40% in the V8 stage (Figure 1).

Part of each experimental unit was used for plant samplings at the R4 stage, for biochemical and biometrical determinations, and the other part was reserved untouched until plants were completely dry, when the harvest was accomplished for final agronomic variables measurement.

2.2. Biometric Determinations in Plants Evaluated at the R4 Stage

Ten plants at random were collected per plot in order to obtain a composite sample, and they were then separated into leaves, stem, kernels, and cob + husks. Leaf area was measured using an Image Analysis System (Delta-T Devices, Cambridge, UK). The dry mass of each sample was determined after drying at $65 \pm 5\,°C$ until constant weight was achieved. The total plant dry mass was obtained as the sum of the dry mass of all plant component parts. Average per plant values, referring to the dry matter of each part and the total plant dry mass, were used for statistical analyses.

2.3. Non-Structural Carbohydrate Determinations in Plants at the R4 Stage

With the aid of a scalpel blade, fresh samples were taken from the central part of the leaf closest to the main ear, as well as from the first internode of the stalk right below the ear, and kernels from the central part of the main ear. In order to obtain representative samples of kernel, stalk, and leaf for biochemical determination, sampling was carried out in 10 plants from each experimental unit, and then mixed to make composite samples of each plant part. The samples were identified and immediately immersed in liquid nitrogen. Then, samples were lyophilized, ground in mortar containing liquid nitrogen, and stored at $-80\,°C$.

For the extraction of reducing sugars and sucrose, 25 mg of the sample was homogenized (Turrax type homogenizer) in a centrifuge tube containing 1 mL of 80% ethanol, incubated at $60\,°C$ for 30 min, and centrifuged ($1200 \times g$, 20 min). Then the supernatant was transferred to a clean tube, and the procedure was repeated two more times. The supernatants were combined and deionized water was added to obtain a volume of 8 mL. This extract was subjected to the determination of reducing sugars and sucrose, while the remaining pellet was reserved for the determination of starch [26]. Reducing sugars were quantified by reaction with DNSA (3,5-dinitrosalicylic acid) and subsequent spectrophotometric determination, as proposed by Miller [27]. The sucrose contents were quantified by reaction with resorcinol and spectrophotometric determination [28]. The starch content was determined by using the remaining pellet, after hydrolysis with α-amylase and amyloglucosidase [29], and the glucose content in the hydrolyzate was determined by the reaction with DNSA (3,5-dinitrosalicylic acid) and subsequent spectrophotometric determination [27], calculating the content of starch as indicated by Brown and Hubber [30].

2.4. Enzymatic Activities Determination in Plants at the R4 Stage

The activities of sucrose synthase (UDP-Gluc: D-fructose 2-glucosyl-transferase, EC 2. 4. 1. 13) and invertases (α-fructofuranosidase, EC 3. 2. 1. 26) were determined in the same samples harvested, prepared and stored as described above for non-structural carbohydrates determinations.

The extraction of soluble enzymes (sucrose synthase and soluble invertase) was performed according to Singletary et al. [31]. Briefly, 100 mg of lyophilized sample was homogenized (Turrax type homogenizer) (30 s at $4\,°C$) in 2 mL of a mixture containing 50 mM HEPES (50 mM N-2-hydroxymethylpiperazine-N′-2-ethanesulfonic acid) buffer (pH 7.5), 5 mM $MgCl_2$, and 1 mM DTT (dithiothreitol), and centrifuged ($20,000 \times g$, $4\,°C$, 20 min). The supernatant was transferred to Spectrapor 4 dialysis tubes and dialyzed in 10 mM HEPES buffer pH 7.2 containing 5 mM $MgCl_2$ and 1 mM DTT for 24 h at $4\,°C$, while the pellet was reserved for the determination of bound invertase activity.

Sucrose synthase was assayed in the direction of sucrose degradation, in a mixture containing 80 mM MES (2(N-morfolino)-ethanesulfonic acid) (pH 6.0), 300 mM sucrose, and 10 mM UDP (uridine diphosphate). Soluble invertase (α-fructofuranosidase, EC 3. 2. 1. 26) activity was assayed following the procedure described by Doehlert and Felker [32]. The mixture contained 200 mM acetate buffer (pH 5.0), 10 mM sucrose, and the dialyzed enzyme extract in a proportion of 500 μL mL^{-1} final volume. Bound invertase activity was determined using the pellet remaining from the first enzyme extraction, which was submitted to a new extraction (3 min, $4\,°C$) with 10 mM HEPES (pH 7.5) containing 5 mM

MgCl$_2$, 1 mM DTT, and 1 M NaCl. The suspension was centrifuged (20,000× *g*, 20 min) and the activity was assayed immediately due to the instability of the enzyme. The activities of sucrose synthase and the invertases were determined after 20 min at 30 °C, and reducing sugars produced were measured using the DNSA method [27].

From the specific activities of the enzymes (activities per unit of tissue dry matter) and the total amounts of dry matter of the respective tissues per plant, the total enzymatic activities per plant were also estimated.

2.5. Agronomic Determinations in Plants Evaluated at Final Harest Time

After the complete drying of the crop in the field, the remaining plants in the useful area of the plots reserved for this purpose were counted, as were the number of ears per plant, and all the ears from the counted plants were subsequently harvested. After the harvest, 20 ears were chosen at random and evaluated to obtain the average number of kernels and the weight of kernels per ear. The average weight of kernels was determined after the evaluation of sub-samples containing 1000 kernels, and the average weight is expressed in mg per kernel. Considering all the plants harvested in the useful area of the plot, the average kernel yield per plant was calculated. The final grain yield was estimated by multiplying the average yield of a single plant per density of plants of each treatment, and the result is expressed in kg ha^{-1}.

2.6. Statistical Analysis

Data were submitted to analysis of variance, using the F test at $p \leq 0.05$ and $p \leq 0.01$. Significant differences ($p \leq 0.05$) were assayed by polynomial regression analysis. As more than one statistical model was always significant (at least $p \leq 0.05$), the model having the highest coefficient of determination (R^2) was chosen. All statistical analyses were performed following indications of Barbosa and Maldonado Jr [33], using the software AgroEstat [33].

3. Results

3.1. Biometric Variables Determined in Plants Evaluated at the R4 Stage

Increasing plant density from 30,000 to 90,000 plants ha^{-1} decreased the kernel production per plant by about 21%, while the stalk decreased in a quadratic manner by 40%. Leaf dry matter varied in a quadratic model, showing a minimum (estimated by the model) at 71,192 plants ha^{-1} (Figure 2a). Values of this variable decreased by 27% from 30,000 to 71,192 plants ha^{-1}, with a little increase from this density to 90,000 plants ha^{-1} (Figure 2a). Plant leaf area decreased by 25% from the lowest to highest density studied (Figure 2b), while plant height increased by 5% (Figure 2c). In its turn, the leaf area index (area of leaves per area of soil cultivated) was increased in a quadratic model by about 54% from 30,000 to 90,000 plants ha^{-1} (Figure 2d).

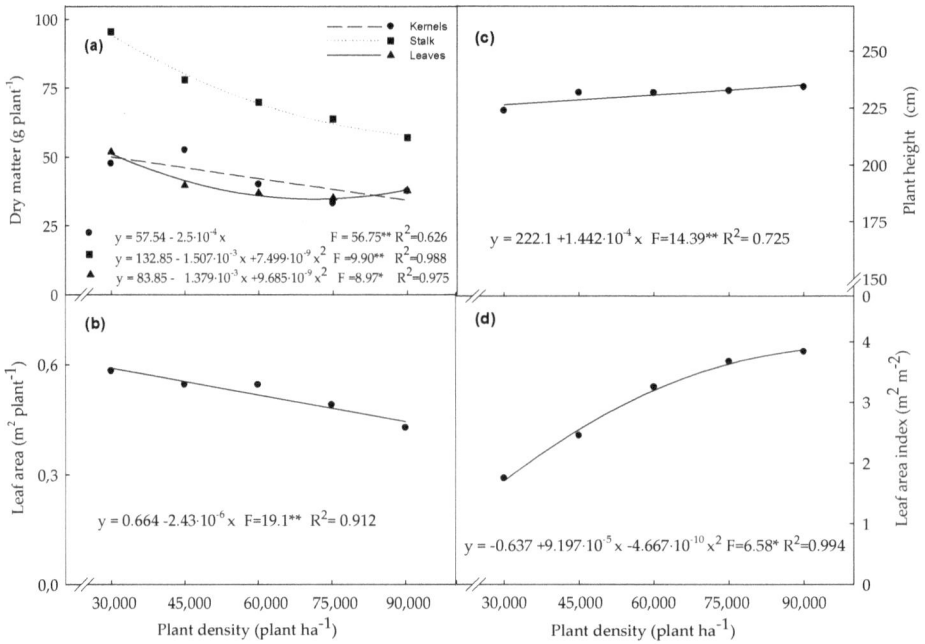

Figure 2. Kernel, leaf, and stalk dry weight per plant (**a**), plant leaf area (**b**), plant height (**c**), and leaf area index (**d**) of corn plants at the R4 stage, as a function of the plant density, ranging from 30,000 to 90,000 plants per hectare. Statistical data were obtained from four field replications. F = F test result of the regression analysis; (*) and (**) indicate that the F test revealed statistical significance at $p \leq 0.05$ and $p \leq 0.01$, respectively; R^2 = determination coefficient.

3.2. Carbohydrates Determined in Plants Evaluated at the R4 Stage

The reducing sugars concentration increased in the stalk by 8% as plant density increased from 30,000 to 90,000 plants ha^{-1}, while this carbohydrate fraction did not vary in kernels and leaves (Figure 3a). Plant density did not affect sucrose concentration in the plant (Figure 3b). Starch concentration remained constant in leaves and stalk, but increased by about 12% in kernels as plant density increased (Figure 3c). Considering that the concentration of carbohydrates did not vary dramatically in plant tissues, the total plant content of reducing sugars (Figure 3d), sucrose (Figure 3e), and starch (Figure 3f) varied mainly as a function of plant growth. Thus, data of the abovementioned carbohydrates presented behavior similar to great extent to those of the kernel, stalk, and leaf dry matter (Figure 2a).

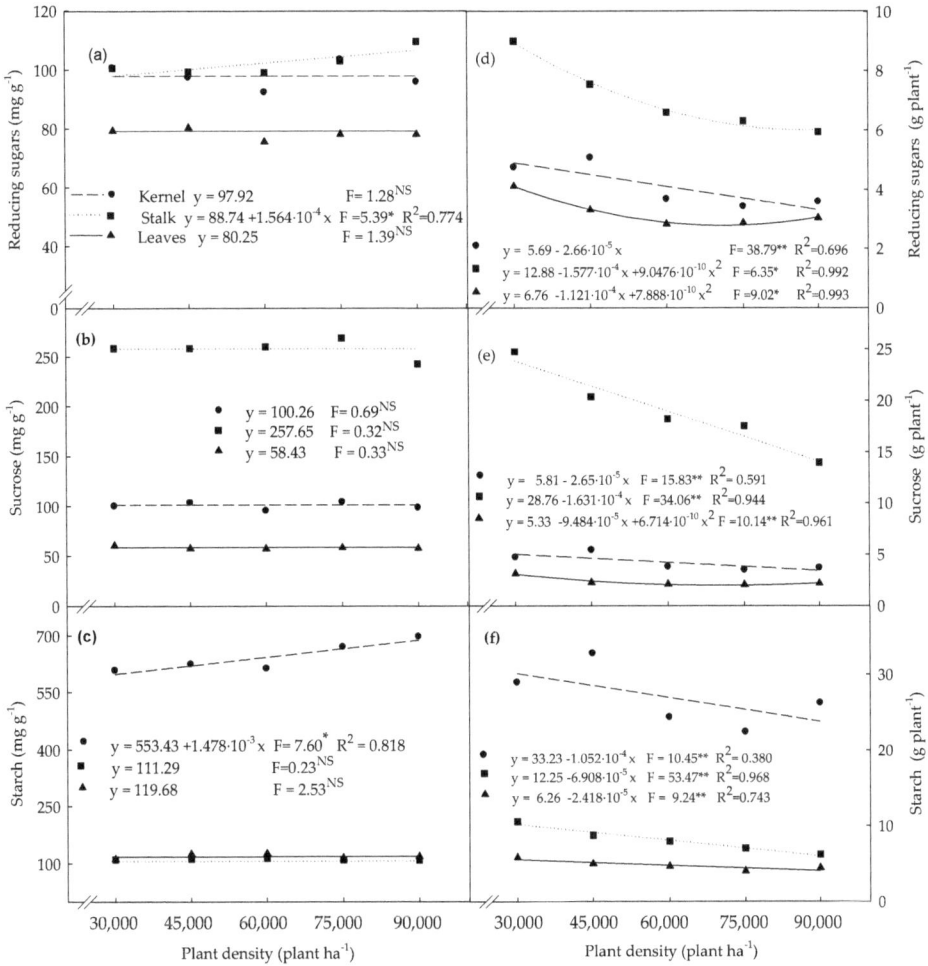

Figure 3. Concentration of reducing sugars (**a**), sucrose (**b**), and starch (**c**), and the amount per plant of reducing sugars (**d**), sucrose (**e**), and starch (**f**) in the kernel, leaf, and stalk of corn plants at the R4 stage, as a function of the plant density, ranging from 30,000 to 90,000 plants per hectare. Statistical data were obtained from four field replications. F = F test result of the regression analysis; (*) and (**) indicate that the F test revealed statistical significance at $p \leq 0.05$ and $p \leq 0.01$, respectively; R^2 = determination coefficient.

3.3. Enzymatic Activities Determined in Plants Evaluated at the R4 Stage

Specific activity (activity per unit of dry matter) of soluble invertase in stalk tissue showed a slight decrease (about 15%) as plant density increased from 30,000 to 90,000 plants ha^{-1}, while no variation was detected in leaves and kernels (Figure 4a). Also, soluble invertase activity in leaves showed values 23% higher than that observed in kernels, and about 10 times higher than that found in stalks (Figure 4a).

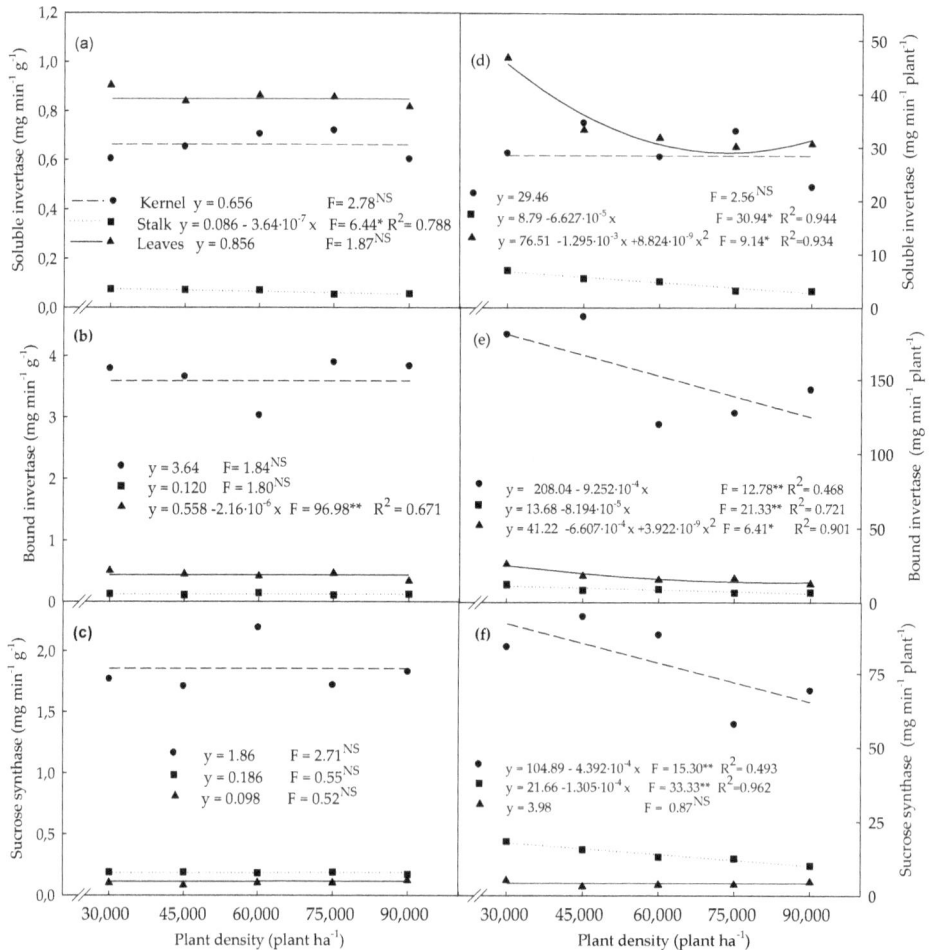

Figure 4. Specific activity of soluble invertase (**a**), bound invertase (**b**), and sucrose synthase (**c**), as well as activity per plant of soluble invertase (**d**), bound invertase (**e**), and sucrose synthase (**f**) in the kernel, leaf, and stalk of corn plants at the R4 stage as a function of the plant density ranging from 30,000 to 90,000 plants per hectare. Statistical data were obtained from four field replications. F = F test result of the regression analysis; (*) and (**) indicate that the F test revealed statistical significance at $p \leq 0.05$ and $p \leq 0.01$, respectively; R^2 = determination coefficient.

Similar to that observed for soluble invertase, bound invertase activity in kernels remained constant as plant density increased. However, the activity of bound invertase in kernels was 5.5 times higher than that of soluble invertase in this tissue (Figure 4b). Bound invertase activity in leaves was reduced by 40% as plant density increased (Figure 4b), and the values were lower (by about half) than those found for soluble invertase (Figure 4a). In its turn, bound invertase in stalk tissue did not respond to plant density, but the activity of this enzyme was 50% higher than that of soluble invertase activity (Figure 4a,b). The activity of bond invertase in kernel tissue was 6.5 times higher than that in leaves, and 30 times higher than that in stalk tissue. As observed in the data of soluble invertase, the activity of bound invertase detected in stalk was much smaller (4.6 times) than that found in leaves (Figure 4b). The activity of sucrose synthase did not change as a result of the increase in plant density (Figure 4c). Stalk tissue showed activity of sucrose synthase 47% higher than that in leaves, and in kernels the activity of this enzyme was about 10 times higher than that found in other studied plant parts (Figure 4c). The values of enzymes activity per plant were obtained by multiplying the specific activity of a tissue by the total dry matter of the respective plant part. So, taking into account that the specific activity of studied enzymes did not vary, or slightly varied in tissues of plants due to treatments (plant density)—in general, the results of soluble invertase (Figure 4d), bound invertase (Figure 4e), and sucrose synthase activity (Figure 4f) were greatly associated with the production of plant dry matter in each treatment (Figure 2a), as also observed for carbohydrates accumulation per plant (Figure 3d–f). In its turn, differences among specific enzymatic activity of tissues tended to be maintained throughout the results of activity per plant, because different tissues showed very distinct activities (Figure 4). The only exemption was verified for soluble invertase in kernels, because data variations in specific activity did not allow for the adjustment of an appropriate model for that variable (Figure 4a), and it seems that this should have reflected the values of activity per plant, leading to a constant response (Figure 4d) instead of a decreasing line as observed for kernel dry weight per plant (Figure 2a).

3.4. Agronomic Variables Determined in Plants Evaluated at Final Harest Time

Final kernel productivity increased with plant density from 30,000 to 90,000 plants ha^{-1} in a quadratic model, as presented in Figure 5a. The maximum productivity was not achieved with the highest plant density tested. However, by applying the mathematic model, it may be predicted to occur at a density of 100,650 plants ha^{-1}. The prolificity (average number of ears per plant) was reduced by 20% with increasing plant density, and this reduction occurred mainly with increasing the density from 30,000 to 60,000 plants ha^{-1} (Figure 5b). The number of kernels per ear (Figure 5c), dry matter of kernels produced per plant (Figure 5d), and kernel weight average (Figure 5e) were reduced proportionally to the increase of plant density, by 28%, 40%, and 12%, respectively, from 30,000 to 60,000 plants ha^{-1}. In its turn, the harvest index increased by 6% from 30,000 to 66,980 plants ha^{-1} (reaching a harvest index maximum estimated as 48.63%), then reduced 15% from densities of 66,980 to 90,000 plants ha^{-1} (Figure 5f).

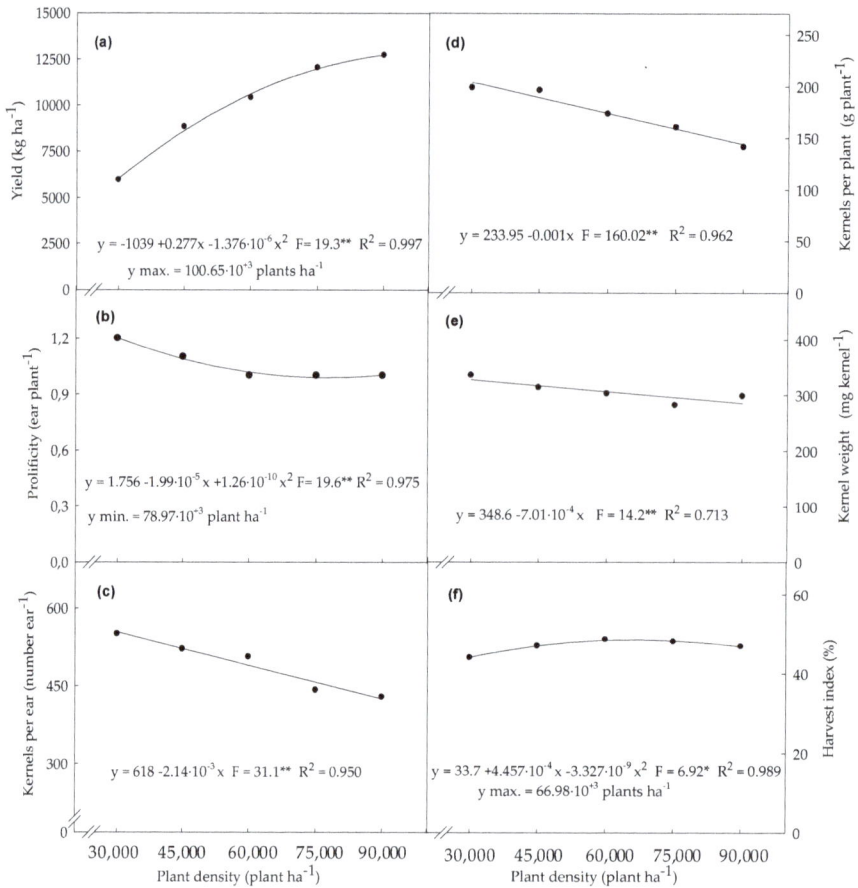

Figure 5. Yield (**a**), prolificity (**b**), number of kernels per ear (**c**), production of kernels per plant (**d**), kernel weight average (**e**), and harvest index (**f**) of corn plants at final harvest time as a function of the plant density. Plant density in the field ranged from 30,000 to 90,000 plants per hectare. Statistical data were obtained from four field replications. F = F test result of the regression analysis; (*) and (**) indicate that the F test revealed statistical significance at $p \leq 0.05$ and $p \leq 0.01$, respectively; R^2 = determination coefficient.

4. Discussion

It is already known that it is possible to improve maize productivity by increasing, to some extent, plant density [9–11]. Also, increasing plant density leads to an increase in plant competition for nutrient, water, light, and causes morphological alterations in corn plants [8,13,14,34]. So, it could be supposed that variables such as non-structural carbohydrates concentration and accumulation, as well as the activity of enzymes involved in carbon metabolism in leaves, stalk, and kernels, would also greatly change to sustain the increases in productivity, reported up to relatively high plant densities [6,16]. In this research, we submitted maize plants to different crop densities, from 30,000 to 90,000 plants ha^{-1}, aiming to study the expected changes in those variables. Biometric measurements were accomplished at the R4 stage to ensure that plants really had been subjected to sufficiently high competition, so that biochemical (metabolites and enzymatic activities) changes could be detected and studied in those plants. Also, agronomic variables were determined in plants evaluated at the final harvest time in order to study the yield, the yield components, and their relationships with the biochemical and biometrical results determined in plants at the R4 stage.

4.1. Biometric Variables Determined in Plants Evaluated at the R4 Stage

The range of plant densities imposed on maize crops in this study effectively caused significant biometrical alterations in plants at the R4 stage (Figure 2). This result is very important, because this allowed us to more clearly study the relationship of plant growth with the results of the biochemical determinations at this same stage. The dry weight of stalk was more sensitive to the effect of increases in plant density than that of leaves and kernels (Figure 2a), with respectively 40%, 27%, and 21% reductions. This increased sensitivity of stalk dry weight compared to the dry weight of other plant parts as a function of plant density was not found in previous critical reviews on this subject [6,16]. This sequence of decreasing effects suggest that plants in the stage of grain filling, and under increasing competition due to the increase of density, directs photoassimilates mainly to the kernels (sink), with the leaves (source) being the second priority and the stalk being the lowest priority, in order to maintain the production of photoassimilates demanded by kernels. Naturally, this fact can lead to the well-known problem of stem weakening and the loss of productivity by plants lodging under high plant densities [6,16]. Other authors verified that the increase in plant density reduces the duration of stalk internode thickening and dry matter accumulation [13], which helps to explain the results observed for stalk tissue (Figure 2a). Parallel to the decrease in dry weight, stalk length increased with plant density (Figure 2c). This was expected since high plant density reduces light intensity in the crop canopy, inhibiting the photodegradation of the hormone auxin, which is responsible for plant shoot elongation [13]. The similar models observed for kernel dry weight (Figure 2a) and leaf area (Figure 2b) per plant may be related to the dependence between kernel production and light interception [34]. Leaf dry weight per plant was reduced in a quadratic model with a minimum estimated at the point of 71,192 plants ha^{-1}; however, it tended to exhibit and increase from this point on (Figure 2a). This may be explained by the fact that distinct leaves of a maize plant subjected to the condition of increasing competition will change shape in order to adapt to this condition [8,9]. However, it is reported that these alterations begin as soon as mild competition is detected, but do not continue any further beyond severe competition [8]. While the leaf area per plant decreased by 5% as plant density varied from 30,000 to 90,000 plants ha^{-1} (Figure 2b), the leaf area index (LAI) increased by 54% (Figure 2d). This difference would justify why maize productivity usually is enhanced by increasing plant density (Figure 5a) even though individual plant growth (Figure 2a) and kernel production per plant is decreased by interplant competition (Figure 5c,d). This statement is supported by other studies suggesting that an increase in maize kernel productivity should be related to the increase in LAI values [8,9,13,14,16,34].

4.2. Non-Structural Carbohydrates Determined in Plants Evaluated at the R4 Stage

Taking into account the evident growth limitations of leaves, stem, and kernels of maize plants subjected to increasing plant density (Figure 2), as well as the reports that stalks may store non-structural carbohydrates under low sink demands and the partition from stalk to sink organs when photosynthesis is not paired with sink demands [2], changes in non-structural carbohydrate concentrations in plants under growing density also would have been expected. However, we found no alterations in the levels of non-structural carbohydrates in the leaf near the developing ear (Figure 3a–c), and no change of sucrose levels was verified in any studied part of the maize plants (Figure 3b). Stalks presented 5-fold more sucrose concentration than leaves, and 2.5-fold more than kernels (Figure 3b). Only slight increases of reducing sugars concentration in stalk tissue (8%) and of starch concentration in kernels (12%) were detected as plant density increased (Figure 3a,c). Previous research [13] reported that plants at stages before silking use the flux of water-soluble carbohydrates produced by the source (mature leaves) chiefly for plant morphological growth. After that, those compounds are used for the mechanical strength formation of stalk tissue, whose phase ends by the silking stage, and after that photoassimilates are driven to the ear under development. Thus, probably due to this competitive relationship between ear and stalk sinks, the continuous flux of water-soluble carbohydrates produced in the leaves did not accumulate in leaves, stalk or kernels. This may explained, at least in part, the lack of alterations verified in this research. So, in our study we could not confirm, for maize plants with increasing density, the general statement that grasses store carbohydrates in stem tissue, whose reserve is used to improve yield stability in grain crops by providing an alternative source to complete grain filling when photosynthesis is not enough or during phases of plant stress [2]. It is interesting to note that there occurred a slight increase in kernel starch concentration with increasing plant density. Schlüter et al. [35] reported that, at least in leaves, starch production was upregulated under stress (low-N conditions), probably to reduce starch turnover, but that the sucrose concentration did not change. So, this plant mechanism is one possibility to explain, at least in part, the increase in starch in kernels of plants under increasing stress (increasing plant density). Another possibility is that the increasing plant density could have accelerated the kernel filling and maturation process [36]; thus, kernels from plants at a higher density were further ahead in finalizing kernel filling and reducing the demand for photoassimilates. This latter theory could also explain the greater starch concentration, and the supposed reduction in kernel sink demand would have induced the reducing sugars to accumulate in the stalk (Figure 3a). Although values of non-structural carbohydrates concentrations were maintained at relatively stable levels (or slightly increased, as in the case of reducing sugars in the stalk and starch in kernels), the total amounts estimated per plant tended to follow the data found for stalk, leaf, and kernels dry matter (Figure 3d–f).

4.3. Enzymatic Activities Determined in Plants Evaluated at the R4 Stage

It is well known that sucrose is the main photosynthetic compound in leaves, which is partitioned from source leaves to the sink organs of maize plants [19–23]. Also, it is already known that invertases are enzymes that irreversibly catalyze the cleavage of sucrose into reducing sugars (glucose + fructose), while sucrose synthase is a cytosolic enzyme that reversibly catalyzes sucrose cleavage, but whose reaction relies on the production of ADP- and UDP-glucose for starch and cell wall polysaccharide synthesis [23]. Thus, alterations in the activity of those enzymes would be expected in plants subjected to the stress of increasing plant density. However, we did not find a clear relationship among the specific activities of soluble invertase (Figure 4a), bound invertase (Figure 4b), and sucrose synthase (Figure 4c) with plant density. Probably, this lack of response is related to the relatively stable contents of soluble sugars observed in the same plants (Figure 2). In a previous study, using the third basal internode of younger maize plants (V6 stage), a decrease in soluble carbohydrate contents was reported as plant density increased [13]. This suggests that in the V6 stage, a difference in the activity of related enzymes would also be found, and it is indicative that the activity of the mentioned enzymes may vary with plant stage. We did not find reports of research on enzymatic activities in adult plants subjected to increased density. Thus, further efforts should be made to better understand the results found in this study, and to elucidate the carbohydrate metabolism of plants under high densities. However, it is already known that genotypes with increased cell wall invertase activities in different tissues and organs, including leaves and developing seeds, exhibit substantially improved grain yield [20] and grain nutrients [22]. This suggests that distinct genotypes may also lead to different responses when grown under increasing plant density.

Notwithstanding the fact that the activity of enzymes did not change with plant density, the specific activities of soluble invertase in stalk, leaves, and kernels (Figure 4a) were opposite to the behavior of the sucrose concentration in the respective plant parts (Figure 3b), while the activity of bound invertase (Figure 4b) was more related to the starch concentration (Figure 3c). Similar to bound invertase, the sucrose synthase activity (Figure 4c) was also greatly related to starch levels in the studied tissues (Figure 3c). These results suggest that soluble invertase activity was the main factor responsible for controlling the transport and accumulation of sucrose in maize plant tissues, while bound invertase was more related to the loading of sucrose into cells of tissues where starch should be deposited.

4.4. Agronomic Variables Determined in Plants Evaluated at Final Harvest Time

The increase in yield with the increase in plant density up to a maximum value at a threshold plant density value, with a decline after this point, has already been extensively reported [6,9–11,37]. So, the obtained result (Figure 5a) was already expected. The only uncertainty concerned the absolute value of kernel productivity, and the respective plant density in which maximum productivity would be obtained. The maximum yield was not achieved within the plant density range tested in this study, but it was predicted by the statistical model (Figure 5a) to occur at densities of 12,901 kg ha^{-1} and 100,650 plants ha^{-1}. This result is in accordance with those obtained by Testa et al. [11], who also proved that a high planting density of up to 100,500 plants ha^{-1} can lead to a sensitive improvement in grain yield when plants are grown in narrow (0.5 m) inter-row spacing. This high limit of plant density was reached in this assay probably as a result of the relatively optimal environmental conditions (Figure 1) and limited nutritional stress (proper fertilization) under which the plants developed. This statement is supported by the fact that the greater the nutritionally and environmentally limitations that a maize crop is submitted to, the lower the threshold plant density in which maximum productivity is reached [6,10]. Also, if plants are grown under severe limitations such as low N fertilization, the yield productivity may even not respond positively to an increase in plant density [9].

While the prolificity (Figure 5b), number of kernels per ear (Figure 5c), dry weight of kernels produced per plant (Figure 5d), and average weight per kernel (Figure 5d) decreased with increasing plant density, the harvest index was enhanced up to a density of 66,982 plants ha^{-1} (density value estimated by the statistical model), after which a decreasing trend became noticeable (Figure 5f). Although there are many discussions about harvest index, and on the variation in the methodology used by distinct researchers to obtain this index [38], the phenomenon of the highest value of harvest index occurring at an intermediate plant density has also been observed in other studies [9,38,39]. In this research it is clear that such a phenomenon occurred because the kernel production per plant reduced in a linear fashion (Figures 3a and 5e), while the values for stalk and leaves (and cob + ear straw, data not shown) decreased under quadratic models. In particular, those quadratic models showed sharp reductions from low to intermediate plant densities, and a very small decrease (in the case of stalk tissue) or even a slight increase (as observed for leaves) in the range from intermediate to high plant densities (Figure 2a). In conjunction, it can be depicted by the data of the biometrical variables that, although the single plant yield potential was reduced by increasing plant density, the increase in plant number per area circumvented this individual reduction and led to an overall increase in kernel productivity.

Future research is needed to determine whether the same lack of response we found by analyzing plants at the R4 stage would occur in other plant stages, when the stalk and tassels are still functioning as sinks and the kernels are not yet under formation. In this study, we analyzed only metabolite concentrations and enzymes activity in a specific part of the stalk (just below the main ear), and in a specific type of leaf (near the main ear). So, studies using distinct parts of the stalk and upper and lower leaves may help to further the understanding of why the metabolic variables evaluated in the studied tissues did not respond to increasing plant density.

5. Conclusions

In summary, we found that the dry weight of stalk tissue was more sensitive to the effect of increases in plant density than those of leaves and kernels. The stalk and the leaf dry weight per single plant decreased in a quadratic model, while the production of grain per plant was reduced linearly with increasing plant density, leading to higher values of harvest index in intermediate plant densities. The sucrose concentration did not change in leaves, stalk, or kernels of plants subjected to increased plant density, evaluated at the R4 stage. The specific activity of soluble invertase, bound invertase, and sucrose synthase did not change in leaves, stalk, or kernels of plants subjected to increased plant density. The productivity was increased with the increase in plant density by using narrow row (0.45 m) spacing in the studied plant density range.

Author Contributions: Both authors were involved in all phases of this research development and manuscript preparation.

Funding: This research was funded by FAPESP—Fundação de Amparo à Pesquisa do Estado de São Paulo.

Acknowledgments: Authors acknowledge the administrative and technical support given by Funep—Fundação de Apoio à Pesquisa, Ensino e Extensão. The first author are grateful to the CNPq (National Council for Scientific and Technological Development) for the granting of research productivity scholarship.

Conflicts of Interest: The authors declare no conflict of interest.

References

1. Gregory, P.J.; George, T.S. Feeding nine bilion: The challenge to sustainable production. *J. Exp. Bot.* **2011**, *62*, 5223–5239. [CrossRef] [PubMed]
2. Slewinski, T.L. Non-structural carbohydrates partitioning in grass stems: A target to increase yield stability, stress tolerance, and biofuel production. *J. Exp. Bot.* **2012**, *63*, 4647–4670. [CrossRef] [PubMed]
3. FAO. FAOSTAT-Agriculture Database. 2018. Available online: http://faostat.fao.org/site/ (accessed on 1 August 2018).
4. White, E.G.; Moose, E.P.; Weil, C.F.; McCann, M.C.; Carpita, N.C.; Below, F.E. Tropical maize: Exploiting maize genetic diversity to develop a novel annual crop for lignocellulosic biomass and sugar production. In *Routs to Cellulosic Ethanol*; Buckeridge, M.S., Goldman, G.G., Eds.; Springer: New York, NY, USA, 2011; pp. 167–179.
5. Lobell, D.B.; Cassman, K.G.; Field, C.B. Crop yield gaps: Their importance, magnitudes, and causes. *Annu. Rev. Environ. Resour.* **2009**, *34*, 179–204. [CrossRef]
6. Sangoi, L. Understanding plant density effects on maize growth and development: An important issue to maximize grain yield. *Ciênc. Rural* **2000**, *31*, 159–168. [CrossRef]
7. Bisht, A.S.; Bhatnagar, A.; Sing, V. Influence of plant density and integrated nutrient management on N, P, and K contents and uptake of quality protein maize. *Madras Agric. J.* **2013**, *100*, 110–112.
8. Song, Y.; Rui, Y.; Bedane, G.; Li, J. Morphological characteristics of maize canopy development as affected by increased plant density. *PLoS ONE* **2016**, *11*, 1–10. [CrossRef] [PubMed]
9. Ciampitti, I.A.; Vyn, T.J. A comprehensive study of plant density consequences on nitrogen uptake dynamics of maize plants from vegetative to reproductive stages. *Field Crops Res.* **2011**, *121*, 2–18. [CrossRef]
10. Tokatlidis, I.S.; Hasb, V.; Melidisc, V.; Hasb, I.; Mylonasa, I.; Evgenidisc, G.; Copandeanb, A.; Ninouc, E.; Fasoulad, V.A. Maize hybrids less dependent on high plant densities improve resource-use efficiency in rainfed and irrigated conditions. *Field Crops Res.* **2011**, *120*, 345–351. [CrossRef]
11. Testa, G.; Reyneri, A.; Blandino, M. Maize grain yield enhancement through high plant density cultivation with different inter-row and intra-row spacings. *Eur. J. Agron.* **2016**, *72*, 28–37. [CrossRef]
12. Yan, P.; Zhang, Q.; Shuai, X.F.; Pan, J.X.; Zhang, W.J.; Shi, J.F.; Wang, M.; Chen, X.P.; Cui, Z.L. Interaction between plant density and nitrogen management strategy in improving maize grain yield and nitrogen use efficiency on the North China Plain. *J. Agric. Sci.* **2016**, *154*, 978–988. [CrossRef]
13. Xue, J.; Zhao, Y.; Gou, L.; Shi, Z.; Yao, M.; Zhang, W. How High Plant Density of Maize Affects Basal Internode Development and Strength Formation. *Crop Sci.* **2016**, *56*, 3295–3306. [CrossRef]
14. Huang, S.; Gaoa, Y.; Lia, Y.; Xub, L.; Taoa, H.; Wanga, P. Influence of plant architecture on maize physiology and yield in the Heilonggang River valley. *Crop J.* **2017**, *5*, 52–62. [CrossRef]
15. Al-Naggar, A.M.M.; Shabana, R.A.; Atta, M.M.M.; Al-Khalil, T.H. Maize response to elevated plant density combined with lowered N-fertilizer rate is genotype-dependent. *Crop J.* **2015**, *3*, 96–109. [CrossRef]
16. Sher, A.; Khan, A.; Cai, L.J.; Ahmad, M.I.; Asharf, U.; Jamoro, S.A. Response of maize grown under high plant density; performance, issues and management—A critical review. *Adv. Crop Sci. Tech.* **2017**, *5*, 1–8. [CrossRef]
17. Xu, Z.; Lai, T.; Li, S.; Si, D.; Zhang, C.; Cui, Z.; Chen, X. Promoting potassium allocation to stalk enhances stalk bending resistance of maize (*Zea mays* L.). *Field Crops Res.* **2018**, *213*, 200–206. [CrossRef]

18. Shiferaw, B.; Prasanna, B.M.; Hellin, J.; Bänziger, M. Crops that feed the world. Past successes and future challenges to the role played by maize in global food security. *Food Sec.* **2011**, *3*, 307. [CrossRef]

19. Cazetta, J.O.; Seebauer, J.R.; Below, F.E. Sucrose and nitrogen supplies regulate growth of maize kernels. *Ann. Bot.* **1999**, *84*, 747–754. [CrossRef]

20. Li, B.; Liu, H.; Zhang, Y.; Kang, T.; Zhang, L.; Tong, J.; Xiao, L.; Zhang, H. Constitutive expression of cell wall invertase genes increases grain yield and starch content in maize. *Plant Biotechnol. J.* **2013**, *11*, 1080–1091. [CrossRef] [PubMed]

21. Jiang, S.; Chi, Y.; Wang, J.; Zhou, J.; Cheng, Y.; Zhang, B.; Ma, A.; Vanitha, J.; Ramachandran, R. Sucrose metabolism gene families and their biological functions. *Nature* **2015**, 17583. [CrossRef] [PubMed]

22. Guo, X.; Duan, X.; Wu, Y.; Cheng, J.; Zhang, J.; Zhang, H.; Li, B. Genetic Engineering of Maize (*Zea mays* L.) with Improved Grain Nutrients. *J. Agric. Food Chem.* **2018**, *66*, 1670–1677. [CrossRef] [PubMed]

23. Tong, X.; Wang, Z.; Ma, B.; Zhang, C.; Zhu, L.; Ma, F.; Li, M. Structure and expression analysis of the sucrose synthase gene family in apple. *J. Integr. Agric.* **2018**, *17*, 847–856. [CrossRef]

24. Van Raij, B.; Andrade, J.C.; Cantarella, H.; Quaggio, J.A. *Análise Química Para Avaliação da Fertilidade de Solos Tropicais*; Instituto Agronômico: Campinas, Brasil, 2001; 285p, ISBN 85-85564-05-9.

25. Van Raij, B.; Cantarella, H. Cereais: Milho para grão e silagem. In *Recomendações de Calagem e Adubação Para o Estado de São Paulo*; van Raij, B., Cantarella, H., Quaggio, J.A., Furlani, A.M.C., Eds.; Instituto Agronômico de Campinas: Campinas, Brasil, 1997; pp. 56–59.

26. Faleiros, R.R.S.; Seebauer, J.R.; Below, F.E. Nutritionally induced changes in endosperm of *shrunken-1* and *brittle-2* maize kernels growth in vitro. *Crop Sci.* **1996**, *36*, 947–954. [CrossRef]

27. Miller, G.L. Use of dinitrosalisylic acid reagent for determination of reducing sugars. *Ann. Chem.* **1959**, *31*, 426–428. [CrossRef]

28. Fieuw, S.; Willenbrink, J. Sucrose synthase and sucrose phosphate synthase in sugar beet plants (*Beta vulgaris* L. ssp. altissima). *J. Plant Physiol.* **1987**, *131*, 153–162. [CrossRef]

29. Hendrix, D.L. Rapid extraction and analysis of nonstructural carbohydrates in plant tissues. *Crop Sci.* **1993**, *33*, 1306–1311. [CrossRef]

30. Brow, C.S.; Huber, S.C. Reserve mobilization and starch formation in soybean (*Glycine max*) cotyledon in relation to seedling growth. *Plant Physiol.* **1988**, *72*, 518–524. [CrossRef]

31. Singletary, G.W.; Doehlert, D.C.; Wilson, C.M.; Muhtich, M.J.; Below, F.E. Response of enzymes and storage proteins of maize endosperm to nitrogen supply. *Plant Physiol.* **1990**, *94*, 858–864. [CrossRef] [PubMed]

32. Doehlert, D.C.; Felker, F.C. Characterization and distribution of invertase activity in developing maize (*Zea mays*) kernels. *Physiol. Plant.* **1987**, *70*, 51–57. [CrossRef]

33. Barbosa, J.C.; Maldonado, W., Jr. *Experimentação Agronômica & AgroEstat: Sistema Para Análise Estatística de Ensaios Agronômicos*; Gráfica Multipress: Jaboticabal, Brazil, 2015; ISBN 978-85-68020-01-02.

34. Balkcom, K.S.; Satterwhiteb, J.L.; Arriagaa, F.J.; Pricea, J.A.; Santenb, E.V. Conventional and glyphosate-resistant maize yields across plant densities in single- and twin-row configurations. *Field Crops Res.* **2011**, *120*, 330–337. [CrossRef]

35. Schlüter, U.; Mascher, M.; Colmsee, C.; Scholz, U.; Bräutigam, A.; Fahnenstich, H.; Sonnewald, U. Maize source leaf adaptation to nitrogen deficiency affects not only nitrogen and carbon metabolism but also control of phosphate homeostasis. *Plant Physiol.* **2012**, *160*, 1384–1406. [CrossRef] [PubMed]

36. Rajcan, I.; Tollenaar, M. Source: Sink ratio and leaf senescence in maize: II. Nitrogen metabolism during grain filling. *Field Crops Res.* **1999**, *60*, 255–265. [CrossRef]

37. Tollenaar, M.; Aguilera, A.; Nissanka, S.P. Grain yield is reduced more by weed interference in an old than in a new maize hybrid. *Agron. J.* **1997**, *89*, 239–246. [CrossRef]

38. Unkovich, M.; Baldock, J.; Forbes, M. Variability in Harvest Index of Grain Crops and Potential Significance for Carbon Accounting: Examples from Australian Agriculture. *Adv. Agron.* **2010**, *105*, 173–219. [CrossRef]

39. Demetrio, C.S.; Fornasieri Filho, D.; Cazetta, J.O.; Cazetta, D.A. Desempenho de híbridos de milho submetidos a diferentes espaçamentos e densidades populacionais. *Pesqui. Agropecu. Bras.* **2008**, *43*, 1691–1697. [CrossRef]

agronomy

Article

Maize Canopy Photosynthetic Efficiency, Plant Growth, and Yield Responses to Tillage Depth

Jiying Sun [1],*[ORCID], Julin Gao [1],*, Zhigang Wang [1], Shuping Hu [2], Fengjie Zhang [1], Haizhu Bao [1] and Yafang Fan [1]

[1] College of Agronomy, Inner Mongolia Agricultural University, No.275, XinJian East Street, Hohhot 010019, China; imauwzg@163.com (Z.W.); zhangfengjie1990@126.com (F.Z.); bhz2009@126.com (H.B.); 15848921389@163.com (Y.F.)

[2] Vocational and Technical College, Inner Mongolia Agricultural University, Baotou 014109, China; bthsp88@163.com

* Correspondence: jiying-sun@imau.edu.cn (J.S.); julin-gao@imau.edu.cn (J.G.); Tel.: +86-139-4713-0409 (J.S.)

Received: 27 October 2018; Accepted: 8 December 2018; Published: 21 December 2018

Abstract: Subsoil tillage loosens compacted soil for better plant growth, but promotes water loss, which is a concern in areas that are commonly irrigated. Therefore, our objective was to determine the physiological responses of high yield spring maize (*Zea mays* L.) to subsoil tillage depth when grown in the Western plain irrigation area of Inner Mongolia, China. Our experiment during 2014 and 2015 used Zhengdan958 (Hybrid of Zheng58 × Chang7-2, produced by Henan academy of agricultural sciences of China, with the characteristics of tight plant type and high yield) and Xianyu335 (Hybrid of PH6WC × PH4CV, produced by Pioneer Corp of USA, with the characteristic of high yield and suitable of machine-harvesting) with three differing subsoil tillage depths (30, 40, or 50 cm) as the trial factor and shallow rotary tillage as a control. The results indicated that subsoil tillage increased shoot dry matter accumulation, leading to a greater shoot/root ratio. Subsoil tillage helped retain a greater leaf area index in each growth stage, increased the leaf area duration, net assimilation rate, and relative growth rate, and effectively delayed the aging of the blade. On average, compared with shallow rotary, the grain yields and water use efficiency increased by 0.7–8.9% and 1.93–18.49% in subsoil tillage treatment, respectively, resulting in the net income being increased by 2.24% to 6.97%. Additionally, the grain yield, water use efficiency, and net income were the highest under the treatment of a subsoil tillage depth of 50 cm. The results provided a theoretical basis for determining the suitable chiseling depth for high-yielding spring corn in the Western irrigation plains of Inner Mongolia.

Keywords: chiseling depth; spring corn; canopy characteristics; photosynthesis quality; Inner Mongolia

---

1. Introduction

Soil is an important carrier of crop growth, and improving soil quality can effectively improve crop yield. Crop yields, in turn, are directly affected by the quality of the ploughing layer. The average effective ploughing layer of the irrigated area in the Inner Mongolia Plain is only 15.1 cm, which is less than China's average value of 16.5 cm and far shallower than that in North America, which has an average effective plough layer of 35 cm. As maize roots are mainly distributed in the soil layer of 3–35 cm [1], the compacted soil below the plough layer not only restricts the development of plant roots, but also hinders the absorption of mineral nutrients and water for plants, and therefore reduces the production capacity of the plant canopy as well as limits the grain yield [2,3]. Notably, appropriate soil tillage measures have great effects for improving the soil's physical and chemical properties, farmland soil quality, and maize photosynthesis efficiency.

Canopy structure greatly influences the leaf photosynthetic rate [4]. Many studies have focused on the canopy structure and physiological characteristics of corn in Inner Mongolia and abroad [5–8].

Leaves are the main organs for photosynthesis, which account for about 95% of the total in maize. Theoretically, yield would be increased by about 1–2% per day if the duration of photosynthetic is extended in maturity [9,10]. The rate of photosynthesis and grain filling in plants are directly affected by the leaf area duration and leaf area index (LAI). The LAI represents the amount of leaf area and is an important quantitative index of canopy structure [11–13]. Leaf senescence is dependent on LAI; the higher the LAI, the less senescence of corn leaves. Within a range from 0 to the optimum maximum leaf area index (e.g., the optimum maximum leaf area index for maize is 5–6), the larger the LAI, the greater the solar utilization efficiency [14–16], and LAI is closely related to the grain number and weight during the filling stage [17,18]. Early research showed that LAI increased with an increased plant density from the jointing stage to the 12-leaves stage, and peaked at the silking stage, which laid a foundation for the accumulation of dry matter in the later period of flowering [19]. Some studies demonstrated that modern maize varieties had longer growth periods, larger leaf areas, and slower leaf senescence, leading to a significant increase of the dry matter accumulation rate compared to early varieties; thus, dry matter accumulation is closely related to leaf senescence [20]. Dry matter accumulation can be improved by increasing the dry matter production rate and duration, which directly affects the corn yield.

However, linkages between the dynamics depth of subsoil tillage and photosynthetic characteristics in different spring corn on the irrigated area of the Inner Mongolia Plain have not been revealed. Therefore, with a comprehensive database, we aimed to (1) investigate the effects of the subsoil tillage depth on the canopy's photosynthetic efficiency, plant growth, and yield of maize, and (2) to fully explore the mechanisms behind the observations.

2. Materials and Methods

2.1. Experimental Site

The field experiment was conducted during 2014 to 2015 in Hulutou village and Zhuergedai village, Salaqi Town, Tumd Right County, Baotou, Inner Mongolia, China. The region, located in the Tumochuan plain has a typical continental semi-arid monsoon climate. Specifically, the annual mean temperature and annual mean precipitation are 7.5 °C and 346 mm, respectively, and maximum temperatures occurred in July (average of 22.9 °C). The region experiences 135 frost-free days and an average of 3095 annual sunshine hours. Drought is the main factor that affected the yield in 2015. The precipitation in 2014 and 2015 varied greatly, with 457.4 mm in 2014 and 230.3 mm in 2015 from May to October, while slight fluctuations of the monthly mean temperature were observed with the highest mean temperature of 23.9 °C in July 2014 and 24.6 °C in July 2015, and the lowest mean temperature of 11.2 °C in October 2014 and 9.7 °C in October 2015 (Figure 1).

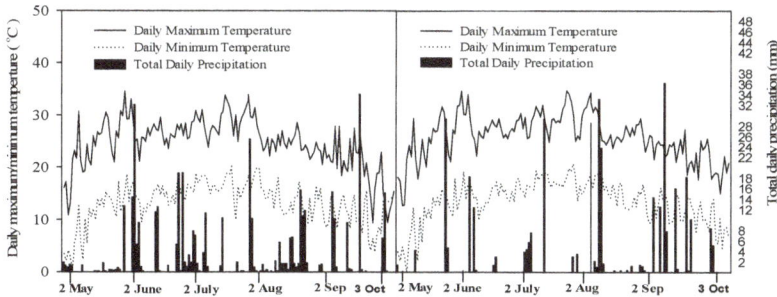

Figure 1. Precipitation and monthly average temperature at the study site during 2014–2015.

Soil properties of the experimental sites are shown in Table 1. The preceding crop (spring maize) in the experimental field, which was subjected to conventional shallow rotary tillage for three years, and the maize stalks were shredded and returned to the field after the corn was harvested.

Table 1. Soil properties in 2014 and 2015 years.

Year	Soil Organic C (g·kg^{-1})	Soil Available N (mg·kg^{-1})	Soil Available P (mg·kg^{-1})	Soil Available K (mg·kg^{-1})
2014 matter	7.30	73.45	15.10	120.40
2015	7.60	77.05	14.05	118.80

2.2. Experimental Treatments and Design

Subsoil tillage depth (30 cm, 40 cm, and 50 cm, designated CH30, CH40, or CH50, respectively) was the trial factor, and was compared to a control of shallow rotary tillage (SR). In 2014, the tested variety was Xianyu335 with a row distance of 50 cm and a plant density of 82,500 plants per ha. Each subplot was an area of 125 m^2, 5 m wide, and 25 m long, with three replications, for a total of 12 plots in a randomized block arrangement. A split-plot design was used for the experiment in 2015, with the subsoil tillage depth (30 cm, 40 cm, 50 cm, and control) as the main factor and varieties (Xianyu335 and Zhengdan958) as the sub-factor. The plant density, row distance, plot size, and the replications were the same with 2014, for a total of 24 plots. All in-crop fertilizer of nitrogen fertilizer, phosphate fertilizer and potash fertilizer was applied at planting (N: 200 kg/ha, P$_2$O$_5$: 105 kg/ha and K$_2$O: 62 kg/ha). Phosphate fertilizer and potash fertilizer were applied as basal fertilizer once before planting and nitrogen fertilizer was applied by 30% (60 kg/ha) at the V6 stage (six leaves with collars visible) and 70% (140 kg/ha) at the V12 stage (12 leaves with collars visible), respectively. Irrigation and other management measures during the whole growth period were similar to local farmer practices. The precipitation and the irrigation rate during the growth stage was recorded.

2.3. Measured Parameters

2.3.1. Leaf Area

Plants were sampled at five growth stages: V6 stage (six leaves with collars visible), V12 stage (12 leaves with collars visible), R1 stage (silking stage), R3 stage (filling stage), and R6 stage (physiological maturity) by three areas in each plot, and in each area, three continuous plants (9 plant per plot) were sampled.

Leaf area was calculated by the leaf length and leaf width at V6, V12, R1, R3, and R6, respectively:

$$A_1 = 0.75 \times \text{Leaf Length} \times \text{Leaf width} \tag{1}$$

$$A_2 = 0.50 \times \text{Leaf Length} \times \text{Leaf width} \tag{2}$$

where A_1 was the area of expanded leaves and A_2 represents the area of unexpanded leaves.

2.3.2. Accumulation and Transport of Dry Matter

At the V6 stage, V12 stage, R1 stage, R3 stage, and R6 stage, three sample areas per plot were chosen, and three uniform plants per area were selected and divided into four parts: Leaf, stem and sheath, female ear, and grain. At the R1 stage, the corresponding roots were examined in the depth of 0–60 cm. Fresh samples were deactivated at 105 °C for 30 min, and dried at 80 °C to a constant weight in an oven, then dry matter of the leaf, stem and sheath, female ear, and grain were weighed.

2.3.3. Photosynthesis Related Parameters

Gas exchange measurements (photosynthesis rate, transpiration rate, stomatal conductance, and intercellular CO$_2$ concentration) were taken between 9:00 am to 11:00 am at the R3 stage on the ear leaf of three uniform plants from three selected sample areas within each plot using a Li-6400XT

Portable Photosynthesis System (Li-Cor Inc., Lincoln, NE, USA). The chamber was adjusted to 25 °C (temperature), 360 μM/mol (CO_2), and 800 μM/m^2/s (photosynthetic photon-flux density).

2.3.4. Production of Photosynthesis

Leaf water use efficiency (LWUE), leaf area duration (LAD), and net assimilation rate (NAR) were calculated by the following formula:

$$\text{LWUE } (\mu molCO_2/mmolH_2O) = Pn/Tr \tag{3}$$

where LWUE is the water use efficiency at the leaf level, Pn is the photosynthesis ratio of the ear leaf at the R3 stage, and Tr is the transpiration ratio of the ear leaf at the R3 stage.

$$\text{LAD } [(m^2 \cdot d)/hm^2] = [(L_1 + L_2)/2] \times (t_2 - t_1) \tag{4}$$

$$\text{NAR } [g/(m^2 \cdot d)] = [(\ln L_2 - \ln L_1) \times (W_2 - W_1)]/[(L_2 - L_1) \times (t_2 - t_1)] \tag{5}$$

where L_1 and L_2 are the leaf area at time t_1 and t_2, respectively; W_1 and W_2 are the dry matter weight at time t_1 and t_2, respectively.

2.3.5. Stover Biomass and Grain Yield

To evaluate the stover biomass and grain yield of maize, plants were sampled at the R6 stage. Sampling consisted of manually excising plants from each plot at R6 (three sample areas for each plot, each sample area had six continuous plants, 18 plants per plot) to determine the stover biomass. The plants at R6 were partitioned into the grain and stover (including husk) components, the total fresh stover was dried to 0% moisture and then weighed. Corn ears were weighed to obtain the grain and cob weight. The grain was removed manually to analyze the moisture content with a seed moisture meter (PM-8188-A, KETT ELECTRIC LABORATORY, Tokyo, Japan), the cob and the grain were dried and weighed again. Dry stover and dry cob weights were summed to calculate the overall R6 stover biomass. 300 randomly selected kernels were weighed to estimate the average individual kernel weight. The kernel number was estimated by dividing the grain yield by the average individual kernel weight of each plot; the kernel number per corn ear was estimated by dividing the kernel number by the number of the corn ear of each plot. All biomass and grain weight measurements were presented on a 0% moisture concentration basis.

Plant stand counts were tallied to confirm plant populations at the R6 plant growth stage. Additionally, ear stand counts were tallied to confirm the ear number per hm^2. The center two rows of each plot were manually harvested to determine the grain yield at physiological maturity, and yield values were presented at 0% moisture concentration too.

2.3.6. Yield Water Use Efficiency (YWUE) and Water Production Efficiency (WPE)

YWUE is defined as the ratio of the grain yield to the water consumed, and it is used to evaluate the plant growth suitability under a water deficit. The WUE was calculated by:

$$\text{YWUE } (kg/ha \cdot mm) = GY/ET \tag{6}$$

where YWUE is the water use efficiency at the yield level, GY is the grain yield, and ET is the maize water consumed.

WPE is defined as the ratio of biomass yield to the water consumed, and it also is an effective indicator for evaluating the plant growth ability. The WPE was calculated by:

$$\text{WPE } (kg/ha \cdot mm) = BY/ET \tag{7}$$

where WPE is the water production efficiency, BY is the biomass yield, and ET is the maize water consumed.

Evapotranspiration is the sum of evaporation from the land surface plus transpiration from plants. The evapotranspiration (ET) was calculated by:

$$ET = P + I + \Delta SWS \tag{8}$$

where ET is the maize water consumed, P is the precipitation during the growth stage, I is the irrigation rate during the growth stage, and ΔSWS is the balance between the soil pondage in the sowing time and the harvest time.

Soil water storage is an important parameter of soil water conservation and field irrigation. The SWS was calculated by:

$$SWS = \text{Soil depth} \times \text{Soil volume weight} \times \text{Soil moisture} \tag{9}$$

At sowing time and harvest time, 0–100 cm depth soils were sampled to measure the soil volume weight and soil moisture from five areas of each plot, using the cutting ring to sample in 0–20 cm, 20–40 cm, 40–60 cm, 60–80 cm, and 80–100 cm soil, with three cutting rings for each soil layer. The fresh weight was weighed, the soil from the cutting ring was excavated, and the soil was put into the oven to dry to 0% moisture. Then it was weighed again, the volume of the cutting ring was measured, and the soil moisture and soil volume weight were calculated.

2.4. Statistical Analysis

Analysis of variance (ANOVA) was used to evaluate treatment effects. Data of the biomass accumulation, leaf area index, ear leaf photosynthetic assimilation, transpiration rate, stomatal conductance and intercellular CO_2 concentration, leaf water use efficiency, dynamics of leaf area duration, net assimilation rate, maize yield, yield components, and economic benefit were analyzed with variance analysis of SAS (Statistical Analysis System) [21]. Pearson's correlation coefficient was used to evaluate the linear association between the grain yield and measured parameters across all treatments and within each rotation, by the correlation analysis procedure of SAS.

3. Results

3.1. Effect of Subsoil Tillage Depth on Dry Matter Accumulation of Maize

With the development of the growth stage, the accumulation of dry matter gradually increased. In 2014, the accumulation of dry matter in the subsoil tillage treatment was significantly higher than that in control during all the growth periods except for the V12 stage. In the V6 stage, V12 stage, R1 stage, R3 stage, and R6 stage, plants in CH50 tillage had higher dry matter accumulation than the control by 78.33%, 16.48%, 13.63%, 56.61%, and 20.79%, respectively; CH40 tillage increased plant growth by 51.46%, 2.56%, 13.11%, 29.76%, and 7.20% compared to the control. Additionally, CH30 was the same as the aforementioned groups, overtopping SR (shallow rotary tillage) by 35.00%, 4.01%, 24.06%, 23.69%, and 8.75% (Table 2). In summary, deeper subsoil tillage led to greater total dry matter than that of the control.

Table 2. Effect of subsoil tillage treatments on the accumulation of Xianyu335 dry matter at different growth stages in 2014. Values are the average ±1 standard error.

Year and Variety	Stage	Treatments	Dry Matter (g Per Plant)				Increased Compared to Control (%)
			Stem	Leaf	Ear	Total	
2014 (Xianyu335)	V6	CH50	† 37.8 ± 1.13 a	† 47.8 ± 6.01 a	—	† 85.6 ± 7.24 a	78.33
		CH40	33.3 ± 4.82 b	39.5 ± 3.03 b	—	72.7 ± 7.85 b	51.46
		CH30	29.2 ± 6.32 b	35.6 ± 4.18 c	—	64.8 ± 5.24 b	35.00
		SR	19.3 ± 1.90 c	28.7 ± 0.38 d	—	48.0 ± 2.28 c	–
	V12	CH50	59.6 ± 4.92 a	45.0 ± 0.96 a	—	104.6 ± 5.88 a	16.48
		CH40	53.5 ± 7.25 b	38.6 ± 4.92 b	—	92.1 ± 6.62 b	2.56
		CH30	52.2 ± 5.98 b	41.2 ± 2.11 b	—	93.4 ± 5.28 b	4.01
		SR	50.9 ± 5.77 c	38.9 ± 3.62 b	—	89.8 ± 8.22 b	–
	R1	CH50	127.1 ± 2.39 a	55.9 ± 4.15 b	† 37.2 ± 5.44 b	220.1 ± 6.54 b	13.63
		CH40	125.2 ± 7.49 b	55.5 ± 7.91 b	38.4 ± 7.03 b	219.1 ± 7.11 b	13.11
		CH30	122.6 ± 2.79 c	66.0 ± 1.68 a	51.7 ± 4.51 a	240.3 ± 4.47 a	24.06
		SR	114.6 ± 5.08 d	50.7 ± 1.96 c	28.4 ± 1.61 c	193.7 ± 3.57 c	–
	R3	CH50	176.0 ± 2.80 a	62.9 ± 1.96 a	192.2 ± 7.82 a	431.0 ± 7.14 a	56.61
		CH40	148.5 ± 5.75 b	56.4 ± 6.82 b	152.1 ± 2.36 b	357.1 ± 9.18 b	29.76
		CH30	136.2 ± 2.89 c	49.3 ± 3.62 b	155.0 ± 4.67 b	340.4 ± 8.29 b	23.69
		SR	112.1 ± 2.99 d	46.2 ± 5.41 c	116.9 ± 4.59 c	275.2 ± 8.40 c	–
	R6	CH50	146.3 ± 5.47 a	49.5 ± 3.94 a	304.0 ± 2.66 a	499.7 ± 4.9.41 a	20.79
		CH40	120.4 ± 6.89 b	44.4 ± 2.30 b	278.7 ± 2.04 b	443.5 ± 10.16 b	7.20
		CH30	116.8 ± 3.16 b	48.4 ± 0.60 a	284.7 ± 0.51 b	449.9 ± 7.29 b	8.75
		SR	106.8 ± 5.02 c	46.1 ± 9.03 b	260.8 ± 6.57 c	413.7 ± 11.24 c	–
2015 (Xianyu335)	V6	CH50	† 37.4 ± 5.26 a	† 40.3 ± 4.44 a	—	† 77.7 ± 4.24 a	50.58
		CH40	26.5 ± 4.67 b	31.4 ± 4.50 b	—	57.9 ± 8.43 b	12.21
		CH30	25.7 ± 5.44 b	29.1 ± 7.05 c	—	54.8 ± 10.49 b	6.20
		SR	23.3 ± 4.20 b	28.4 ± 4.81 c	—	51.6 ± 9.01 b	–
	V12	CH50	53.7 ± 8.34 a	46.2 ± 2.49 a	—	99.9 ± 8.83 a	21.98
		CH40	49.9 ± 3.97 b	39.1 ± 3.45 b	—	88.9 ± 6.45 b	8.55
		CH30	47.6 ± 4.94 b	41.8 ± 1.34 b	—	89.4 ± 5.28 b	9.16
		SR	43.5 ± 1.37 c	38.4 ± 1.88 b	—	81.9 ± 3.25 c	–
	R1	CH50	86.5 ± 4.85 a	43.5 ± 2.89 a	† 48.3 ± 3.00 a	178.3 ± 7.74 a	19.91
		CH40	79.5 ± 2.79 b	42.8 ± 6.14 a	43.3 ± 9.08 b	165.4 ± 8.93 b	11.23
		CH30	73.7 ± 7.04 c	36.7 ± 2.52 b	30.3 ± 3.31 c	140.8 ± 4.04 c	–
		SR	73.5 ± 1.57 c	36.8 ± 5.13 b	38.4 ± 7.88 b	148.7 ± 6.74 c	–
	R3	CH50	91.1 ± 8.50 a	49.7 ± 5.65 a	111.3 ± 7.82 a	252.1 ± 7.14 a	23.70
		CH40	82.3 ± 3.38 b	41.3 ± 4.45 b	101.8 ± 2.60 b	225.3 ± 7.83 b	10.55
		CH30	91.0 ± 2.96 a	44.6 ± 6.30 b	103.7 ± 9.70 b	239.3 ± 9.26 b	17.42
		SR	73.8 ± 7.21 c	36.6 ± 8.85 c	93.4 ± 4.87 c	203.8 ± 7.26 c	–
	R6	CH50	71.6 ± 6.61 a	41.6 ± 7.00 a	254.1 ± 3.03 a	367.2 ± 4.64 a	5.09
		CH40	62.1 ± 5.96 c	41.7 ± 3.20 a	243.7 ± 5.75 b	347.5 ± 9.16 b	–
		CH30	68.6 ± 3.16 b	40.8 ± 8.80 a	252.5 ± 4.13 a	361.8 ± 7.29 a	3.55
		SR	68.6 ± 1.47 b	37.1 ± 5.77 b	243.7 ± 7.01 b	349.4 ± 7.24 b	–
2015 (Zhengdan958)	V6	CH50	† 34.9 ± 4.76 b	† 38.7 ± 6.04 b	—	† 73.6 ± 5.84 b	40.99
		CH40	39.3 ± 5.68 a	42.5 ± 4.54 a	—	81.8 ± 9.45 a	56.70
		CH30	32.8 ± 9.87 b	35.0 ± 6.37 b	—	67.9 ± 6.49 b	30.08
		SR	24.9 ± 4.87 c	27.2 ± 4.53 c	—	52.2 ± 7.55 c	–
	V12	CH50	85.9 ± 7.43 a	43.5 ± 7.14 a	—	129.4 ± 7.14 a	37.37
		CH40	63.6 ± 5.27 b	46.2 ± 1.80 a	—	109.8 ± 6.07 b	16.56
		CH30	55.1 ± 2.69 c	46.7 ± 6.05 a	—	101.8 ± 8.74 b	8.07
		SR	53.7 ± 6.72 c	40.5 ± 6.75 b	—	94.2 ± 7.72 c	–
	R1	CH50	75.6 ± 5.96 a	41.1 ± 5.69 a	† 48.7 ± 3.94 a	165.4 ± 8.69 a	24.74
		CH40	70.4 ± 5.41 a	40.7 ± 7.35 a	46.1 ± 5.56 a	157.1 ± 6.56 b	18.48
		CH30	72.7 ± 3.52 a	37.4 ± 3.35 b	48.5 ± 3.59 a	158.6 ± 6.87 b	19.61
		SR	63.0 ± 5.87 b	30.5 ± 1.58 c	39.1 ± 5.29 b	132.6 ± 6.77 c	–
	R3	CH50	75.2 ± 8.80 a	37.7 ± 3.11 b	115.9 ± 2.90 a	228.7 ± 6.01 a	9.95
		CH40	74.4 ± 2.21 a	39.4 ± 5.54 b	104.9 ± 3.46 b	218.7 ± 5.67 a	5.14
		CH30	72.6 ± 4.09 a	43.7 ± 5.01 a	104.6 ± 3.64 b	220.9 ± 7.73 a	6.20
		SR	68.2 ± 2.60 b	37.0 ± 5.62 b	102.8 ± 7.07 b	208.0 ± 8.22 b	–
	R6	CH50	66.1 ± 1.44 a	43.6 ± 8.41 a	247.0 ± 9.07 b	356.7 ± 6.85 b	10.19
		CH40	64.5 ± 6.60 a	45.6 ± 8.18 a	255.5 ± 3.56 a	365.7 ± 9.78 a	12.97
		CH30	67.7 ± 9.30 a	37.2 ± 3.86 b	252.6 ± 3.39 a	357.5 ± 6.39 b	10.44
		SR	62.4 ± 7.49 a	35.3 ± 4.81 b	226.0 ± 7.35 c	323.7 ± 7.48 c	–

† Means within a column and growth stage followed by the same letter are not significantly different at $p \leq 0.05$, and the different letters are significantly different at $p \leq 0.05$.

In 2015, the growth performance trend of the accumulation of dry matter was consistent with that of 2014. Deeper subsoil tillage led to higher total dry matter accumulation than that of the control. The accumulation of dry matter in the tillage of subsoil tillage treatment was significantly higher than that in the control for most of the growth periods.

In the V6 stage, V12 stage, R1 stage, R3 stage, and R6 stage, Xianyu335 treated with CH50 tillage had higher dry matter accumulation than the control by 50.58%, 21.98%, 19.91%, 23.70%, and 5.09%, respectively; CH40 tillage increased plant growth by 12.21%, 8.55%, 11.23%, and 10.55% in the V6 stage, V12 stage, R1 stage, and R3 stage compared to the control; however, no significant difference was between the control in the R6 stage. CH30 was the same as the aforementioned groups, overtopping SR by 6.20%, 9.16%, 17.42%, and 3.55% in the V6 stage, V12 stage, R3 stage, and R6 stage, while no significant difference between the control in the R1 stage was observed (Table 2).

Compared with the control, the dry matter accumulation of Zhengdan958 under the treatments of CH50 was improved by 40.99%, 37.37%, 24.77%, 9.95%, and 10.19% in the V6 stage, V12 stage, R1 stage, R3 stage, and R6 stage, respectively; dry matter accumulation of Zhengdan958 under the treatments of CH40 was improved by 56.70%, 16.56%, 18.48%, 5.14%, and 12.97% in the V6 stage, V12 stage, R1 stage, R3 stage, and R6 stage, respectively; dry matter accumulation of Zhengdan958 under the treatments of CH30 was improved by 30.08%, 8.07%, 19.61%, 6.20%, and 10.44% in the V6 stage, V12 stage, R1 stage, R3 stage, and R6 stage, respectively (Table 2).

The results of the two years of experiment showed that subsoil tillage could increase dry matter accumulation and lay a foundation for greater grain fill and, therefore, increase yield. The response of the dry matter accumulation by Xianyu335 to subsoil tillage depth was more sensitive than the response by Zhengdan958 (Table 2).

3.2. Effects of Subsoil Tillage Depth on Leaf Area Index of Maize

The size of the green leaf area directly affects the photosynthetic ability of plants and is an important index that determines the yield. Our results indicated that all plants reached maximum LAI at the R1 stage, and gradually showed a decreasing trend after the R1stage (Figure 2).

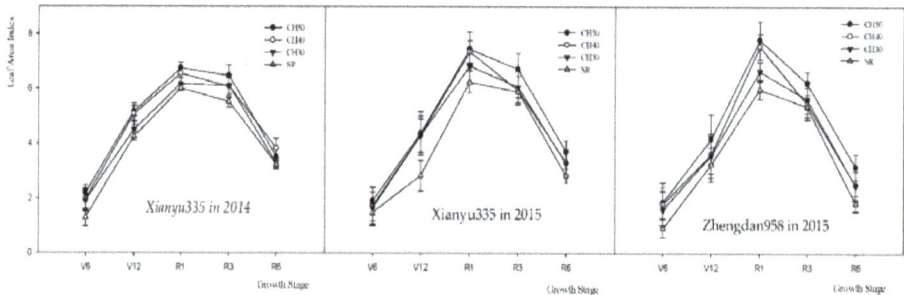

Figure 2. The change of the leaf area index due to the subsoil tillage depth. Values are the average of three replications. Bars indicate ±1 standard error of the mean.

The results over the two years showed that CH50 led to a greater LAI than the control. The CH40 treatment led to a greater LAI than the control over the whole growth period except in the R3 stage in 2015. The CH30 treatment lead to a greater LAI than the control, except in the V12 stage and R1 stage in 2014. At other growth stages, CH30 treatment led to a greater LAI than the control. The results showed that subsoil tillage depth could lead to maintenance of a relatively high LAI and more prolonged LAI at different stages, which provided the possibility for plants to capture more light for photosynthesis.

The LAI of Zhengdan958 was greater than that of Xianyu335 from the V3 stage to the R1 stage, but less than that of Xianyu335 after the R1 stage, indicating that the leaf senescence of Zhengdan958 was faster than that of Xianyu335 after the R1 stage.

3.3. Photosynthesis Related Parameters of Ear Leaf

Photosynthetic assimilation (Pn), transpiration rate (Tr), stomatal conductance (Gs), intercellular CO_2 concentration (Ci), and leaf water use efficiency of the Maize cultivars, Xianyu335 and Zhengdan958, were obviously affected by the subsoil tillage depth in different years (Figure 3A–E).

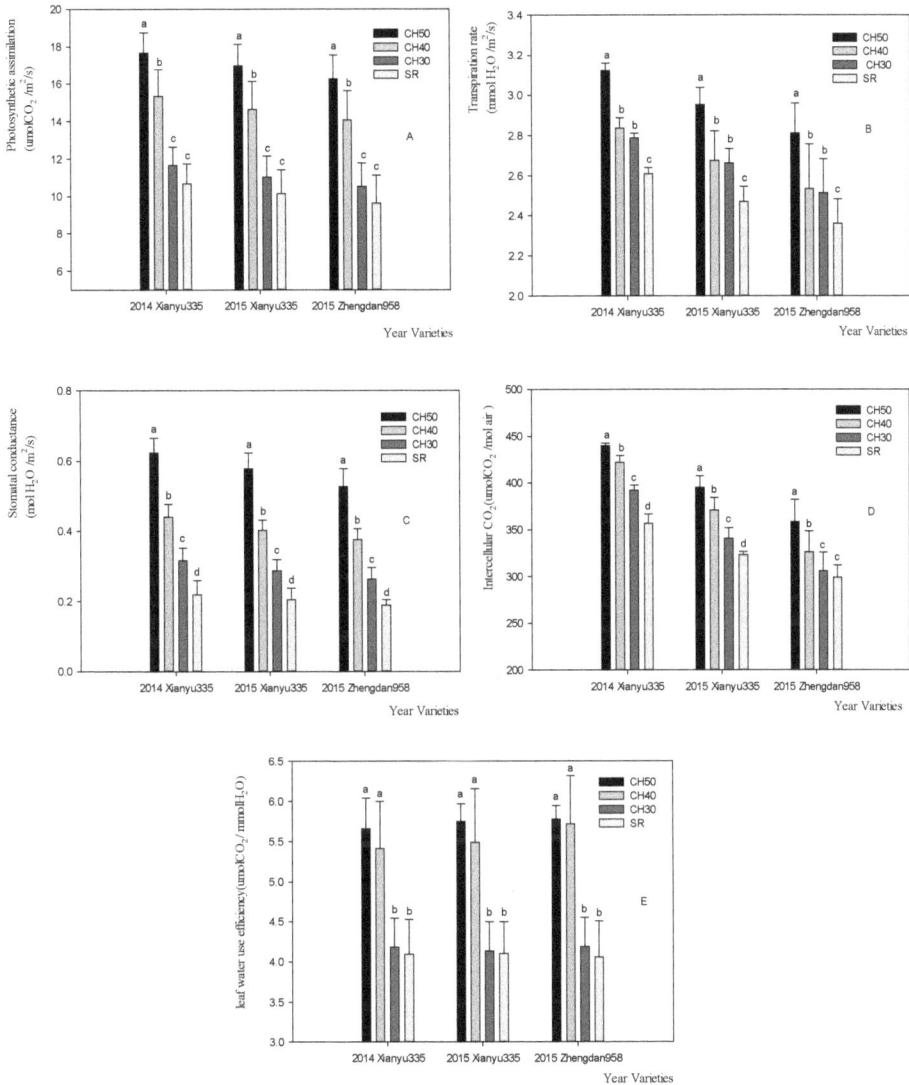

Figure 3. Photosynthetic assimilation (**A**), transpiration rate (**B**), stomatal conductance (**C**), intercellular CO_2 concentration (**D**), and water use efficiency (**E**) of maize due to subsoil tillage depth. Values are means ± SE of three replicates, bars represent ± 1 standard error of the mean, the same letters labeled upon the bars are not significantly different at $p \leq 0.05$, and the different letters are significantly different at $p \leq 0.05$.

Pn of the maize ear leaf in different varieties and different years treated by CH50 and CH40 was significantly higher than that of CH30 and the control, and there were no significant differences between CH30 and the control. Pn of the maize ear leaf treated by CH50 and CH40 increased by 61.25–69.20% and 41.46–49.88% more than the control. Tr of the maize ear leaf in different varieties and different years of subsoil tillage was significantly higher than that of the control. Among the treatments, the Tr of the maize ear leaf treated by CH50 was the highest, and there was no significant difference between the treatment of CH40 and CH30. Tr of the maize ear leaf treated by CH50, CH40, and CH30 increased by 19.14–19.67%, 7.29–8.69%, and 6.43–7.87% more than the control. As for stomatal conductance (Gs) and intercellular CO^2 concentration (Ci), the subsoil tillage resulted in a significant influence in Gs compared to the control in different varieties and years. Among the four-trial treatment, the Gs increased from the control to the CH30, and then the CH40, and the highest Gs was the CH50. The difference was significant between SR, CH30, CH40, and CH50. Ci of the ear leaf of Xianyu335 in 2014 and 2015, and were significantly different between the subsoil tillage and the control. The Ci of the ear leaf of Zhengdan958 treated by CH50 and CH40 was significantly higher than that of CH30 and the control, and there was no significant difference between CH30 and the control. Leaf water use efficiency treated by CH50 and CH40 was significantly higher than that of CH30 and the control, and there was no significant difference between CH50 and CH40, and no significant difference between CH30 and the control either. Leaf water use efficiency of maize treated by CH50 and CH40 increased by 38.29–42.41% and 32.21–40.87% more than the control.

3.4. Effects of Subsoil Tillage Depth on Dynamics of Leaf Area Duration

Photosynthetic productivity was not only related to LAI, but also related to LAD. The duration of the leaf area reflects the photosynthesis time of maize, so it has great influence on the yield because the grain is primarily composed of starch. Table 3 shows that a significantly higher LAD was observed after silking, in descending order, followed by CH50, CH40, CH30, and the control. LAD in the CH50 treatment was significantly greater than that of CH40 or CH30; and there was a smaller difference of LAD between the CH40 and CH30 treatments. The LAD from the silking stage to the filling stage sustained the best during the growth stage, then it began to decrease, and the decrease from the filling stage to maturity was more pronounced, with no difference between the two varieties in LAD response to the subsoil tillage depth (Table 3).

Table 3. Dynamics of leaf area duration under different chiseling depths for two maize varieties grown in 2014 and/or 2015. Values are the average ±1 standard error.

| Year Varieties | Treatment | Dynamics of Leaf Area Duration [$10^4 m^2/(d \cdot ha)$] | | | |
		V6 Stage–V12 Stage	V12 Stage–R1 Stage	R1 Stage–R3 Stage	R3g Stage–R6 Stage
2014 Xianyu335	CH50	† 3.1 ± 0.19 a	† 13.0 ± 1.09 a	† 17.6 ± 1.14 a	† 12.7 ±0.56 a
	CH40	3.0 ± 0.13 a	12.7 ± 1.01 a	16.9 ± 0.79 b	12.6 ± 1.19 a
	CH30	2.7 ± 0.14 a	11.6 ± 0.63 a	16.4 ± 0.58 b	11.9 ± 0.89 b
	SR	2.4 ± 0.19 a	11.2 ± 1.18 a	15.4 ± 0.49 c	11.1 ± 0.75 b
2015 Xianyu335	CH50	2.7 ± 0.22 a	12.9 ± 0.59 a	18.9 ± 1.03 a	13.3 ± 0.79 a
	CH40	2.6 ± 0.19 a	12.7 ± 0.19 a	17.8 ± 1.21 b	11.8 ± 1.13 b
	CH30	2.5 ± 0.21 a	12.2 ± 0.89 a	17.2 ± 1.37 b	11.9 ± 1.07 b
	SR	1.8 ± 0.29 b	9.9 ± 0.75 b	16.2 ± 0.89 c	11.1 ± 0.87 b
2015 Zhengdan958	CH50	3.1 ± 0.16 a	13.8 ± 0.97 a	19.3 ± 1.10 a	13.2 ± 0.56 a
	CH40	2.9 ± 0.20 a	13.0 ± 0.89 a	18.1 ± 1.15 b	11.6 ± 0.67 b
	CH30	2.8 ± 0.09 a	12.1 ± 1.19 b	17.2 ± 0.97 b	11.7 ± 0.71 b
	SR	2.4 ± 0.09 a	11.2 ± 1.06 b	16.1 ± 0.94 c	10.7 ± 0.55 c

† Means within a column and year variety followed by the same letter are not significantly different at $p \leq 0.05$, and the different letters are significantly different at $p \leq 0.05$.

3.5. Effects of Subsoil Tillage Depth on Net Assimilation Rate

From Table 4, compared with the control, there was a significant difference in NAR between the V6 stage to the V12 stage due to the subsoil tillage depth treatment, and the order of NAR under different tillage depth treatments was as follows: CH50 < CH40 < CH30 < SR, which reflected the rapid recovery of plant growth in this period due to the restriction of early growth.

Table 4. Dynamics of net assimilation rate between plant stages when grown under different chiseling depths. Values are the mean of three replications ±1 standard error.

Year Varieties	Treatment	Dynamics of Net Assimilation Rate [g/(m²·d)]			
		V6 to V12 Stage	V12 stage to R1 Stage	R1 Stage to R3 Stage	R3 Stage to R6 Stage
2014 Xianyu335	CH50	† 13.3 ± 1.19 c	† 5.8 ± 0.19 a	† 5.4 ± 0.59 a	† 5.6 ± 0.44 c
	CH40	14.2 ± 1.01 c	6.9 ± 0.13 a	5.4 ± 0.19 a	7.0 ± 0.28 c
	CH30	19.0 ± 0.63 b	6.3 ± 0.14 a	4.3 ± 0.49 b	9.5 ± 0.31 b
	SR	29.4 ± 1.18 a	5.8 ± 0.19 a	4.6 ± 0.38 b	12.7 ± 0.52 a
2015 Xianyu335	CH50	8.8 ± 0.59 c	6.2 ± 0.22 a	4.9 ± 0.44 a	8.9 ± 0.42 c
	CH40	13.0 ± 0.79 b	6.2 ± 0.19 a	4.4 ± 0.58 a	10.6 ± 0.57 b
	CH30	14.8 ± 0.75 b	7.3 ± 0.31 a	5.7 ± 0.42 a	10.6 ± 0.36 b
	SR	17.1 ± 0.89 a	7.9 ± 0.35 a	3.4 ± 0.39 b	13.7 ± 0.31 a
2015 Zhengdan958	CH50	11.1 ± 0.97 b	3.1 ± 0.49 a	3.3 ± 0.41 a	10.0 ± 0.46 b
	CH40	9.9 ± 0.89 c	3.8 ± 0.52 a	3.4 ± 0.56 a	13.1 ± 0.29 a
	CH30	12.4 ± 1.12 b	4.8 ± 0.55 a	3.6 ± 0.35 a	12.0 ± 0.19 a
	SR	18.2 ± 1.06 a	4.4 ± 0.25 a	2.8 ± 0.39 b	13.3 ± 0.27 a

† Means within a column and year variety followed by the same letter are not significantly different at $p \leq 0.05$, and the different letters are significantly different at $p \leq 0.05$.

As for the plant NAR, no significant difference between the treatments from the V12 stage to the R1 stage was observed, which indicated that the vegetative growth of the treatment was stable. Meanwhile, the NAR from the R1 stage to the R3 stage was clearly higher than that in the control. Additionally, the subsoil tillage treatment showed greater advantages than the control. Our results indicated that during the reproductive stages, the NAR in all the subsoil tillage depth treatments increased because of the increase of the filling rate. The performance in the subsoil tillage depth treatment from the R3 stage to the R6 stage was significantly lower than that in the control, which indicated that the subsoil tillage treatment improved the NAR of the maize plant in the stage from R1 to R3, and with enough assimilation of photosynthesis, the NAR of the maize plant treated by the subsoil tillage declined to be lower than that of the control. The two varieties in 2015 had similar responses to NAR for the tillage depth.

3.6. Effects of Subsoil Tillage Depth on Shoot-Root Ratio

Table 5 shows that the dry weight of stem, leaf, female ear, and root in the subsoil tillage increased significantly compared with that in the shallow rotary tillage treatment in 2014. The dry weights of the stem, leaf, female ear, and root of Xianyu335 plants receiving the CH50 and CH40 treatments in 2015, except the CH30 treatment, increased significantly compared to those receiving the shallow rotary tillage treatment.

However, the dry weight of the stem, leaf, female ear, and root of Zhengdan958 were significantly increased with tillage compared to the shallow rotary tillage treatment. The ratio of shoot-root of plants receiving the subsoil tillage depth treatments was greater than that from the shallow rotary tillage treatment, which indicated that the growth of shoot aboveground was more than that of the root system, and therefore resulted in the accumulation of greater shoot aboveground and an increase in yield. The ratio of shoot -root in 2015 was lower than that in 2014, which was due to the drought in

2015. In this condition, the crop growth was under water stress, thus the relative biomass of the root system increased, and the ratio of shoot- root decreased [22].

Table 5. Dry matter distribution in different organs and shoot root ratio in maize at the silking stage as affected by tillage treatment. Values are the mean of three replications ±1 standard error.

Year Varieties	Treatment	Dry Matter (g Per Plant)					Shoot/Root
		† Stem	† Leaf	† Ear	† Shoot	† Root	
2014 XianYu 335	CH50	121.5 ± 14.37 b	54.5 ± 3.34 a	60.3 ± 1.82 a	236.3 ± 18.31 a	18.4 ± 0.73 a	12.84
	CH40	138.3 ± 10.5 a	54.3 ± 4.14 a	47.4 ± 7.62 b	240.0 ± 16.71 a	17.6 ± 0.25 a	13.64
	CH30	123.9 ± 14.27 b	49.2 ± 4.55 b	48.2 ± 6.40 b	221.3 ± 6.34 b	16.2 ± 0.67 b	13.66
	SR	102.0 ± 16.04 c	39.9 ± 4.69 c	37.4 ± 2.40 c	179.3 ± 30.03 c	16.1 ± 0.33 b	11.14
2015 XianYu 335	CH50	85.1 ± 2.59 a	33.5 ± 2.89 b	44.9 ± 4.81 a	163.6 ± 16.29 a	18.5 ± 0.50 a	8.84
	CH40	79.5 ± 12.79 a	42.8 ± 6.14 a	43.3 ± 3.09 a	165.6 ± 22.32 a	17.8 ± 0.42 ab	9.30
	CH30	63.7 ± 9.12 b	36.7 ± 2.52 ab	37.0 ± 6.11 b	137.5 ± 17.58 b	17.2 ± 0.92 bc	7.99
	SR	63.5 ± 7.73 b	36.8 ± 5.13 ab	35.1 ± 2.11 b	135.3 ± 5.22 b	16.3 ± 0.22 c	8.30
2015 Zheng Dan958	CH50	75.6 ± 5.96 a	41.1 ± 5.69 a	48.7 ± 3.94 a	165.4 ± 23.94 a	19.5 ± 0.34 a	8.48
	CH40	70.4 ± 5.41 ab	40.7 ± 7.35 a	39.7 ± 5.56 b	150.8 ± 18.17 b	18.6 ± 0.75 ab	8.11
	CH30	72.6 ± 3.52 b	37.4 ± 3.35 a	48.5 ± 2.59 a	158.6 ± 16.19 ab	18.5 ± 0.48 b	8.57
	SR	63.0 ± 5.88 c	30.5 ± 1.58 b	39.1 ± 5.29 b	132.6 ± 12.93 c	17.4 ± 0.58 c	7.62

† Means within a column and year variety followed by the same letter are not significantly different at $p \leq 0.05$, and the different letters are significantly different at $p \leq 0.05$.

3.7. Effects of Subsoil Tillage Depth on Maize Yield and Economic Benefit

3.7.1. Effects of Subsoil Tillage Depth on Maize Yield and Its Components

Table 6 shows that there were significant differences in the biomass accumulation and yield due to the tillage depth. In 2014, the subsoil tillage of CH50, CH40, and CH30 increased yields more than that of the control (SR) by 6.9%, 3.5%, and 3.5%, respectively, with an average increase of 4.6%.

Table 6. Biomass, grain yield, and yield components under different subsoiling treatments. Values are the mean of three replications ±1 standard error.

Year Varieties	Treatment	t/ha		10^4/ha	Per Ear	g
		Biomass	Grain Yield	Ear number	Kernel Number	100-Kernel Weight
2014 Xianyu 335	CH50	† 41.6 ± 0.11 a	† 15.4 ± 0.26 a	† 7.67 ± 0.03 a	† 646 ± 24.12 a	† 35.4 ± 0.15 a
	CH40	38.4 ± 0.72 b	14.9 ± 0.14 b	7.60 ± 0.05 a	637 ± 33.41 a	35.2 ± 0.11 a
	CH30	37.9 ± 0.72 b	14.9 ± 0.11 b	7.50 ± 0.16 a	651 ± 12.13 a	34.9 ± 0.12 ab
	SR	37.6 ± 0.54 b	14.4 ± 0.16 c	7.62 ± 0.19 a	656 ± 5.76 a	34.5 ± 0.34 b
2015 Xianyu 335	CH50	29.7 ± 0.68 a	14.7 ± 0.11 a	8.80 ± 0.01 a	618 ± 16.94 a	31.7 ± 0.09 a
	CH40	29.2 ± 0.66 a	14.2 ± 0.17 b	8.44 ± 0.25 a	610 ± 11.15 a	31.9 ± 0.13 a
	CH30	29.6 ± 0.64 a	14.0 ± 0.29 b	8.51 ± 0.18 a	607 ± 11.32 a	32.8 ± 0.12 a
	SR	28.1 ± 0.45 b	13.5 ± 0.13 c	8.43 ± 0.19 a	610 ± 10.67 a	30.9 ± 0.17 b
2015 Zhengdan 958	CH50	30.0 ± 0.49 a	13.6 ± 0.11 a	8.65 ± 0.06 a	589 ± 12.15 a	31.3 ± 0.14 a
	CH40	29.3 ± 0.14 a	13.2 ± 0.14 a	8.49 ± 0.18 a	574 ± 9.98 a	31.8 ± 0.18 a
	CH30	29.4 ± 0.23 a	13.1 ± 0.09 a	8.44 ± 0.16 a	573 ± 19.56 a	32.2 ± 0.10 a
	SR	28.3 ± 0.54 b	13.0 ± 0.1 b	8.43 ± 0.14 a	575 ± 13.12 a	31.6 ± 0.16 a

† Means within a column and year variety followed by the same letter are not significantly different at $p \leq 0.05$, and the different letters are significantly different at $p \leq 0.05$.

The biomass from the CH50 tillage was 10.6% higher than that in the control. The CH50 tillage led to a significantly greater biomass than that in SR. However, there was no significant difference in the biomass between the CH40, CH30, and SR treatments. In 2015, the yield of the two varieties showed the same trend, with the subsoil tillage depth treatments generating greater yields than those receiving the shallow rotary tillage treatment. Subsoil tillage increased Xianyu335 yields by 3.7% to 8.9%, but there was no significant difference between the CH40 and CH30 treatments. The biomass in the subsoil tillage depth treatment was significantly higher than that of SR, with an average increase of

5.0%. As for Zhengdan958, plants in the CH50, CH40, and CH30 tillage treatment increased the yield, on average, by 4.0%, 1.5%, and 0.7%, respectively. In addition, the biomass of Zhengdan958 increased by 6.0%, 3.9%, and 3.7% compared to the control. There was no significant difference among the three treatments in biomass.

For the production components, the 100-grain weight of Xianyu335 was significantly increased by the subsoil tillage depth treatment, but not for Zhengdan958. It indicated that the treatment of deepening the subsoil depth could lead to a significant increase in the 100-grain weight of Xianyu335, but no significant increase of Zhengdan958.

3.7.2. Correlation Analysis between Canopy Characteristics and Yield

Correlation analysis between the canopy characteristics and yield showed that the leaf area duration, stover biomass dry matter, and leaf area index were significantly positively correlated with the yield at the 0.01 probability level. Photosynthesis assimilation, transpiration ratio, stomatal conductance, and intercellular CO_2 concentration were significantly positively correlated with the yield at the 0.05 probability level (Table 7). The net assimilation rate, photosynthesis rate, and intercellular CO_2 concentration were significantly positively correlated with leaf water use efficiency at the 0.05 probability level. The correlation coefficients of the leaf area duration, stover biomass dry matter, and leaf area index with the yield were 0.997, 0.972, and 0.952, and the correlation coefficients of the photosynthesis ratio, transpiration ratio, stomatal conductance, and intercellular CO_2 concentration were 0.948, 0.979, 0.980, and 0.976, respectively. The results showed that the leaf area duration and dry matter were the main factors affecting the grain yield. The correlation coefficients of the net assimilation rate, photosynthesis rate, and intercellular CO_2 concentration with the leaf water use efficiency were 0.963, 0.981, and 0.928, respectively. The results showed that the ear leaf photosynthesis rate and the plant net assimilation rate were the main factors affecting the leaf water use efficiency.

Table 7. Pearson correlation coefficients and associated significance level for final grain yield between selected corn canopy parameters as influenced by subsoil tillage depth.

	NAR	LAD	Pn	Tr	Gs	Ci	DM	LAI	GY	LWUE
NAR	1.000									
LAD	0.732	1.000								
Pn	0.909	0.940	1.000							
Tr	0.630	0.965 *	0.896	1.000						
Gs	0.817	0.967 *	0.982 *	0.962 *	1.000					
Ci	0.864	0.973 *	0.993 **	0.929	0.989 *	1.000				
DM	0.612	0.943	0.883	0.997 **	0.955 *	0.912	1.000			
LAI	0.642	0.989 *	0.882	0.949 *	0.921	0.930	0.902 **	1.000		
GY	0.736	0.997 **	0.948 *	0.979 *	0.980 *	0.976 *	0.972 **	0.952 **	1.000	
LWUE	0.963 *	0.888	0.981 *	0.801	0.928	0.965 *	0.779	0.825	0.888	1.000

** Significant at the 0.01 probability level, * Significant at the 0.05 probability level. NAR: Net assimilation rate, LAD: leaf area duration, Pn: photosynthesis assimilation, Tr: transpiration ratio, Gs: stomatal conductance, Ci: intercellular CO_2 concentration, DM: dry Matter, LAI: leaf area index, GY: grain yield, LWUE: leaf water use efficiency at leaf level.

3.7.3. Effect of Subsoil Tillage Depth on Water Use Efficiency

Regardless of drought or rainy years, plant WUE was significantly improved by the subsoil tillage depth treatment (Table 8). In 2014, the treatments of CH50, CH40, and CH30 increased WUE by 14.62%, 8.29%, and 6.92%, respectively. In 2015, CH50, CH40, and CH30 increased WUE of Xianyu335 by 18.49%, 8.74%, and 9.53% respectively, and WUE of Zhengdan958 by 6.41%, 1.93%, and 2.28% respectively, compared with the control. The two years of data showed that the CH50 treatment led to higher WUE under the condition of the lower water availability, followed by CH40 and CH30. Notably, there was no significant difference in WUE between the CH40 and CH30 treatments.

Table 8. Water use efficiency of spring maize under different tillage depths. Values are the mean of three replications.

Year Varieties	Treatment	mm	kg/ha		kg/(ha·mm)	(kg/ha·mm)
		Water Consume	Seed Yield	Biological Yield	Water Use Efficiency	Water Productivity
2014 Xianyu335	CH50	† 770.9 bc	† 15,447 a	† 41,614 a	† 20.04 a	† 53.98 a
	CH40	786.6 b	14,891 b	38,405 b	18.93 b	48.83 b
	CH30	796.2 b	14,883 b	37,876 b	18.69 b	47.57 b
	SR	823.6 a	14,398 c	37,551 b	17.48 c	45.59 c
2015 Xianyu335	CH50	540.7 c	14,660 a	29,710 a	27.11 a	54.95 a
	CH40	562.1 b	13,985 bc	29,196 a	24.88 b	51.94 b
	CH30	574.9 b	14,404 ab	29,638 a	25.06 b	51.56 b
	SR	591.5 a	13,531 c	28,100 b	22.88 c	47.51 c
2015 Zhengdan958	CH50	560.5 b	13,576 a	29,993 a	24.22 a	53.51 a
	CH40	568.7 b	13,192 a	29,319 a	23.20 b	51.55 b
	CH30	568.6 b	13,234 a	29,374 a	23.28 b	51.66 b
	SR	573.3 a	13,048 b	28,284 b	22.76 c	49.34 c

† Means within a column and year variety followed by the same letter are not significantly different at $p \leq 0.05$, and the different letters are significantly different at $p \leq 0.05$.

In 2014, CH50, CH40, and CH30 increased WPE more than the control by 18.40%, 7.11%, and 4.34%, respectively; in 2015, the WPE of Xianyu335 increased by 15.66%, 9.32%, and 8.52%, and 8.45%, 4.48%, and 4.70% of Zhengdan958, respectively. The CH50 tillage increased WPE by 3.80% and 3.58% compared to CH40 and CH30, respectively, but WPE showed no significant difference between the CH40 and CH30 treatments.

3.7.4. Economic Benefit Analysis

The depth of the subsoil tillage is an important factor for farmers when considering the cost. Therefore, our experiment analyzed the economic input-output ratio for growing maize using the different tillage depths.

The results are presented in Table 9. On average, from 2014 to 2015, the net income increased due to increasing depths of the subsoil tillage by 2.24% to 6.97% more than that of the shallow rotary tillage treatment. Among them, the subsoil tillage depth of 50 cm led to the highest returns, followed by tillage to a depth of 30 cm. There was no significant difference between the yields from tilling to 40 cm versus 30 cm depths. The test results showed that the most economic advantage was the subsoil tillage to a depth of 50 cm. Thus, the results of this study could offer a reference for farmers to choose the subsoil tillage depth and to increase income.

Table 9. Inputs and outputs of maize production for different treatments in the year, 2014.

Year			CH50	CH40	CH30	SR
2014	Inputs	Seeds (RMB/ha)	1050	1050	1050	1050
		Fertilizer (RMB/ha)	1600	1600	1600	1600
		Pesticides (RMB/ha)	1500	1500	1500	1500
		Irrigation (RMB/ha)	1200	1200	1200	1200
		Mechanical work (RMB/ha)	900	750	600	450
		Total (RMB/ha)	6250	6100	5950	5800
	Outputs	Yield (kg/ha)	15,447	14,891	14,883	14,398
		Price (RMByuan/kg)	1	1	1	1
		Income (RMB/ha)	15,447	14,891	14,883	14,398
		Net income (RMB/ha)	9197	8791	8933	8598
		Increase (%)	6.97	2.24	3.90	

Table 9. *Cont.*

Year			CH50	CH40	CH30	SR
2015	Inputs	Seeds (RMB/ha)	1050	1050	1050	1050
		Fertilizer (RMB/ha)	1600	1600	1600	1600
		Pesticides (RMB/ha)	1500	1500	1500	1500
		Irrigation (RMB/ha)	1200	1200	1200	1200
		Mechanical work (RMB/ha)	900	750	600	450
		Total (RMB/ha)	6250	6100	5950	5800
	Outputs	Yield (kg/ha)	† 14,118	† 13,589	† 13,819	† 13,290
		Price (RMByuan/kg)	1	1	1	1
		Income (RMB/ha)	14,118	13,589	13,819	13,290
		Net income (RMB/ha)	7868	7489	7869	7490
		Increase (%)	5.05	–	5.06	

† Means within a column and subsoil tillage depth treatment and the control are average of Xianyu335 and Zhengdan958 in 2015.

4. Discussion

4.1. The Effect of Subsoil Tillage on Corn Canopy

As the leaf area of the maize plants and Pn, Tr of the maize ear leaf treated with subsoil tillage increased, the total dry matter accumulation amount and the rate increased; especially at the later stage, the yield was significantly increased [23–26]. The yield of spring corn increased by 14.6% through subsoil tillage [27], which may have loosened soil, improved permeability, and promoted dry matter accumulation of winter wheat and summer corn [28,29], and in so doing, significantly improved the grain yield and water utilization efficiency of crops [30–32]. This study indicated that subsoil tillage could result in a high photosynthetic assimilation, transpiration rate, stomatal conductance and intercellular CO^2 concentration at the R3 stage, maintain a relatively high LAI in different growth periods, and increase LAD. The deeper the subsoil tillage, the longer it maintained plant vitality. The subsoil tillage effectively delayed leaf senescence, which provided the possibility for plants to capture more light for photosynthesis. The net assimilation rate in the late silking period was obviously increased by the subsoiling tillage compared to the control, with CH50 > CH40 > CH30. These results indicated that subsoil tillage was beneficial to the accumulation of dry matter in the early growth stage and laid a foundation for the formation of yield in the late growth period.

Many studies have demonstrated that subsoil tillage increased soil porosity, water infiltration, as well as root penetration [33]. Similar initial decreases in soil density and penetration resistance compared to no-tillage plots [33]. Subsoil tillage practices can improve the content of water and nutrient in soil [34], increase soil structure [35] and promote crop yields [36]. Specifically, compared to rotary tillage, subsoil tillage led decrease in water consumption by 1.5%, increase in soil water content by 0.1%, WUE by 2.5% and maize yield by 29.4 kg ha^{-1} in Northern Huang–Huai–Hai Valley [37]. Subsoil tillage is typical cultivation method applied to promote crop yields in arid areas [38,39], such as the dryland region of northwest China [40]. Similarly, in abroad, using subsoil tillage fracture dense layers in a loamy sand soil and reduction of penetration resistance was found and yield increased in wheat (Triticum aestivum L), soybean [Glycine max L. (Merr.)] and maize [33].

4.2. Response of Different Corn Varieties to Subsoil Tillage Depth

The LAI of Zhengdan958 was higher than Xianyu335 from the V6 stage to the R1 stage, but lower than Xianyu335 after the R1 stage, indicating that the leaf senescence rate of Zhengdan958 was faster than that of Xianyu335 after the R1 stage. The WUE of Xianyu335 and Zhengdan958 with the subsoil tillage increased by 12.25% and 3.54% more compared to the control, respectively, which indicated that Xianyu335 was more sensitive to subsoil tillage depth than Zhengdan958. The results of this study indicated that different varieties had different responses to the subsoil tillage depth, but how the different varieties respond to different soil types and climatic conditions remains to be further studied [41].

4.3. The Effect of Subsoil Tillage on Economic Efficiency

The intensity of the subsoiling tillage should be suitable to avoid economic efficiency decreasing [42]. Cai Hongguang [43] found that a subsoil tillage of 50 cm was superior to that of 30 cm or no chiseling. Our study indicated that subsoiling tillage of 50 cm was optimal, and there was no significant economic difference between the 40 cm and 30 cm subsoil tillages. Compared to compacted soil by shallow rotary tillage for many years, subsoil tillage maximized the energy gain, while, in contrast, reduced tillage or no tillage minimized energy intensity for corn–soybean in eastern Nebraska [44]. Compacted soils reduced the plant height of field corn, and decreased the aboveground biomass in potato, snap bean, cucumber, and cabbage [45]. Subsoil tillage has been recommended to use across the United States to alleviate the negative effect of a compacted layer on potato (*Solanum tuberosum* L.) productivity [46]. Compared with no-till plots, subsoil tillage dramatically decreased foliar symptoms of sudden death syndrome for soybeans [47].

4.4. Preliminary Discuss on Area Suitable to Subsoil Tillage

The ratio of yield and WUE were improved by subsoil tillage compared to the control in 2014, and 2015 showed that the result of subsoil tillage was effected by the precipitation or the irrigation rate. The result can be better expressed in the condition of more precipitation or irrigation rate, due to strong moisture conservation by loosened soil. On the contrary, even if the soil possesses a large storage ability of moisture through treatment by subsoil tillage, without enough water supply, the results of moisture conservation will not be shown. A lower soil bulk density, greater soil porosity, and decreased soil moisture was observed in subsoiled plots [47]. For example, in the semi-arid Segarra region in Spain, no-tillage is regarded as the best system for executing fallow only, if residues of the preceding crops are left spread over the soil [48]. Similarly, residue management and tillage effects on soil-water storage and grain yields of dryland wheat and sorghum [*Sorghum bicolor L. (Moench)*] for a clay loam in Texas. No-tillage increased the average soil water storage compared to stubble mulch-tillage. Therefore, compared to subsoil tillage, no-tillage residue management was more favorable for dryland crop production [33]. Various tradeoffs indicate that farmers should alternate between subsoil tillage and no-tillage to enhance the soil quality, and to decrease disease and yield problems, which may occur with continuous minimum tillage [49]. After all, the subsoil tillage is a method to maintain the moisture, not the method to produce the moisture, so the subsoil tillage can be used in a rainfed area with certain precipitation. The range of the precipitation suitable for soil moisture conservation by subsoil tillage is a topic for future research.

5. Conclusions

Subsoil tillage increased the ear leaf photosynthetic assimilation, transpiration rate, stomatal conductance, and intercellular CO^2 concentration; maintained relatively high LAI; and extended LAD. In this experiment, the deeper the subsoil tillage, the longer it lasted, and the senescence of leaves was effectively delayed, which made it possible to prolong the photosynthetic time of plants. Compared with the control, subsoil tillage obviously increased NAR after the R1 stage, ordered from high to low values: CH50, CH40, and CH30.

There was a significant difference in yield among treatments, and the yield under the subsoil tillage treatment was significantly higher (0.7% to 8.9%) than that of the control (SR). In terms of yield components, subsoil tillage significantly increased the 100-grain weight of Xianyu335, while other factors had no significant difference. Correlation analysis between the canopy parameters and yield indicated that LAD and dry matter were the main factors affecting the final yield. Considering the economic benefits, the net income of the CH50 was higher than that of the shallow rotary tillage treatment, thus, the best tillage system was the 50 cm subsoil tillage treatment.

Agronomy **2019**, *9*, 3

Author Contributions: J.S. and J.G. conceived and designed the experiments; F.Z. and Y.F. performed the experiments; H.B. performed the statistical analysis; Z.W. and S.H. analyzed the results; J.S. wrote the paper.

Funding: This study was funded by the National Natural Science Foundation of China (31360304), National Key Research and Development Program of China [2017YFD0300802,2016YFD0300103], the Maize Industrial Technology System Construction of Modem Agriculture of China (CARS-02-63) and the Fund of Crop Cultivation Scientific Observation Experimental Station in North China Loess Plateau of China (25204120).

Acknowledgments: We would like to thank the Maize High-yield and High-efficiency Cultivation Team for field and data collection, and especially Juliann Seebauer for manuscript revisions.

Conflicts of Interest: The authors declare no conflict of interest.

References

1. Wang, Z.; Wang, C.S.; Zhang, J.B. Study of Soil Deep Loosening Influencing Crop Growth. *Heilongjiang Agric. Sci.* **2009**, *4*, 33–35.
2. Zhao, Y.L.; Xue, Z.W.; Guo, H.B.; Mu, X.Y.; Li, C.H. Effects of Tillage and Straw Returning on Water Consumption Characteristics and Water Use Efficiency in the Winter Wheat and Summer Maize Rotation System. *Sci. Agric. Sin.* **2014**, *47*, 3359–3371. [CrossRef]
3. Ghosh, P.K.; Mohanty, M.; Bandyopadhyay, K.K.; Painuli, D.K.; Misra, A.K. Growth, Competition, Yield Advantage and Economics in Soybean/pigeonpea Intercropping System in Semi-arid Tropics of India. *Field Crops Res.* **2005**, *96*, 89. [CrossRef]
4. Guo, J.; Xiao, K.; Guo, X.; Zhao, C. Review on Maize Canopy Structure, Light Distributing and Canopy Photosynthesis. *Maize Sci.* **2005**, *13*, 55–59. [CrossRef]
5. Borras, L.; Maddonni, G.A.; Otegui, M.E. Leaf Senescence in Maize Hybrids: Plant Population, Rowspacing and Kernel Set Effects. *Field Crops Res.* **2003**, *82*, 13–14. [CrossRef]
6. He, P.; Osaki, M.; Takebe, M.; Shinano, T. Changes of Photosynthetic Characteristics in Relation to Leaf Senescence in Two Maize Hybrids with Different Senescent Appearance. *Photosynthetica* **2002**, *40*, 547–552. [CrossRef]
7. Dong, S.T.; Gao, R.Q.; Hu, C.H.; Wang, Q.; Wang, K. Study of Canopy Photosynthesis Property and High Yield Potential after Anthesis in Maize. *Acta Agron. Sin.* **1997**, *23*, 318–325.
8. Dong, X.W.; Liu, S.T. A study of Canopy Apparent Photosynthesis Property in Summer Maize with Superhigh Yield. *Acta Agric. Boreali Sin.* **1999**, *14*, 36–41.
9. Zheng, P.R. Photosynthetic Physiology. In *Introduction to Crop Physiology*, 5th ed.; Hu, C.H., Yu, Z.W., Eds.; China Agricultural University Press: Beijing, China, 1992; Volume 5, pp. 222–275, ISBN 7-S1002-259-8/O.260.
10. Bai, J.F. Effect of Different Farming Measure on Canopy and Root of High Yield Spring Maize. Master's Thesis, Inner Mongolia Agricultural University, Hohhot, China, July 2012.
11. Huang, Z.X.; Wang, Y.J.; Wang, K.J.; Li, D.H.; Zhao, M.; Liu, J.G.; Dong, S.T.; Wang, H.J.; Wang, J.H.; Yang, J.S. Photosynthetic Characteristics During Grain Filling Stage of Summer Maize Hybrids with High Yield Potential of 15 000 kg·ha^{-1}. *Sci. Agric. Sin.* **2007**, *40*, 1898–1906.
12. Lin, Z.H.; Xiang, Y.Q.; Mo, X.G.; Li, J.; Wang, L. Normalized Leaf Area Index Model for Summer Maize. *Chin. J. Eco-Agric.* **2003**, *11*, 69–72.
13. Sun, R.; Zhu, P.; Wang, Z.M.; Cong, Y.; Gou, L.; Fang, L.; Zhao, M. Effect of Plant Density on Dynamic Characteristics of Leaf Area Index in Development of Spring Maize. *Acta Agron. Sin.* **2009**, *35*, 1097–1105. [CrossRef]
14. Chanh, T.T.; Roeske, C.A.; Eaglesham, A.R. Changes in Maize-Stalk Proteins during Ear Development. *Physiol. Plant.* **1993**, *87*, 21–24. [CrossRef]
15. Guo, S.Y.; Zhang, X.; Zhang, Q.J.; Wang, Z.H.; Li, Y.Z.; Gu, S.F.; Jiao, N.Y.; Yin, F.; Fu, G.Z. Effects of Straw Mulching and Water-retaining Agent on Ear Leaf Senescence after Anthesis and Yield of Summer Maize. *Maize Sci.* **2012**, *20*, 104–107. [CrossRef]
16. Tong, S.Y.; Song, F.B.; Xu, H.W. Differences of Morphological Senescence of Leaves in Various Maize Varieties during Mature Period of Seed. *Acta Agric. Boreali Sin.* **2009**, *24*, 11–15.
17. Paponov, I.A.; Sambo, P.; Schulte auf'm, G.E.; Presterl, T.; Geiger, H.H.; Engels, C. Kernel Set in Maize Genotypes Differing in Nitrogen Use Efficiency in Response to Resource Availability around Flowering. *Plant Soil* **2005**, *272*, 101–110. [CrossRef]

18. Paponov, I.A.; Sambo, P.; Schulteaufm, G.E.; Presterl, T.; Geiger, H.H.; Engels, C. Grain Yield and Kernel Weight of Two Maize Genotypes Differing in Nitrogen Use Efficiency at Various Levels of Nitrogen and Carbohydrate Availability during Flowering and Grain Filling. *Plant Soil* **2005**, *272*, 111–123. [CrossRef]

19. Song, B.; Wu, S.L.; Yu, S.H.; Chen, F.; Xu, L.; Fan, W.B. A Study on Population Quality Indexes of Maize under Different Ecological Condition. *Tillage Cultiv.* **1998**, 23–28. [CrossRef]

20. Li, S.; Peng, Y.F.; Yu, P.; Zhang, Y.; Fang, Z.; Li, C.J. Accumulation and Distribution of Dry Matter and Potassium in Maize Varieties Released in Different Years. *Plant Nutr. Fertil. Sci.* **2011**, *17*, 325–332.

21. *Statistical Analysis System (SAS) Version SAS/STAT 9.0*; SAS Institute Inc.: Cary, NC, USA, 2004.

22. Song, H.X.; Li, S.X. Effects of Root Growing Space of on Maize Its Absorbing Characteristics. *Sci. Agric. Sin.* **2003**, *36*, 899–904.

23. Li, D.M.; Guo, H.; Zhu, H.Y.; Liu, M.; Chen, T.; Qi, H. Effect of Different Tillageon the Development, Root Distribution and Yield in Maize. *Maize Sci.* **2014**, *22*, 115–119. [CrossRef]

24. Song, R.; Wu, C.S.; Mu, J.M.; Xu, K.Z. The Effect of Breaking the Bottom of the Plow on The Growth and Development of Corn Roots. *Tillage Cultiv.* **2000**, *5*, 6–7. [CrossRef]

25. Zou, H.T.; Zhang, Y.L.; Huang, Y.; Huang, Y.; Song, H.; Yu, N.; Zhang, Y.; Sun, Z. Effect of Deep Tillage on Maize Growth in the Semi-arid Region of Liaoning Northwest Area. *Shenyang Agric. Univ.* **2009**, *40*, 475–477.

26. Wang, T.C.; Li, X.M.; Sui, R.T.; Liu, D.J. A Primary Study on the Technical Effects of Subsoiling in Row Space at Corn Seedling Growth Stage. *Chin. Agric. Sci. Bull.* **2003**, *19*, 40–43.

27. Shang, J.X.; Li, J.; Jia, Z.K.; Zhang, L.H. Soil Water Conservation Effect, Yield and Income Increments of Conservation Tillage Measures on Dryland Wheat Field. *Sci. Agric. Sin.* **2010**, *43*, 2668–2678. [CrossRef]

28. Serrano, J.M.; Shahidian, S.; Da Silva, J.M.; Carvalho, M. Monitoring of Soil Organic Carbon over 10 Years in a Mediterranean Silvo-pastoral System: Potential Evaluation for Differential Management. *Precis. Agric.* **2016**, *17*, 274–295. [CrossRef]

29. Zhao, J.B.; Mei, X.R.; Xue, J.H.; Zhong, Z.Z. The Effect of Straw Mulch on Crop Water Use Efficiency in Dryland. *Sci. Agric. Sin.* **1996**, *29*, 59–66.

30. Fu, G.Z.; Li, C.H.; Wang, J.Z.; Wang, Z.L.; Cao, H.M.; Jiao, N.Y.; Chen, M.C. Effects of Stubble Mulch and Tillage Managements on Soil Physical Properties and Water Use Efficiency of Summer Maize. *Agric. Eng.* **2005**, *21*, 52–56.

31. Wang, X.B.; Wu, H.J.; Dai, K.; Zhang, D.; Feng, Z.; Zhao, Q.; Wu, X.; Jin, K.; Cai, D.; Oenema, O.; et al. Tillage and Crop Residue Effects on Rainfed Wheat and Maize Production in Northern China. *Field Crops Res.* **2012**, *132*, 106–116. [CrossRef]

32. Zhu, P.F.; Yu, Z.W.; Wang, D.; Zhang, Y. Effects of Tillage on Water Consumption Characteristics and Grain Yield of Wheat. *Sci. Agric. Sin.* **2010**, *43*, 3954–3964. [CrossRef]

33. Baumhardt, R.L.; Jones, O.R. Residue Management and Tillage Effects on Soil-water Storage and Grain Yield of Dryland Wheat and Sorghum for a Clay Loam in Texas. *Soil Tillage Res.* **2002**, *68*, 71–82. [CrossRef]

34. Moraru, P.I.; Rusu, T. Soil Tillage Conservation and its Effect on Soil Organic Matter, Water Management and Carbon Sequestration. *J. Food Agric. Environ.* **2010**, *8*, 309–312.

35. Rusu, T.; Gus, P.; Bogdan, I.; Moraru, P.I.; Pop, A.I.; Clapa, D.; Marin, D.I.; Oroian, I.; Pop, L.I. Implications of Minimum Tillage Systems on Sustainability of Agricultural Production and Soil Conservation. *J. Food Agric. Environ.* **2002**, *216*, 335–338.

36. Chen, Y.; Liu, S.; Li, H.; Li, X.F.; Song, C.Y.; Cruse, R.M.; Zhang, X.Y. Effects of Conservation Tillage on Corn and Soybean Yield in the Humid Continental Climate Region of Northeast China. *Soil Tillage Res.* **2011**, *115*, 56–61. [CrossRef]

37. Tao, Z.; Li, C.; Li, J.; Ding, Z.; Xu, J.; Sun, X.; Zhou, P.; Zhao, M. Tillage and Straw Mulching Impacts on Grain Yield and Water Use Efficiency of Spring Maize in Northern Huang-Huai-Hai Valley. *Crop J.* **2015**, *3*, 445–450. [CrossRef]

38. Hou, X.Q.; Li, R.; Jia, Z.K.; Han, Q.F.; Yang, B.P.; Nie, J.F. Effects of Rotational Tillage Practices on Soil Structure, Organic Carbon Concentration and Crop Yields in Semi-arid Areas of Northwest China. *Soil Use Manag.* **2012**, *28*, 551–558. [CrossRef]

39. Verhulst, N.; Nelissen, V.; Jespers, N.; Haven, H.; Sayre, K.D.; Raes, D.; Deckers, J.; Govaerts, B. Soil Water Content, Maize Yield and its Stability as Affected by Tillage and Crop Residue Management in Rainfed Semi-arid Highlands. *Plant Soil* **2011**, *344*, 73–85. [CrossRef]

40. Liu, Y.; Gao, M.; Wu, W.; Tanveer, S.K.; Wen, X.; Liao, Y. The Effects of Conservation Tillage Practices on the Soil Water-holding Capacity of a non-irrigated Apple Orchard in the Loess Plateau, China. *Soil Tillage Res.* **2013**, *130*, 7–12. [CrossRef]

41. Feng, Y.; Zhang, Y.X.; Wang, C.L.; Zhang, J.H. An Influence of Different Subsoiling Depth on Corn Root Activity and Output. *Inner Mong. Univ. Natl.* **2013**, *28*, 196–199. [CrossRef]

42. He, J.; Li, H.W.; Gao, H.W. Subsoiling Effect and Economic Benefit under Conservation Tillage Mode in Northern China. *Trans. Chin. Soc. Agric. Eng.* **2006**, *22*, 62–67.

43. Cai, H.G.; Ma, W.; Zhang, X.Z.; Ping, J.; Yan, X.; Liu, J.; Yuan, J.; Wang, L.; Ren, J. Effect of Subsoil Tillage Depth on Nutrient Accumulation, Root Distribution, and Grain Yield in Spring Maize. *Crop J.* **2014**, *2*, 297–307. [CrossRef]

44. Rathke, G.W.; Wienhold, B.J.; Wilheim, W.W.; Diepenbrock, W. Tillage and Rotation Effect on Corn-soybean Energy Balances in Eastern Nebraska. *Soil Tillage Res.* **2007**, *97*, 60–70. [CrossRef]

45. Alva, A.K.; Hodges, T.; Boydston, R.A.; Collins, H.P. Effects of Irrigation and Tillage Practices on Yield of Potato under High Production Conditions in the Pacific Northwest. *Commun. Soil Sci. Plant Anal.* **2002**, *33*, 1451–1460. [CrossRef]

46. Copas, M.E.; Bussan, A.J.; Drilias, M.J.; Wolkowski, R.P. Potato Yield and Quality Response to Subsoil Tillage and Compaction. *Agron. J.* **2009**, *101*, 82–90. [CrossRef]

47. Vick, C.M.; Chong, S.K.; Bond, J.P.; Russin, J.S. Response of Soybean Sudden Death Syndrome to Subsoil tillage. *Plant Dis.* **2003**, *87*, 629–632. [CrossRef]

48. Lampurlanes, J.; Angas, P.; Cantero-Martinez, C. Tillage Effects on Water Storage During Fallow, and on Barley Root Growth and Yield in Two Contrasting Soils of the Semi-arid Segarra Region in Spain. *Soil Tillage Res.* **2002**, *65*, 207–220. [CrossRef]

49. Jackson, L.E.; Ramirez, I.; Yokota, R.; Fennimore, S.A.; Koike, S.T.; Henderson, D.M.; Chaney, W.E.; Calderón, F.J.; Klonsky, K. On-farm Assessment of Organic Matter and Tillage Management on Vegetable Yield, Soil, Weeds, Pests, and Economics in California. *Agric. Ecosyst. Environ.* **2004**, *103*, 443–463. [CrossRef]

![agronomy logo] *agronomy*

MDPI

Article

Polyaspartic Acid Improves Maize (*Zea mays* L.) Seedling Nitrogen Assimilation Mainly by Enhancing Nitrate Reductase Activity

Qingyan Wang [ORCID]**, Huihui Tang, Guangyan Li, Hui Dong, Xuerui Dong, Yanli Xu and Zhiqiang Dong** *

Key Laboratory of Crop Ecophysiology and Cultivation, Center for Crop Management & Farming System, Institute of Crop Sciences, Chinese Academy of Agricultural Sciences, Beijing 100081, China; wangqyan1201160@163.com (Q.W.); tanghuihui0609@163.com (H.T.); guangyan5112@126.com (G.L.); donghui2013@163.com (H.D.); sjtu_008@163.com (X.D.); xuyanli412@163.com (Y.X.)
* Correspondence: dongzhiqiang@caas.cn; Tel.: +86-010-8210-6043

Received: 7 August 2018; Accepted: 11 September 2018; Published: 13 September 2018

Abstract: Improvement of nitrogen use efficiency is of great importance in maize (*Zea mays* L.) production. In the present study, an eco-friendly growth substance, polyaspartic acid (PASP), was applied to maize seedlings grown with different nitrate (NO_3^-) doses by foliar spraying, aimed at evaluating its effects on maize nitrogen assimilation at both the physiological and molecular level. The results showed that PASP promoted biomass and nitrogen accumulation in maize seedlings, especially under low NO_3^- doses. Among different NO_3^- conditions, the most noticeable increase in plant biomass by PASP addition was observed in seedlings grown with 1 mmol L^{-1} NO_3^-, which was a little less than the optimum concentration (2 mmol L^{-1}) for plant growth. Furthermore, the total nitrogen accumulation increased greatly with additions of PASP to plants grown under suboptimal NO_3^- conditions. The promotion of nitrogen assimilation was mostly due to the increase of nitrate reductase (NR) activities. The NR activities in seedlings grown under low NO_3^- doses (0.5 and 1.0 mmol L^{-1}) were extremely increased by PASP, while the activities of glutamine synthetase (GS), aspartate aminotransferase (AspAT), and alanine aminotransferase (AlaAT) were slightly changed. Moreover, the regulation of PASP on NR activity was most probably due to the promotion of the protein accumulation rather than gene expression. Accumulation of NR protein was similarly affected as NR activity, which was markedly increased by PASP treatment. In conclusion, the present study provides insights into the promotion by PASP of nitrogen assimilation and identifies candidate regulatory enzymatic mechanisms, which warrant further investigation with the use of PASP in promoting nitrogen utilization in crops.

Keywords: polyaspartic acid; nitrate reductase; nitrogen metabolism; enzymatic activity; gene expression; protein accumulation

---

1. Introduction

Nitrogen is one of the most important nutrients that strikingly affects plant growth, development, and production. China is one of the world's largest nitrogen fertilizer producers and consumers, accounting for about 61% of the worldwide increase in nitrogen fertilizer production and 52% of the increase in nitrogen fertilizer consumption that occurred between 1990 and 2009 [1–3]. The excessive use of nitrogen fertilizer has contributed to serious damage to the environment, including soil acidification [4] as well as water and air pollution [5,6]. Therefore, there is an urgent need to find strategies to improve the nitrogen use efficiency of field crops, especially crops that are widely cultivated, such as maize (*Zea mays* L.), to simultaneously ensure food security and environmental quality.

Despite the importance of improving nitrogen use efficiency in maize, a number of previous studies had focused on various agronomic strategies to optimize nitrogen application and its biological mechanisms, such as the tillage type, rate and timing of nitrogen fertilizer application, and better sources of nitrogen fertilizer [7,8]. For maize, on the one hand, cultivars with high nitrogen use efficiency have been proven to be a great option for increasing grain yield under low nitrogen conditions while also maintaining the health of the environment [9]; on the other hand, the development of highly efficient nitrogen fertilizers is another effective way to resolve these problems [10]. Controlled-release urea has been demonstrated to significantly improve not only grain yields but also the nitrogen use efficiency of maize [11]. Recently, an eco-friendly polymer, polyaspartic acid (PASP), has been studied as a superabsorbent material and a promoter of fertilizer absorption [12–14] due to its free carboxylic and amide groups [15]. Polyaspartic acid is a hydrophilic, nontoxic, and biodegradable polymer of aspartic acid, with good dispersibility, chelating ability, and adsorption capacity [16]. Polyaspartic acid is found naturally in snails and mollusks, but for industrial production, it is commonly obtained through mild alkaline hydrolysis of polysuccinimide with high yield and low cost [17]. Considerable attention has been received for PASP in the medicine, cosmetic, and food industries [18]. In agriculture, PASP is usually used as a fertilizer absorption promoter and has been studied in nitrogen and potassium utilization [19]. Fertilizers containing PASP, especially PASP urea, have been gradually developed and applied in crop production. However, in previous studies, PASP was usually supplied together with a fertilizer or nutrient solution. Therefore, its promotion of fertilizer absorption was most probably due to its strong absorbency for ions, which reduces nutrient loss and improves the nutrient level of the soil [10–12]. However, information about the direct effects of PASP on plant growth and nitrogen assimilation, especially its physiological and molecular mechanism, remains limited. Thus, PASP was applied by foliar spraying to avoid the interaction of PASP and soil-based nutrients and to investigate the direct influence of PASP on maize growth and nitrogen assimilation, especially at enzymatic levels, and the genetic basis of the key enzyme involved.

Nitrate (NO_3^-) is the predominant form of nitrogen nutrition in most agricultural systems [20]. The pathway for nitrate assimilation in crops has been well documented [21]. Briefly, after uptake by roots, NO_3^- is first reduced by nitrate reductase (NR) to NO_2^- [22] and further reduced by nitrite reductase (NiR) to NH_4^+ [23]. Then, the NH4+ is assimilated into glutamine and glutamate by glutamine synthetase (GS) and glutamate synthase (GOGAT) [24–27]. The amino group of glutamate can be further transferred into other amino acids by various amino transferases, such as alanine aminotransferase (AlaAT) and aspartate aminotransferase (AspAT) [28–31]. Among these enzymes, NR is a primary rate-limiting enzyme for nitrogen assimilation [32,33]. Glutamine synthetase (GS, EC 6.3.1.2) is a key enzyme in nitrogen assimilation and remobilization [34]. AspAT and AlaAT can serve as markers of nitrogen use efficiency [35].

In this study, a commercial variety of maize, Zhengdan 958, was used, which is widely cultivated in China. The seedlings were cultivated under different doses of NO_3^- and were treated with PASP by foliar spraying. The objectives of the present study were to: (1) determine the effect of PASP on seedling biomass production and nitrogen assimilation in maize; (2) analyze the enzymatic mechanism of PASP regulation on maize nitrogen assimilation; and (3) investigate the genetic basis of the key enzyme involved in the regulation of PASP on nitrogen assimilation. Our study may provide information on the theoretical and practical bases for optimizing the use of PASP in promoting nitrogen utilization and plant growth in maize.

2. Materials and Methods

2.1. Plant Material and Growth Conditions

A commercial variety of maize (*Zea mays* L., cv. Zhengdan 958) was used in the experiments. The seeds were sterilized with 10% (*v/v*) H_2O_2 for 15 min, washed with distilled water, and germinated for 2 days in the dark on a moist filter paper at 30 °C. Then, the germinated seeds were transferred to silica sand to grow. Uniform seedlings with two visible leaves were selected and transferred to vessels containing 1/2 modified Hoagland solution with the following nutrients: 2 mmol L^{-1} of KNO_3, 1 mmol L^{-1} of $CaCl_2$, 0.5 mmol L^{-1} of $MgSO_4$, 0.1 mmol L^{-1} of KH_2PO_4, 0.1 mmol L^{-1} of EDTA-FeNa, 0.03 mmol L^{-1} of H_3BO_3, 0.0008 mmol L^{-1} of $CuSO_4 \cdot 5H_2O$, 0.005 mmol L^{-1} of $MnSO_4 \cdot H_2O$, 0.00003 mmol L^{-1} of $(NH_4)_6Mo_7O_{24} \cdot 4H_2O$, and 0.0025 mmol L^{-1} of $ZnSO_4 \cdot 7H_2O$. The pH of the solution was adjusted to 6.0. The nutrient solution was continuously aerated using an electric pump and renewed every 4 days. Each pot (7 L) contained 30 plants. When the second leaves were fully expanded, the seedlings were transferred to vessels (7 L) containing full-strength modified Hoagland solution with different concentrations of NO_3^- (0, 0.5, 1.0, 2.0, and 4.0 mmol L^{-1}, referred to as N0, N0.5, N1, N2, and N4, respectively). The concentrations of NO_3^-, K^+, and Ca^{2+} were balanced by varying the supply of KNO_3, $Ca(NO_3)_2$, KCl, and $CaCl_2$. The vessels were placed in a growth chamber controlled at 28 °C with a 16-h/8-h light/dark cycle. A photosynthetic photon flux density of 400 µmol m^{-2} s^{-1} was provided during the 16-h light period. The relative humidity was approximately 65%.

Two days after treatment with different concentrations of NO_3^-, PASP was applied to the plant of each treatment (PASP treatment to seedlings grown under N0, N0.5, N1, N2, and N4 conditions referred to as N0P, N0.5P, N1P, N2P, and N4P, respectively) by foliar spraying, and an equal amount of water was applied to the control treatment. Polyaspartic acid was prepared from polysuccinimide (AR, obtained from Desai Chemical Engineering Company, Shijiazhuang, China) with a molecular mass of 3000–5000 Da. Polysuccinimide (53.19 g) was dissolved in 100 mL of H_2O with 40.96 g of KOH to make the PASP solution [36]. Citric acid was added to adjust the solution pH to 8.0. Then, 0.2 mL of the above PASP solution was added to 1 L of water containing 0.1% (*v/v*) Tween-20 as a surfactant to make the final PASP concentration (approximately 73.55 mg of polysuccinimide L^{-1}). A compression sprayer (capacity, 1 L) was used for this purpose to ensure the even distribution of PASP on all leaves. Spraying was performed in the morning (between 9:00 and 10:00 a.m.). The PASP solution was sprayed on the shoot until complete leaf wetting (approximately 3 mL of solution per plant). The experiment was performed four times.

2.2. Measurement of Biomass and Nitrogen Accumulation

Seven days after PASP treatment, 15 plants per treatment were separated into two parts (aboveground and underground) and oven-dried at 80 °C until a constant weight was reached to measure the respective dry weights. Then, the dry samples were ground and used for total nitrogen accumulation determination using the Kjeldahl method.

2.3. Measurement of Nitrate Reductase (NR, EC 1.6.6.1) Activity

The method was adapted from the in vivo NR assay of Majláth et al. [37]. Briefly, approximately 200 mg fresh weight of samples (the latest fully expanded leaf and roots, respectively) were cut into small sections and incubated in 1 mL of 100 mmol L^{-1} of Na-phosphate buffer (pH 7.5) containing 200 mmol L^{-1} of KNO_3 at 37 °C in the dark for 1 h. Next, 0.4 mL of 30% (*m/v*) trichloroacetic acid was added to stop the conversion of NO_3^- to NO_2^-. Then, nitrite production was detected calorimetrically by adding 2 mL of 0.2% 1-naphthylamine and 2 mL of 1% sulphanilamide (dissolved in 30% acetic acid) to the reaction mixture. After 30 min, the optical density of solutions was measured at 540 nm. The incubation buffer was used as a blank. NR activity was calculated in µmol nitrite produced per gram of fresh tissue using a standard curve, based on known nitrite dilutions. The experiment was conducted at 9:00 a.m. on the 1st, 3rd, and 7th day after PASP treatment.

2.4. Measurement of Glutamine Synthetase (GS, EC 6.3.1.2) Activity

Approximately 200 mg fresh weight of samples (the latest fully expanded leaf and roots, respectively) were homogenized at 4 °C with 2 mL of extraction buffer (0.5 mol L^{-1} of Tris-HCl, 2 mmol L^{-1} of $MgCl_2$, 2 mmol L^{-1} of DTT, and 0.4 mol L^{-1} of sucrose; pH 8.0). Then, the homogenates were centrifuged at $12,000 \times g$ at 4 °C for 15 min, and the supernatant was recovered. GS activity was determined in the supernatant by transferase assay [38].

2.5. Measurement of Alanine Aminotransferase (AlaAT EC 2.6.1.2) and Aspartate Aminotransferase (AspAT EC 2.6.1.1) Activity

AlaAT and AspAT activity were assayed in conditions adapted from the study conducted by Gibon et al. [39]. Approximately 200 mg fresh weight of samples (the latest fully expanded leaf and roots, respectively) were homogenized at 4 °C with 2 mL of extraction buffer (0.05 mol L^{-1} of Tris-HCl) and centrifuged at $12,000 \times g$, 4 °C for 15 min. Then, the supernatant was recovered for enzyme activity assays. AlaAT activity was assayed using a solution containing 0.1 mol L^{-1} of Na-phosphate buffer (pH 7.2), 0.2 mol L^{-1} of l-alanine, and 2 mmol L^{-1} of α-ketoglutarate. Plant sample protein extract (75 μL) was added to the 125-μL assay solution and incubated at 37 °C for 30 min. The reaction was stopped by adding 125 μL of 2,4-dinitrophenylhydrazine. Pyruvic acid production was detected calorimetrically by adding 1.25 mL of 0.4 mol L^{-1} of NaOH to the reaction mixture at 5 min after the addition of 2,4-dinitrophenylhydrazine. After 30 min, the optical density of solutions was measured at 500 nm. The incubation buffer was used as a blank. Pyruvic acid standards (0–0.4 μmol in extraction buffer) were run in parallel. The activities were expressed as μmol pyruvic acid produced per gram of fresh tissue.

AspAT activity was determined using an assay solution containing 0.1 mol L^{-1} of Na-phosphate buffer (pH 7.2), 0.2 mol L^{-1} of aspartic acid, and 2 mmol L^{-1} of α-ketoglutarate. Otherwise, the enzyme activity assays and activity calculations were the same as that for AlaAT.

2.6. Real-Time Quantitative PCR (qPCR)

Fresh samples of 15 seedlings from individual treatments were collected in liquid nitrogen for the isolation of RNA. Total RNA was extracted using a total Plant RNA kit (Gene Mark, Taiwan). Reverse transcription was performed with 1–2 μg of purified total RNA using TaqMan® Reverse Transcription Reagents (Invitrogen™, Carlsbad, CA, USA) according to the manufacturer's protocol. The qPCR was performed by a 7500 Real-Time PCR system (Applied Biosystems, Foster City, CA, USA) using a PowerUp™ SYBR® Green Master Mix (Applied Biosystems™, Foster City, CA, USA). The following protocol was applied in the qRT-PCR reaction: denaturation at 95 °C for 10 min, followed by 41 cycles of denaturation at 95 °C for 15 s, and annealing extension at 60 °C for 1 min. The relative gene expression levels were calculated using the $2^{-\Delta\Delta Ct}$ method [40], with *actin* as an internal control, and three repetitions were performed for each sample. The sequences of the gene primers are shown in Table 1.

Table 1. Primers for the qPCR.

Gene	Gene ID	Forward Primer (5'-3')	Reverse Primer (5'-3')
ZmNR1	GRMZM2G568636	ATGATCCAGTTCGCCATCTC	GTCCGTGGTACGTCGTAGGT
ZmNR2	GRMZM2G428027	AGCAAGTCTTGAGGGAGCAC	CGCCTTGCATGACATTCGTT [41]
ZmNR3	GRMZM5G878558	ACTGGTGCTGGTGCTTCTGGTCC	ATGCCGATCTCGCCCTTGTGC [42]
ZmNR4	GRMZM2G076723	GCGTGCAGTTTCAATTCGGT	AGCTATTCCCCGTTGCCATC
actin	XM_008656735	GATTCCTGGGATTGCCGAT	TCTGCTGCTGAAAAGTGCTGAG [43]

2.7. Nitrate Reductease Protein Extraction and Quantification by Enzyme-Linked Immunosorbent Assay (ELISA)

Fresh samples of 15 seedlings from individual treatments were collected in liquid nitrogen for the isolation of NR protein. The samples were collected at 9:00 a.m. on the 1st, 3rd, and 7th day after PASP treatment. NR protein extraction and quantification was performed using a Plant Nitrate reductase (NR) ELISA Kit and was carried out by Beijing Fangcheng Jiahong Science and Technology Co. (Beijing, China).

2.8. Statistical Analysis of Data

Data were analyzed by one-way ANOVA using SPSS 17.0 software (SPSS Inc., Chicago, IL, USA, 2002). The treatment means were separated using Duncan's Multiple Range Test or Student's test. Statistical comparisons were considered significant at $p < 0.05$.

3. Results

3.1. Changes in Plant Biomass Accumulation in Maize Seedlings

Biomass accumulation in maize seedlings differed due to different NO_3^- doses. As shown in Figure 1A, the shoot biomass had a bell-shaped curve pattern, which peaked at 2 mmol L^{-1} of NO_3^-, while the root biomass was significantly ($p < 0.05$) reduced with the increase in NO_3^- doses (Figure 1B). Consequently, the root/shoot ratio shrunk along with the increase in NO_3^- dose (Figure 1C). Furthermore, the response of the total biomass accumulation per plant to NO_3^- dose was the same with the shoot biomass accumulation, which indicated 2 mmol L^{-1} of NO_3^- as the optimum concentration for plant biomass production.

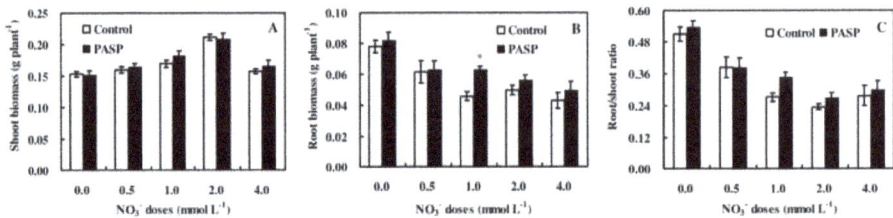

Figure 1. The biomass accumulation in the shoots (**A**) and roots (**B**), as well as the root/shoot ratio (**C**), of maize differed on the 7th day after nitrogen and polyaspartic acid (PASP) treatment. The PASP and control indicate seedlings with and without PASP addition, respectively. Data was presented as the mean ± standard error of 15 plants in each treatment. *, statistically significant differences between PASP treatment and control within each nitrogen level, according to Student's test at $p < 0.05$.

The PASP treatment mainly improved biomass accumulation, especially in seedlings supplied with 1 mmol L^{-1} of NO_3^- (Figure 1A). The root biomass in N1P was significantly ($p < 0.05$) increased by 37.8% from applications of PASP to plants receiving the N1 treatment, while no significant ($p < 0.05$) differences were observed from PASP applications with the rest of the nitrogen levels (0, 0.5, 2.0, and 4.0 mmol L^{-1}). The increase of the shoot biomass from PASP applications was slight, and no significant ($p < 0.05$) difference was observed (Figure 1A). However, the total biomass per plant grown in N1P was significantly ($p < 0.05$) higher than those grown in N1 by 13.9%, which was consistent with the response of the root biomass (Figure 1B). The root/shoot ratio was mildly boosted by PASP, especially in the N1P treatment, which was 28.5% greater than plants grown in N1 (Figure 1C). These results suggest that the effect of PASP in seedlings supplied with 1 mmol L^{-1} of NO_3^- was most significant among different nitrogen doses, and the improvement by PASP on root biomass was greater than that on shoots.

3.2. Changes in Nitrogen Accumulation in Maize Seedlings

Nitrogen and PASP treatments markedly affected the total nitrogen accumulation in seedlings (Figure 2). Similar to shoot biomass accumulation, nitrogen content in both the roots and shoots had a bell-shaped curve pattern, which peaked with 2 mmol L^{-1} of NO_3^- supply. Additions of PASP mostly improved the nitrogen content in both the shoots (Figure 2A) and roots (Figure 2B). Among the different NO_3^- doses, the increase of plant nitrogen content by PASP in conjunction with the N0.5P and N1P treatments were the most remarkable. N0.5P and N1P had 13.2% and 12.6% greater total nitrogen content in the whole plant when compared to N0.5 and N1, respectively. This increase was followed by the N2P treatment, in which the total nitrogen accumulation per plant increased by 4.8% when compared to N2. For seedlings with a high NO_3^- dose (4 mmol L^{-1}), only the shoot nitrogen content was increased by PASP application (Figure 2A). Overall, these findings indicate that PASP positively affects nitrogen assimilation in maize, especially in seedlings grown under low nitrogen levels.

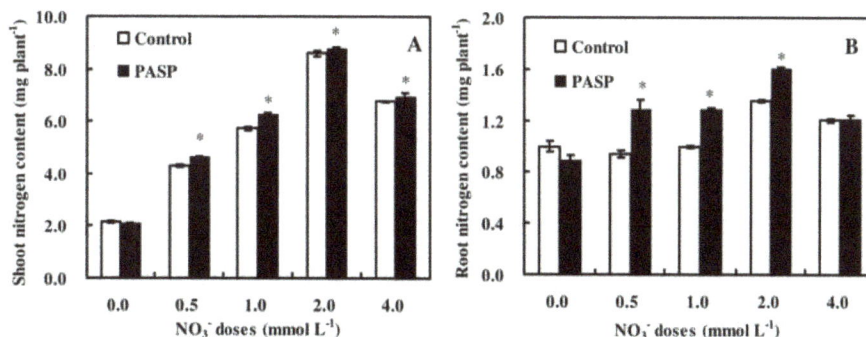

Figure 2. Total nitrogen accumulations in the shoots (**A**) and roots (**B**) of maize on the 7th day after nitrogen and PASP treatment. The PASP and control indicate seedlings with and without PASP addition, respectively. Data rewash presented as the mean ± standard error of 15 plants in each treatment. *, Statistically significant differences between PASP treatment and control within each nitrogen level, according to Student's test at $p < 0.05$.

3.3. Changes in Enzyme Activities Correlated to Nitrogen Metabolism in Leaves and Roots of Maize Seedlings

In order to further investigate the physiological mechanism in which PASP affects nitrogen assimilation, enzyme activities involved with nitrogen metabolism, such as NR, GS, AspAT, and AlaAT, were estimated. Multiple analyses showed that the activity of NR was most affected by PASP (Table 2).

Table 2. Multiple analyses of enzyme activities correlated to nitrogen metabolism in maize seedlings leaves and roots at the 1st, 3rd, and 7th day after PASP treatment.

Parts	Source of Variation	Days after PASP Treatment											
		1d				3d				7d			
		NR	GS	AspAT	AlaAT	NR	GS	AspAT	AlaAT	NR	GS	AspAT	AlaAT
Leaves	NO_3^-	***	**	**	*	***	**	ns	ns	***	***	**	*
	PASP	***	ns	**	ns	**	ns	ns	ns	**	ns	ns	ns
	NO_3^- × PASP	***	ns	ns	ns	**	ns	*	*	***	ns	ns	ns
Roots	NO_3^-	***	***	ns	**	***	***	*	ns	***	***	***	***
	PASP	***	ns	ns	*	ns	ns	ns	ns	***	**	ns	*
	NO_3^- × PASP	***	ns	ns	ns	ns	ns	ns	**	***	ns	ns	*

Note: ns, no significant difference, * $p < 0.05$, ** $p < 0.01$, *** $p < 0.001$.

NO_3^- doses significantly ($p < 0.05$) affected most of the four enzyme activities measured during the experimental period in both the leaves and roots, except the root AspAT activity on the 1st day, the leaf AspAT activity on the 3rd day, and both of the leaf and root AlaAT activities on the 3rd day (Table 2). Supplementing plants with PASP, however, almost had no significant ($p < 0.05$) effects on all of the GS, AspAT, and AlaAT enzyme activities, with only changes of the leaf AspAT activity ($p < 0.01$) and root AlaAT activity ($p < 0.05$) on the 1st day as well as the root GS ($p < 0.01$) and AlaAT ($p < 0.05$) activities on the 7th day being significant ($p < 0.05$) (Table 2).

The effect of PASP on NR activity, however, was definitely remarkable. The NR activities in both the roots and leaves on all 3 days were significantly ($p < 0.001$) affected by PASP additions (Table 2). Moreover, the interaction of PASP and nitrogen treatment on NR activity was significant ($p < 0.001$) too (Table 2). In leaves, PASP treatment strikingly stimulated the NR activities at low

NO_3^- doses, i.e., 0.5 and 1.0 mmol L^{-1} (Figure 3A–C). On the 1st day after PASP treatment, leaf NR activities increased 3.4-fold and 2.6-fold over the levels found in N0.5 and N1, respectively (Figure 3A). On the 3rd day, a similar increase in leaf NR activities of 2.4-fold and 1.6-fold was observed in N0.5P and N1P in comparison to N0.5 and N1, respectively (Figure 3B). On the 7th day, PASP supplementation significantly ($p < 0.05$) upregulated the NR activities in leaves by 1.7-fold and 4.6-fold when compared to N0.5 and N1, respectively (Figure 3C). However, in seedlings grown under high NO_3^- doses, the changes in NR activities caused by PASP additions were quite small or even negative (Figure 3A–C). In roots, NR activity responded to foliarly applied PASP after a short period of time. On the 1st day after treatment, PASP only induced a significant ($p < 0.05$) increase (by 2.9-fold) in NR activity in seedlings supplied with 1 mmol L^{-1} of NO_3^- when compared to control (Figure 4A). Furthermore, no significant ($p < 0.05$) effect was observed on the 3rd day between these treatments (Figure 4B). However, on the 7th day, root NR activities in all N0.5P, N1P, N2P, and N4P treatments were significantly ($p < 0.05$) upregulated by 1.8-fold, 2.6-fold, 5.4-fold, and 1.6-fold, respectively, when compared to N0.5, N1, N2, and N4, respectively (Figure 4C).

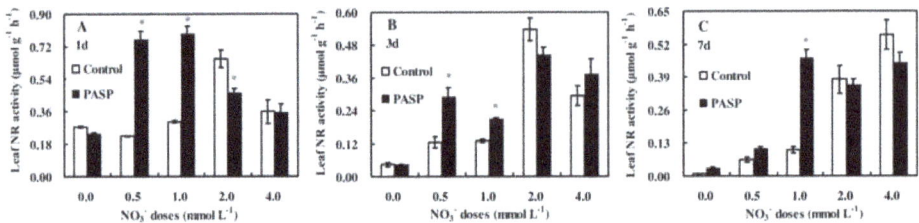

Figure 3. Changes in leaf nitrate reductase activities on the 1st (**A**), 3rd (**B**), and 7th (**C**) day after PASP treatment. The PASP and control indicate seedlings with and without PASP addition, respectively. Data were presented as the mean ± standard error of 15 plants in each treatment. *, Statistically significant differences between PASP treatment and control within each nitrogen level, according to Student's test at $p < 0.05$.

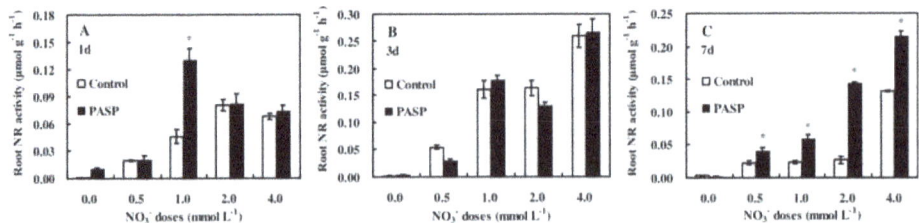

Figure 4. Changes in root nitrate reductase activities on the 1st (**A**), 3rd (**B**), and 7th (**C**) day after PASP treatment. The PASP and control indicate seedlings with and without PASP addition, respectively. Data were presented as the mean ± standard error of 15 plants in each treatment. *, Statistically significant differences between PASP treatment and control within each nitrogen level, according to Student's test at $p < 0.05$.

Overall, these results show that NR reacted more positively to PASP level than GS, AspAT, and AlaAT, especially in seedlings cultured under low nitrogen conditions. Thus, NR is probably the key enzyme involved in the promotion of nitrogen assimilation by PASP.

3.4. Changes in NR Gene Expression Levels in Maize Seedlings

To further investigate the mechanism of PASP regulation on NR activities, qPCR analysis was performed to examine the NR gene expression levels in leaves and roots after nitrogen and PASP treatment. Nitrate dose level significantly ($p < 0.001$) affected the expression levels of all four NR genes in both the leaves (Figure 5) and the roots (Figure 6). The PASP treatment, however, induced almost no significant ($p < 0.05$) changes on the expression of *ZmNR1* to *4*, especially in the leaves (Figure 5). In roots, only the *ZmNR1* expression on the 3rd day after PASP treatment was significantly ($p < 0.05$) affected by PASP (Figure 6). However, the interaction of PASP addition and NO_3^- doses were significant ($p < 0.001$) (Figures 5 and 6). In the leaves of seedlings grown under 1–4 mmol L^{-1} NO_3^- doses, the expression patterns of *ZmNR1* and *ZmNR3* reacted in a similar way to NR activity (Figure 3) to PASP addition, which were generally upregulated by PASP under low NO_3^- levels but downregulated under high NO_3^- levels (Figure 5). This indicated that *ZmNR1* and *ZmNR3* might be the candidate genes in response to PASP addition. However, the changes of the gene expression levels were much smaller than that of the NR activities (Figure 3). On the 1st day, PASP application significantly ($p < 0.05$) upregulated the expression of *ZmNR1* and *ZmNR3* by 1.8-fold and 1.4-fold when compared to N1 (Figure 5). On the 3rd day, a similar increase in leaf *ZmNR1* and *ZmNR3* expression of 1.2-fold and 1.3-fold was observed in N1P in comparison to N1, respectively (Figure 5). Moreover, the expression pattern of *ZmNR1* and *ZmNR3* in seedlings grown with 0.5 mmol L^{-1} NO_3^-, which was scarcely affected by PASP addition (Figure 5), was not consistent with the NR activity in response to PASP addition (Figure 3). In roots, few changes of the *ZmNR1* and *ZmNR3* expression were induced by PASP. Noticeable decreases were only observed in *ZmNR1* expression in N2P and N4P in comparison to N2 and N4, respectively, and in *ZmNR3* expression in N2P compared to N2 on the 3rd day (Figure 6). On the 1st day, the expression of *ZmNR2* and *ZmNR4* in roots was markedly regulated by PASP addition, but the general expression pattern of these two genes on the 3 days were not consistent with NR activities in roots in response to PASP addition. Thus, *ZmNR1* and *ZmNR3* seem to be candidate genes in leaves in response to PASP addition, but generally, the gene expression does not appear to be the main approach where PASP upregulates the NR activity in maize under low nitrogen conditions.

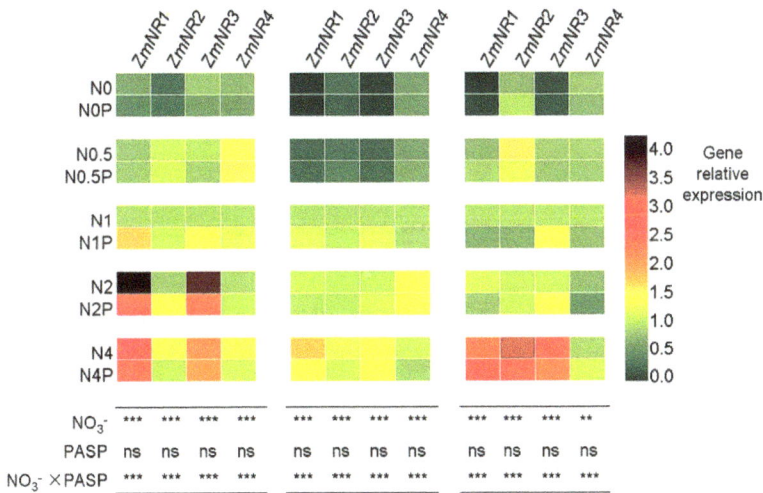

Figure 5. Relative expression of *ZmNR1*, *ZmNR2*, *ZmNR3*, and *ZmNR4* in the leaves on the 1st, 3rd, and 7th day after PASP treatment. Data were presented as the mean of 15 plants in each treatment. ns, *, **, and ***, Statistically significant differences according to multiple analyses at $p > 0.05$, $p < 0.05$, $p < 0.01$, and $p < 0.001$. N0, N0.5, N1, N2, and N4 indicate the seedlings treated with 0, 0.5, 1.0, 2.0, and 4.0 mmol L^{-1} of NO_3^- without PASP treatment, respectively. N0P, N0.5P, N1P, N2P, and N4P indicate the seedlings treated with 0, 0.5, 1.0, 2.0, and 4.0 mmol L^{-1} of NO_3^- plus PASP addition, respectively.

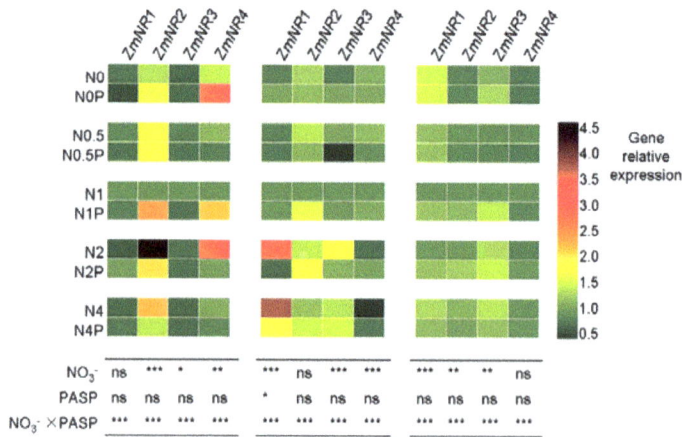

Figure 6. Relative expression of *ZmNR1*, *ZmNR2*, *ZmNR3*, and *ZmNR4* in the roots on the 1st, 3rd, and 7th day after PASP treatment. Data were presented as the mean of 15 plants in each treatment. ns, *, **, and ***, Statistically significant differences according to multiple analyses at $p > 0.05$, $p < 0.05$, $p < 0.01$, and $p < 0.001$. N0, N0.5, N1, N2, and N4 indicate the seedlings treated with 0, 0.5, 1.0, 2.0, and 4.0 mmol L^{-1} of NO$_3$$^{-}$ without PASP treatment, respectively. N0P, N0.5P, N1P, N2P, and N4P indicate the seedlings treated with 0, 0.5, 1.0, 2.0, and 4.0 mmol L^{-1} of NO$_3$$^{-}$ plus PASP addition, respectively.

3.5. Changes in Nitrate Reductase Protein Accumulation in Maize Seedlings

In addition to the gene expression, NR protein accumulation was analyzed by ELISA to better appreciate the regulation of PASP on NR activity in maize. Nitrate reductase protein accumulation in seedlings grown in combination with low NO$_3$$^{-}$ doses was generally increased by PASP (Figures 7 and 8), which was similar to the NR activity (Figures 3 and 4) changes in response to PASP supplementation. On the 1st day after PASP treatment, NR protein accumulations in leaves in the N0.5P and N1P treatments were much greater, by 1.6-fold and 2.1-fold, compared to the levels found from the N0.5 and N1 treatments, respectively (Figure 7A). On the 3rd day after PASP application, similar increases of 1.5-fold and 1.4-fold in leaf NR protein content were observed when compared to N0.5 and N1, respectively (Figure 7B). On the 7th day, PASP increased the leaf NR protein content of N1-treated plants by 1.3-fold (Figure 7C). However, by the 7th day, there was no significant ($p < 0.05$) difference due to PASP treatment in leaf NR protein levels in plants grown with 0.5 mmol L^{-1} NO$_3$$^{-}$. Conversely, in seedlings grown under high NO$_3$$^{-}$ doses (2 and 4 mmol L^{-1}), the NR protein content in leaves was generally deceased by PASP treatment (Figure 7A–C), which was similar to the response of NR activity to PASP treatment (Figure 3A–C).

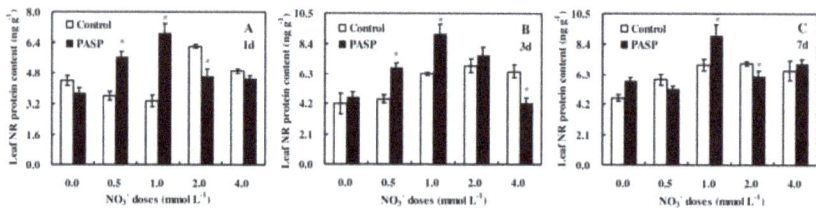

Figure 7. Leaf nitrate reductase (NR) protein content analysis by ELISA on the 1st (**A**), 3rd (**B**), and 7th (**C**) day after PASP treatment. The PASP and control indicate seedlings with and without PASP addition, respectively. Data were presented as the mean ± standard error of 15 plants in each treatment. *, Statistically significant differences between PASP treatment and control within each nitrogen level, according to Student's test at $p < 0.05$.

In roots, the response of NR protein accumulation to PASP (Figure 8) was in accordance with that of the NR enzymatic activity (Figure 4). On the 1st day, PASP markedly increased the root NR protein content in both N0.5P and N1P by 1.7-fold when compared to N0.5 and N1, respectively (Figure 8A). On the 3rd day, no significant ($p < 0.05$) difference was observed in root NR protein content from PASP treatment under all the experiment NO_3^- doses (Figure 8B). On the 7th day, root NR protein accumulation was generally increased by 1.3–1.6 fold by PASP when NO_3^- was present (Figure 8C).

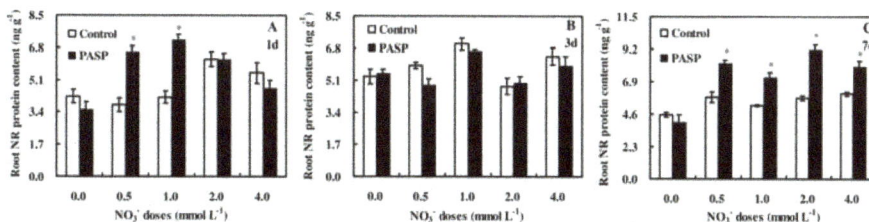

Figure 8. Root NR protein content analysis by ELISA on the 1st (**A**), 3rd (**B**), and 7th (**C**) day after PASP treatment. The PASP and control indicate seedlings with and without PASP addition, respectively. Data were presented as the mean ± standard error of 15 plants in each treatment. *, Statistically significant differences between PASP treatment and control within each nitrogen level, according to Student's test at $p < 0.05$.

4. Discussion

4.1. PASP Promoted Seedling Growth and Nitrogen Accumulation in Maize under Low Nitrogen Conditions

Plant growth reaction to a nitrogen source supply has been widely considered to be positive, increasing leaf area, plant height, and biomass production as the nitrogen supply rises [44,45]. In the present study, shoot biomass accumulation gradually increased along with the increase in NO_3^- dose from 0 to 2 mmol L^{-1} (Figure 1A). By contrast, root biomass production was strikingly inhibited through increasing NO_3^- doses (Figure 1B). This was in accordance with a previous study conducted by Tian et al. [46]. In fact, many recent studies have pointed out that the influence of nitrogen supply to plant growth may be less than expected or even detrimental to plant development in some conditions and areas [47,48]. Similarly, biomass accumulation in both the shoots (Figure 1A) and roots (Figure 1B) remarkably decreased in seedlings that were grown under the high nitrogen condition (N4) when compared to N2. The total biomass accumulation per plant in N2 was markedly higher than that in other treatments without PASP treatment, which suggests that the 2 mmol L^{-1} of NO_3^- dose was the optimum concentration for plant growth.

Polyaspartic acid, which has a large number of carboxylic and amide groups [15], has been studied for promoting plant growth and nutrient use efficiency in crops [10,12,13,19]. In the present study, PASP generally increased the biomass production of maize seedlings (Figure 1A,B). This result is consistent with the findings reported by Du et al. [13], in which PASP was shown to promote seedling and root growth in rice. The most noticeable increase in plant biomass production after PASP treatment was observed in N1P (Figure 1B), which was under NO_3^- doses that were slightly less than the optimum concentration for plant growth. Larger changes in biomass accumulation were observed in the root than in the shoot from the PASP treatment. Therefore, the root/shoot ratio (Figure 1C) exhibited a slight increase after PASP treatment.

In terms of the significant relationship among experimental conditions, nitrogen content in the roots was observed to significantly ($p < 0.05$) increase along with the increase in NO_3^- doses (Figure 2). These results are in accordance with the reports of previous studies, which indicated the dependence of nitrogen accumulation and redistribution of crops on nitrogen rate [49]. However, nitrogen accumulation in the shoots (Figure 2A) had a bell-shaped curve pattern, which peaked from the N2 treatment. A decrease in shoot nitrogen accumulation was observed by supplementing with more nitrogen in N4 when compared to N2, which was consistent with the shoot biomass production (Figure 1A). Similar to the plant biomass, nitrogen accumulation was generally increased by PASP additions in both the roots and shoots (Figure 2). In particular, the changes between N0.5P and N0.5, as well as N1P and N1, were pronouncedly great. However, nitrogen content per gram of dry matter did not significantly ($p < 0.05$) differ as a result of these treatments, as calculated in Figures 1 and 2. These results suggest that the promotion of nitrogen assimilation by PASP is probably simultaneous with other types of metabolism, such as carbon assimilation [19], which consequently results in the extreme expansion in total nitrogen accumulation per plant, with a small influence in nitrogen concentration per gram of dry matter.

Overall, these results showed that the N2 treatment was the optimum NO_3^- dose for seedling growth in these experiment conditions. Moreover, PASP treatment most probably enhanced nitrogen assimilation as well as seedling growth under low nitrogen levels (0.5 and 1.0 mmol L^{-1} of NO_3^-).

4.2. The Improvement of PASP on Nitrogen Accumulation in Maize Was Primarily Attributed to Changes in NR Activities

A few comprehensive reviews [31,34] have discussed the progress of nitrogen assimilation. In these reviews, various enzymes have been reported to participate in nitrate assimilation, such as NR, GS, AspAT, and AlaAT. In the current study, the most noticeable change after PASP treatment was the large increase in NR activity in seedlings when grown under low NO_3^- doses, especially 0.5 and 1.0 mmol L^{-1} (Figures 3 and 4). This finding was in accordance with the increase of nitrogen accumulation after PASP treatment, which suggests the key role of NR in maize nitrate assimilation in response to PASP. However, in seedlings grown under high NO_3^- doses (2 and 4 mmol L^{-1}), NR activity rarely or even negatively responded to PASP on the 1st and 3rd day after treatment (Figures 3 and 4). This response may be due to the abundance of NO_3^- available. Regarding the response of NR to the increasing NO_3^- dose, both control and PASP treatment exhibited a bell-shape curve. The response of NR activity to PASP under high NO_3^- conditions was similar to that response to high NO_3^- doses when a decrease in NR activity (Figures 3 and 4) and nitrogen (Figure 2) and biomass (Figure 1) accumulation was observed in N4 when compared to N2. Furthermore, previous studies have demonstrated that the activity of NR is positively induced by exogenous nitrate only at low nitrogen conditions [50]. Moreover, the cellular compartmentation of nitrate, rather than the exogenous nitrate, was considered to be the main factor that regulates NR activity [50,51]. The PASP treatment may minimize the threshold concentration of the supplied nitrate, which induces NR activity by enhancing nitrate uptake under low nitrogen conditions and elevates the internal nitrate concentration in the metabolic pool. Furthermore, the enhanced NR activity was simultaneously followed by a high production of nitrite [52], which may inhibit NR activity through a feedback inhibition mechanism. Different from that on the 1st (Figure 4A) and 3rd (Figure 4B) days after supplementation, root NR activity on the 7th day was remarkably increased by PASP (Figure 4C). These differences may be due to the variation of nitrogen demand in different growth processes [53,54]. Thus, it was hypothesized that in seedlings with nitrogen availability less than the growth demand, NR activity was upregulated by PASP, while in seedlings supplied with sufficient nitrogen, this activity was reduced.

In addition to NR, GS is another critical enzyme of nitrate assimilation which catalyzes the assimilation of ammonium into glutamine [55]. AlaAT and AspAT are two important aminotransferases in plants which catalyze the reversible transfer of the amino group from glutamate to oxalocaetate and pyruvate, respectively [28–30]. GS and AlaAT were positively coregulated with NR [56,57]. However, in the present study, the activities of all GS, AlaAT, and AspAT enzymes were barely affected by PASP treatment (Table 2) and appeared not to be the dominant processes that allowed PASP to promote nitrate assimilation.

Overall, these results revealed that NR predominantly contributed to the promotion of PASP in nitrogen accumulation in maize by PASP, especially under low nitrogen conditions.

4.3. The Regulation of PASP on NR Activity Was Mainly Due to the Increasing Accumulation of Protein Rather than Gene Transcription

As the primary rate-limiting enzyme of nitrogen nutrition [32,33], NR has been widely studied. These studies revealed that NR activity in higher plants was regulated at both the transcriptional and post-translational levels [42,58,59]. In order to further investigate the genetic basis of PASP regulation on NR, the gene expression and protein accumulation of NR was measured in the present study.

The present results revealed that there were limited correlations between the gene transcript level and enzymatic activity of NR in response to PASP supplementation. In particular, on the 7th day after PASP treatment, none of the four NR genes was consistently transcribed with that of the NR activity (Figures 5 and 6) in response to PASP treatment. On the 1st and 3rd days after treatment, PASP affected the transcript of *ZmNR1* and *ZmNR3* (Figure 5) in leaves in a similar way to that of NR activity (Figure 3A), but the changes in the gene transcription were much less than that of NR activity. In roots, the expression of *ZmNR1* and *ZmNR3* (Figure 6) was only significantly regulated by PASP on the 3rd day in plants grown with high NO_3^- dose, which was not consistent with NR activity in response to PASP addition. The transcription levels of *ZmNR2* and *ZmNR4* (Figure 6) in roots on the 1st day were markedly affected by PASP addition, but on the 3rd and 7th days, the expression patterns of these two genes were not consistent with the NR activity (Figure 4) in response to PASP addition. These results indicate that *ZmNR1* and *ZmNR3* may be candidate genes in response to PASP addition in leaves, while *ZmNR2* and *ZmNR4* may take a role in the response of roots to PASP application, but all the transcriptions of these genes are not the main approach where PASP upregulates the NR activity in maize under low nitrogen conditions. This was not unexpected. In fact, numerous studies have found similar results, in which transcript levels and their relevant enzyme activities were not tightly correlated [39,60,61]. Moreover, on the 1st day after treatment, the transcription levels of *ZmNR2* and *ZmNR4* (Figure 6) in the roots of seedlings grown without NO_3^- application were markedly upregulated by PASP addition, while no significant differences were observed in the NR protein content (Figure 8), NR activity (Figure 4), nitrogen accumulation (Figure 2), or plant dry weight (Figure 1). This indicates that there needs to be nitrogen available in the first place to observe physiological changes.

Different from gene transcription, the protein accumulation of NR (Figures 7 and 8) was affected similar to the NR activity (Figures 3 and 4) by PASP treatment, especially in the leaves. The response of the NR protein content to PASP could be interpreted in two ways. First, PASP might increase the translation of NR mRNA to protein, but this still needs evidence. Second, a decrease in degradation in NR protein may be induced by PASP. NR has a short half-life of several hours. In previous reviews, NR phosphorylation and 14-3-3-binding, as well as sugar signals, were considered to be involved in NR degradations [62,63]. It appears that the activated NR protein was more stable. In fact, the changes in protein accumulation were less than that of NR activity after PASP treatment in the present study (indicated by Figures 3, 4, 7 and 8). On the 7th day after treatment, especially in the roots, the NR protein content (Figure 8C) and NR activity (Figure 4C) did not exhibit a completely consistent response to PASP treatment. Thus, it appears that PASP most probably regulates NR activity according to protein accumulation as well as post-translational control.

Overall, the transcription and protein accumulation results imply that the regulation of PASP on NR activity is predominantly attributed to protein accumulation and may be a post-translational regulation rather than gene transcription.

These results imply that the application of PASP by foliar spraying generally promotes plant growth and nitrogen assimilation, especially the NR activity. In previous studies, PASP was usually supplied together with a fertilizer or nutrient solution, and its promotion of fertilizer absorption was usually considered to be due to its strong absorbency for ions [10,12,13]. In this study, PASP was supplied by foliar spraying, which avoided the interaction of PASP and fertilizer. Thus, we infer that PASP may be absorbed in the leaves and acts in a similar way to a plant growth regulator. In fact, Xu et al. [64] has studied PASP as a plant growth regulator together with Kinetin and 1-Naphthaleneacetic acid. However, the absorption of PASP in leaves and the mechanism by which it works still need further investigation.

Agronomy **2018**, *8*, 188

5. Conclusions

Application of PASP mainly promoted seedling growth, the increase of the root/shoot ratio, and the accumulation of total nitrogen in maize plants, especially under low nitrogen conditions. The promotion by PASP of maize nitrogen accumulation was primarily due to the increase in NR activity. The transcription of *ZmNR1* to 4 was significantly ($p < 0.05$) affected by NO_3^-, but few changes were induced by PASP treatment. However, the accumulation of NR protein was strikingly increased by PASP treatment, which was consistent with the changes in NR activities. Thus, it appears that the regulation by PASP of NR activity was most probably due to the promotion of accumulating protein rather than gene expression. The present study provides useful insight into the action of PASP on maize nitrogen assimilation. Therefore, PASP could be used for enhancing nitrogen utilization in maize.

Author Contributions: Q.W. and Z.D. conceived and designed the experiments; Q.W., H.D., X.D., and Y.X. performed the experiments; Q.W. and H.T. analyzed the data; G.L. and H.D. contributed reagents/materials/analysis tools; Q.W. wrote the paper.

Funding: This research was funded by the National Key R&D Program of China (Grant No. 2018YFD0200608) and the National Natural Science Foundation of China (Grant No. 31470087).

Acknowledgments: We thank the reviewers and editors for their exceptionally helpful comments about the manuscript.

Conflicts of Interest: The authors declare no conflict of interest.

References

1. Cao, Q.; Miao, Y.X.; Feng, G.H.; Gao, X.W.; Liu, B.; Liu, Y.Q.; Li, F.; Khosla, R.; Mulla, D.J.; Zhang, F.S. Improving nitrogen use efficiency with minimal environmental risks using an active canopy sensor in a wheat-maize cropping system. *Field Crops Res.* **2017**, *214*, 365–372. [CrossRef]

2. Norse, D.; Ju, X. Environmental costs of China's food security. *Agric. Ecosyst. Environ.* **2015**, *209*, 5–14. [CrossRef]

3. Zhang, W.F.; Dou, Z.X.; He, P.; Ju, X.T.; Powlson, D.; Chadwick, D.; Norse, D.; Lu, Y.L.; Zhang, Y.; Wu, L.; et al. New technologies reduce greenhouse gas emissions from nitrogenous fertilizer in China. *Proc. Nat. Acad. Sci. USA* **2013**, *110*, 8375–8380. [CrossRef] [PubMed]

4. Guo, J.H.; Liu, X.J.; Zhang, Y.; Shen, J.L.; Han, W.X.; Zhang, W.F.; Christie, P.; Goulding, K.W.T.; Vitousek, P.M.; Zhang, F.S. Significant acidification in major Chinese croplands. *Science* **2010**, *327*, 1008–1010. [CrossRef] [PubMed]

5. Galloway, J.N.; Townsend, A.R.; Erisman, J.W.; Bekunda, M.; Cai, Z.; Freney, J.R.; Martinelli, L.A.; Seitzinger, S.P.; Sutton, M.A. Transformation of the nitrogen cycle: Recent trends, questions, and potential solutions. *Science* **2008**, *320*, 889–892. [CrossRef] [PubMed]

6. Liu, X.J.; Zhang, Y.; Han, W.X.; Tang, A.H.; Shen, J.L.; Cui, Z.L.; Vitousek, P.; Erisman, J.W.; Goulding, K.; Christie, P.; et al. Enhanced nitrogen deposition over China. *Nature* **2013**, *494*, 459–462. [CrossRef] [PubMed]

7. Berenguer, P.; Santiveri, F.; Boixadera, J.; Lloveras, J. Nitrogen fertilization of irrigated maize under Mediterranean conditions. *Eur. J. Agron.* **2009**, *30*, 163–171. [CrossRef]

8. Good, A.G.; Beatty, P.H. Fertilizing Nature: A Tragedy of Excess in the Commons. *PLoS Biol.* **2011**, *9*, e1001124. [CrossRef] [PubMed]

9. Hirel, B.; Gallais, A. Nitrogen use efficiency—Physiological, molecular and genetic investigations towards crop improvement. In *Advances in Maize*; Prioul, J.L., Thévenot, C., Molnar, T., Eds.; Society for Experimental Biology: London, UK, 2011; Volume 3, pp. 285–310.

10. Deng, F.; Wang, L.; Ren, W.J.; Mei, X.F.; Li, S.X. Optimized nitrogen managements and polyaspartic acid urea improved dry matter production and yield of indica hybrid rice. *Soil Tillage Res.* **2015**, *145*, 1–9. [CrossRef]

11. Hu, H.Y.; Ning, T.Y.; Li, Z.J.; Han, H.F.; Zhang, Z.Z.; Qin, S.J.; Zheng, Y.H. Coupling effects of urea types and subsoiling on nitrogen—Water use and yield of different varieties of maize in northern China. *Field Crops Res.* **2013**, *142*, 85–94. [CrossRef]

12. Deng, F.; Wang, L.; Ren, W.J.; Mei, X.F.; Li, S.X. Enhancing nitrogen utilization and soil nitrogen balance in paddy fields by optimizing nitrogen management and using polyaspartic acid urea. *Field Crops Res.* **2014**, *169*, 30–38. [CrossRef]

13. Du, Z.J.; Yang, H.; Wang, Y.Z.; Luo, H.Y.; Xu, L.; Wang, K.L.; Wang, B. Effects on yield and phosphorus nutrition absorbing for rice using homologous polypeptides of polyaspartic acids. *Mod. Agric. Sci. Technol.* **2012**, *18*, 12–13, 15, (In Chinese with English abstract).

14. Xu, Y.; Zhao, L.L.; Wang, L.N.; Xu, S.Y.; Cui, Y.C. Synthesis of polyaspartic acid–melamine grafted copolymer and evaluation of its scale inhibition performance and dispersion capacity for ferric oxide. *Desalination* **2012**, *286*, 285–289. [CrossRef]

15. Tomida, M.; Nakato, T.; Matsunami, S.; Kakuchi, T. Convenient synthesis of high molecular weight poly (succinimide) by acid-catalysed polycondensation of L-aspartic acid. *Polymer* **1997**, *38*, 4733–4736. [CrossRef]

16. Vega-Chacón, J.; Amaya Arbeláez, M.I.; Jorge, J.H.; Marques, R.F.C.; Jafelicci, M., Jr. pH-responsive poly(aspartic acid) hydrogel-coated magnetite nanoparticles for biomedical applications. *Mater. Sci. Eng. C* **2017**, *77*, 366–373. [CrossRef] [PubMed]

17. Kumar, A. Polyaspartic Acid—A Versatile Green Chemical. *Chem. Sci. Rev. Lett.* **2012**, *1*, 162–167.

18. Chiriac, A.P.; Nita, L.E.; Neamtu, I. Poly(ethylene glycol) functionalized by polycondensing procedure with poly(succinimide). *Polymer* **2010**, *55*, 641–645.

19. Jiang, W.; Zhou, D.B.; Zhang, H.S.; Zhang, Y.S. The effect of polyaspartic acid on maize growth at seedling stage under different fertilizer applied condition. *J. Maize Sci.* **2007**, *15*, 121–124. (In Chinese with English Abstract)

20. Miller, A.J.; Fan, X.; Orsel, M.; Smith, S.J.; Wells, D.M. Nitrate transport and signalling. *J. Exp. Bot.* **2007**, *58*, 2297–2306. [CrossRef] [PubMed]

21. Plett, D.; Baumann, U.; Schreiber, A.W.; Holtham, L.; Kalashyan, E.; Toubia, J.; Nau, J.; Beatty, M.; Rafalski, A.; Dhugga, K.S.; et al. Maize maintains growth in response to decreased nitrate supply through a highly dynamic and developmental stage-speciic transcriptional response. *Plant Biotechnol. J.* **2016**, *14*, 342–353. [CrossRef] [PubMed]

22. Lea, U.S.; Leydecker, M.T.; Quillere, I.; Meyer, C.; Lillo, C. Post-translational regulation of nitrate reductase strongly affects the levels of free amino acids and nitrate, whereas transcriptional regulation has only minor inluence. *Plant Physiol.* **2006**, *140*, 1085–1094. [CrossRef] [PubMed]

23. Takahashi, M.; Sasaki, Y.; Ida, S.; Morikawa, H. Nitrite reductase gene enrichment improves assimilation of NO2 in Arabidopsis. *Plant Physiol.* **2001**, *126*, 731–741. [CrossRef] [PubMed]

24. Bernard, S.; Moller, A.L.B.; Dionisio, G.; Kichey, T.; Jahn, T.P.; Dubois, F.; Baudo, M.; Lopes, M.S.; Terce-Laforgue, T.; Foyer, C.H.; et al. Gene expression, cellular localisation and function of glutamine synthetase isozymes in wheat (Triticum aestivum L.). *Plant Mol. Biol.* **2008**, *67*, 89–105. [CrossRef] [PubMed]

25. Martin, A.; Lee, J.; Kichey, T.; Gerentes, D.; Zivy, M.; Tatout, C.; Dubois, F.; Balliau, T.; Valot, B.; Davanture, M.; et al. Two cytosolic glutamine synthetase isoforms of maize are specifically involved in the control of grain production. *Plant Cell* **2006**, *18*, 3252–3274. [CrossRef] [PubMed]

26. Swarbreck, S.M.; Defoin-Platel, M.; Hindle, M.; Saqi, M.; Habash, D.Z. New perspectives on glutamine synthetase in grasses. *J. Exp. Bot.* **2011**, *62*, 1511–1522. [CrossRef] [PubMed]

27. Yamaya, T.; Kusano, M. Evidence supporting distinct functions of three cytosolic glutamine synthetases and two NADH-glutamate synthases in rice. *J. Exp. Bot.* **2014**, *65*, 5519–5525. [CrossRef] [PubMed]

28. De la Torre, F.; Cañas, R.A.; Pascual, M.B.; Avila, C.; Cánovas, F.M. Plastidic aspartate aminotransferases and the biosynthesis of essential amino acids in plants. *J. Exp. Bot.* **2014**, *65*, 5527–5534. [CrossRef] [PubMed]

29. De la Torre, F.; El-Azaz, J.; Ávila, C.; Cánovas, F.M. Deciphering the role of aspartate and prephenate aminotransferase activities in plastid nitrogen metabolism. *Plant Physiol.* **2014**, *164*, 92–104. [CrossRef] [PubMed]

30. Beatty, P.H.; Shrawat, A.K.; Carroll, R.T.; Zhu, T.; Good, A.G. Transcriptome analysis of nitrogen-efficient rice over-expressing alanine aminotransferase. *Plant Biotechnol. J.* **2009**, *7*, 562–576. [CrossRef] [PubMed]

31. McAllister, C.H.; Beatty, P.H.; Good, A.G. Engineering nitrogen use efficient crop plants: The current status. *Plant Biotechnol. J.* **2012**, *10*, 1011–1025. [CrossRef] [PubMed]

32. Campbell, W.H. Nitrate reductase structure, function and regulation: Bridging the gap between biochemistry and physiology. *Annu. Rev. Plant Physiol. Plant Mol. Biol.* **1999**, *50*, 277–303. [CrossRef] [PubMed]

33. Xia, B.X.; Sun, Z.G.; Wang, L.H.; Zhou, Q.; Huang, X.H. Analysis of the combined effects of lanthanum and acid rain, and their mechanisms, on nitrate reductase transcription in plants. *Ecotoxicol. Environ. Saf.* **2017**, *138*, 170–178. [CrossRef] [PubMed]

34. Li, H.; Hu, B.; Chu, C.C. Nitrogen use efficiency in crops: Lessons from Arabidopsis and rice. *J. Exp. Bot.* **2017**, *4*, 2–12. [CrossRef] [PubMed]

35. Cañas, R.A.; Quilleré, I.; Lea, P.J.; Hirel, B. Analysis of amino acid metabolism in the ear of maize mutants deficient in two cytosolic glutamine synthetase isoenzymes highlights the importance of asparagine for nitrogen translocation within sink organs. *Plant Biotechnol. J.* **2010**, *8*, 966–978. [CrossRef] [PubMed]

36. Mei, Q.H. Study on the Application of Environmental Friendly Polymer PASP in Agriculture. Master's Thesis, East China University of Science and Technology, Shanghai, China, 2005.

37. Majláth, I.; Darko, E.; Palla, B.; Nagy, Z.; Janda, T.; Szalai, G. Reduced light and moderate water deficiency sustain nitrogen assimilation and sucrose degradation at low temperature in durum wheat. *J Plant Physiol.* **2016**, *191*, 149–158. [CrossRef] [PubMed]

38. Teixeira, J.; Pereira, S.; Canovas, F.; Salema, R. Glutamine synthetase of potato (Solanum tuberosum L. cv. Desiree) plants: Cell- and organ-specific expression and differential developmental regulation reveal specific roles in nitrogen assimilation and mobilization. *J. Exp. Bot.* **2015**, *56*, 663–671. [CrossRef] [PubMed]

39. Gibon, Y.; Blaesing, O.E.; Hannemann, J.; Carillo, P.; Hohne, M.; Hendriks, J.H.M.; Palacios, N.; Cross, J.; Selbig, J.; Mark, S.M. A Robot-based platform to measure multiple enzyme activities in arabidopsis using a set of cycling assays: Comparison of changes of enzyme activities and transcript levels during diurnal cycles and in prolonged darkness. *Plant Cell* **2004**, *16*, 3304–3325. [CrossRef] [PubMed]

40. Livak, K.J.; Schmittgen, T.D. Analysis of relative gene expression data using real-time quantitative PCR and the $2^{-\triangle\triangle Ct}$ method. *Methods* **2001**, *25*, 402–408. [CrossRef] [PubMed]

41. Yu, J.J.; Han, J.N.; Wang, R.F.; Li, X.X. Down-regulation of nitrogen/carbon metabolism coupled with coordinative hormone modulation contributes to developmental inhibition of the maize ear under nitrogen limitation. *Planta* **2016**, *244*, 111–124. [CrossRef] [PubMed]

42. Liseron-Monfils, C.; Bi, Y.M.; Downs, G.; Wu, W.Q.; Signorelli, T.; Lu, G.W.; Chen, X.; Bondo, E.; Zhu, T.; Lukens, L.N.; et al. Nitrogen transporter and assimilation genes exhibit developmental stage-selective expression in maize (*Zea mays* L.) associated with distinct cis-acting promoter motifs. *Plant Signal. Behav.* **2013**, *8*, e26056. [CrossRef] [PubMed]

43. Li, Y.J.; Wang, M.L.; Zhang, F.X.; Xu, Y.D.; Chen, X.H.; Qin, X.L.; Wen, X.X. Effect of post-silking drought on nitrogen partitioning and gene expression patterns of glutamine synthetase and asparagine synthetase in two maize (*Zea mays* L.) varieties. *Plant Physiol. Biochem.* **2016**, *102*, 62–69. [CrossRef] [PubMed]

44. Pandey, R.K.; Maranville, J.W.; Chetima, M.M. Deficit irrigation and nitrogen effects on maize in a Sahelian environment II. Shoot growth, nitrogen uptake and water extraction. *Agric. Water Manag.* **2000**, *46*, 15–27. [CrossRef]

45. Weligama, C.; Tang, C.; Sale, P.W.G.; Conyers, M.K.; Liu, D.L. Application of nitrogen in NO3- form increases rhizosphere alkalisation in the subsurface soil layers in an acid soil. *Plant Soil* **2010**, *333*, 403–416. [CrossRef]

46. Tian, Q.; Chen, F.; Liu, J.; Zhang, F.; Mi, G. Inhibition of maize root growth by high nitrate supply is correlated with reduced IAA levels in roots. *J Plant Physiol.* **2008**, *165*, 942–951. [CrossRef] [PubMed]

47. Danalatos, N.G.; Archontoulis, S.V. Growth and biomass productivity of kenaf (*Hibiscus cannabinus* L.) under different agricultural inputs and management practices incentral Greece. *Ind. Crops Prod.* **2010**, *32*, 231–240. [CrossRef]

48. Valentinuz, O.R.; Tollenaar, M. Effect of genotype, nitrogen, plant density, and row spacing on the area-per-leaf profile in maize. *Agron. J.* **2006**, *98*, 94–99. [CrossRef]

49. Qiao, J.; Yang, L.Z.; Yan, T.M.; Xue, F.; Zhao, D. Rice dry matter and nitrogen accumulation, soil mineral Naround root and N leaching, with increasing application rates of fertilizer. *Eur. J. Agron.* **2013**, *49*, 93–103. [CrossRef]

50. Chen, B.M.; Wang, Z.H.; Li, S.X.; Wan, G.X.; Song, H.X.; Wang, X.N. Effects of nitrate supply on plant growth, nitrate accumulation, metabolic nitrate concentration and nitrate reductase activity in three leafy vegetables. *Plant Sci.* **2004**, *167*, 635–643. [CrossRef]

51. Ferrari, T.E.; Yoder, O.C.; Filner, D. Anaerobic nitrite production by plant cells and tissues: Evidence for two nitrate pools. *Plant Physiol.* **1973**, *51*, 423–431. [CrossRef] [PubMed]

52. Saiz-Fernández, I.; De Diego, N.; Brzobohatý, B.; Muñoz-Ruedad, A.; Lacuesta, M. The imbalance between C and N metabolism during high nitrate supply inhibits photosynthesis and overall growth in maize (*Zea mays* L.). *Plant Physiol. Biochem.* **2017**, *120*, 213–222. [CrossRef] [PubMed]

53. Garnett, T.; Conn, V.; Plett, D.; Conn, S.; Zanghellini, J.; Mackenzie, N.; Enju, A.; Francis, K.; Holtham, L.; Roessner, U.; et al. The response of the maize nitrate transport system to nitrogen demand and supply across the lifecycle. *New Phytol.* **2013**, *198*, 82–94. [CrossRef] [PubMed]

54. Sabermanesh, K.; Holtham, L.R.; George, J.; Roessner, U.; Boughton, B.A.; Heuer, S.; Tester, M.; Plett, D.C.; Garnett, T.P. Transition from a maternal to external nitrogen source in maize seedlings. *J. Integr. Plant Biol.* **2017**, *59*, 261–274. [CrossRef] [PubMed]

55. Cren, M.; Hirel, B. Glutamine synthetase in higher plants: Regulation of gene and protein expression from the organ to the cell. *Plant Cell Physiol.* **1999**, *40*, 1187–1193. [CrossRef]

56. Silva, I.T.; Abbaraju, H.K.R.; Fallis, L.P.; Liu, H.J.; Lee, M.; Dhugga, K.S. Biochemical and genetic analyses of N metabolism in maize testcross seedlings: 1. Leaves. *Theor. Appl. Genet.* **2017**, *130*, 1453–1466. [CrossRef] [PubMed]

57. Zhang, N.Y.; Gibon, Y.; Gur, A.; Chen, C.; Lepak, N.; Hohne, M.; Zhang, Z.W.; Kroon, D.; Tschoep, H.; Stitt, M.; et al. Fine quantitative trait loci mapping of carbon and nitrogen metabolism enzyme activities and seedling biomass in the maize IBM mapping population. *Plant Physiol.* **2010**, *154*, 1753–1765. [CrossRef] [PubMed]

58. Yanagisawa, S. Transcription factors involved in controlling the expression of nitrate reductase genes in higher plants. *Plant Sci.* **2014**, *229*, 167–171. [CrossRef] [PubMed]

59. Lillo, C.; Meyer, C.; Lea, U.S.; Provan, F.; Oltedal, S. Mechanism and importance of post-translational regulation of nitrate reductase. *J. Exp. Bot.* **2004**, *55*, 1275–1282. [CrossRef] [PubMed]

60. Amiour, N.; Imbaud, S.; Clement, G.; Agier, N.; Zivy, M.; Valot, B.; Balliau, T.; Armengaud, P.; Quillere, I.; Canas, R.; et al. The use of metabolomics integrated with transcriptomic and proteomic studies for identifying key steps involved in the control of nitrogen metabolism in crops such as maize. *J. Exp. Bot.* **2012**, *63*, 5017–5033. [CrossRef] [PubMed]

61. Fernie, A.R.; Stitt, M. On the discordance of metabolomics with proteomics and transcriptomics: Coping with increasing complexity in logic, chemistry, and network interactions. *Plant Physiol.* **2012**, *158*, 1139–1145. [CrossRef] [PubMed]

62. Kaiser, W.M.; Weiner, H.; Kandlbinder, A.; Tsai, C.B.; Rockel, P.; Sonoda, M.; Planchet, E. Modulation of nitrate reductase: Some new insights, an unusual case and a potentially important side reaction. *J. Exp. Bot.* **2002**, *53*, 875–882. [CrossRef] [PubMed]

63. Kaiser, W.M.; Huber, S.C. Post-translational regulation of nitrate reductase: Mechanism, physiological relevance and environmental triggers. *J. Exp. Bot.* **2001**, *52*, 1981–1989. [CrossRef] [PubMed]

64. Xu, T.J.; Dong, Z.Q.; Lan, H.L.; Pei, Z.C.; Gao, J.; Xie, Z.X. Effects of PASP-KT-NAA on photosynthesis and antioxidantenzyme activities of maize seedlings under low temperature stress. *Acta Agron. Sin.* **2012**, *38*, 352–359. (In Chinese with English Abstract) [CrossRef]

MDPI

St. Alban-Anlage 66

4052 Basel

Switzerland

Tel. +41 61 683 77 34

Fax +41 61 302 89 18

www.mdpi.com

Agronomy Editorial Office

E-mail: agronomy@mdpi.com

www.mdpi.com/journal/agronomy

www.ingramcontent.com/pod-product-compliance
Lightning Source LLC
Chambersburg PA
CBHW051314020426
42333CB00028B/3336